PHOSPHORIMETRY: *Theory, Instrumentation, and Applications*

PHOSPHORIMETRY: *Theory, Instrumentation, and Applications*

Robert J. Hurtubise
Chemistry Department
University of Wyoming
Laramie, WY 82071-3838

Robert J. Hurtubise
Department of Chemistry
University of Wyoming
Laramie, Wyoming 82071-3838

Library of Congress Cataloging-in-Publication Data

Hurtubise, Robert J., 1941-
  Phosphorimetry : theory, instrumentation, and applications /
Robert J. Hurtubise.
      p.   cm.
  Includes bibliographical references
  ISBN 0-89573-749-3
  1. Phosphorimetry.   I. Title.
QD79.P4H87 1990
543'.0852--dc20   90-12284
                                                                         CIP

British Library Cataloguing in Publication Data

Hurtubise, Robert J.
  Phosphorimetry : theory, instrumentation and applications.
  1. Chemical analysis. Spectroscopy fluorescence &
  spectroscopy phosphorescence
  I. Title
  543.0858

    ISBN 0-89573-749-3

Printed in the United States of America.
ISBN 0-89573-749-3 VCH Publishers
ISBN 3-52727-861-3 VCH Verlagsgesellschaft

Printing History:
10 9 8 7 6 5 4 3 2 1

Published jointly by:

VCH Publishers Inc.   VCH Verlagsgesellschaft mbH   VCH Publishers (UK) Ltd.
220 East 23rd Street   P.O. Box 10 11 61   8 Wellington Court
Suite 909   D-6940 Weinheim   Cambridge CB1 1HW
New York, New York 10010   Federal Republic of Germany   United Kingdom

TABLE OF CONTENTS

To Paula, Tim, Dave, Suzanne, and Gretta

PREFACE

Low-temperature phosphorimetry from solutions was considered for years as the primary means of obtaining phosphorescence signals. However, with the advent of several new developments in phosphorimetry, it is possible to obtain phosphorescence signals under a variety of experimental conditions at room temperature. For example, by simply adsorbing a phosphor on a solid material under dry conditions, it is feasible to obtain strong room-temperature phosphorescence from numerous compounds. Also, micelle-stabilized phosphors have been employed to obtain room-temperature phosphorescence, and phosphor-sensitized room-temperature phosphorescence in solution has been developed. From these and other new advances in phosphorimetry, a number of novel phosphorescence applications have been developed. In the areas of protein research and polymer research, several reports have appeared in which both room-temperature phosphorescence and low-temperature phosphorescence have been employed to characterize the structural features of these important materials.

With the new developments in obtaining phosphorescence, one has seen significant improvements in phosphorescence instrumentation. Modifications to commercial instruments have been relatively minor or very sophisticated depending on individual needs. Also, entire instrumental systems have been built which are computerized and permit time-resolved phosphorescence measurements to be made.

The primary purpose in writing this book was to give a rather detailed survey of phosphorimetry from its inception to modern times. However, emphasis has been placed on the more modern developments in phosphorimetry and its use in organic trace analysis. The book will be of interest to those working in luminescence spectrometry and to those who have more specialized interests in phosphorimetry. The book should find use as a reference text or as a supplemental text for courses involved with optical methods of analysis.

In Chapter 1, a historical survey is presented, whereas in Chapter 2 several of the photophysical aspects of luminescence are discussed. The instrumentation for phosphorimetry is detailed in Chapter 3, although various instrumental aspects are also considered in other parts of the book. In Chapter 4, a variety of important analytical concepts such as the relationship between phosphorescence intensity and concentration is presented.

Chapter 5 is devoted to low-temperature phosphorescence, where instrumentation and a variety of low-temperature phosphorescence applications are considered. One example presented in this chapter is phosphorescence line-narrowing spectrometry. In Chapter 6, sample preparation, types of phosphorescence data, sample holders, and instrumentation for solid-surface room-temperature phosphorescence are explored. Chapter 7 gives a rather detailed discussion of the important physicochemical interactions in solid-surface room-temperature phosphorescence, whereas several applications for solid-surface RTP are detailed in Chapter 8.

Sensitized and quenched phosphorescence in solution at room temperature are considered from the general theoretical and application viewpoints in Chapter 9. A similar approach is used in Chapter 10 for micelle-stabilized, cyclodextrin, and colloidal/microcrystalline room-temperature phosphorescence, but with emphasis on micelle-stablized room-temperature phosphorescence. Chapter 11 presents a brief consideration of the phosphorescence of proteins, polypeptides, and peptides. The use of phosphorescence for the structural characterization of polymers is stressed in Chapter 12. Finally, in Chapter 13, the future trends in phosphorimetry are discussed. To the author's knowledge this book is the only up-to-date treatment that surveys the entire area of phosphorimetry.

Appreciation is extended to Professor James D. Winefordner for prepublication material on phosphorimetry. Professor Winefordner has been a great source of inspiration, and he and his colleagues, and other researchers, have developed many of the new paths in analytical phosphorimetry.

Finally, it is a pleasure to acknowledge Ms. Sandy Pease for her expert assistance in the early stages of the preparation of this book, and Ms. Janet Zemanek for her expertise in typing and in the final formatting of the book.

CHAPTER 1

INTRODUCTION

1.1. General Comments

Luminescence is defined generally as radiation emitted by atoms or molecules when they undergo a radiative transition from an excited energy level to a lower energy level. The emission from an excited energy level, which has been produced by absorption of incident radiation, is referred to as photoluminescence. Photoluminescence includes both fluorescence and phosphorescence. The phenomena of fluorescence and phosphorescence have fundamentally different properties, and these will be discussed more fully in Chapter 2. However, in molecular fluorescence, the lifetime of fluorescence is usually much shorter than the lifetime of molecular phosphorescence, and the emitted wavelengths of fluorescence are normally shorter than the emitted wavelengths of phosphorescence.

A variety of instrumental systems and modern instruments is used for fluorescence and phosphorescence (luminescence) analysis work. These instruments and systems can be used from simple collection of luminescence data to sophisticated data processing and handling. Applications in luminescence analysis are copious. One finds these in the following types of analysis: in clinical, drug, air and water pollution, biological and medical, enzyme, chromatographic, industrial, forensic, agricultural, immunochemical, polymer, and so forth (1). Luminescence analysis is widely used primarily because of the sensitivity and selectivity that can be achieved in trace analysis.

Phosphorescence analysis is not as widely used as fluorescence analysis. This is partly related to the fact that in earlier work phosphorescence signals were obtained at liquid nitrogen temperature, which required the use of cryogenic solvents. However, one can obtain excellent sensitivity and selectivity with low-temperature phosphorescence techniques. New advances in analytical phosphorimetry have shown that useful phosphorescence signals can be obtained at room temperature from phosphors adsorbed on solid matrices (2,3), by micelle-stabilized phosphors in solution (4), and by phosphor-sensitized room-temperature phosphorescence in solution (5). Two particular areas that demonstrate the usefulness of phosphorescence are in the characterization of proteins and polymers (6,7). However, there are several other important areas in which phosphorescence has been employed, and most of these areas will be considered throughout the book. The new advances in phosphorescence will significantly advance the applicability of phosphorimetry.

1.2.   Historical Survey

Edmond Becquerel developed the first phosphoroscope in 1858 (8). In 1861 he established the exponential law of decay of phosphorescence. E. Wiedemann observed the phosphorescence of aniline dyes in solid solution and in gelatin in 1887 (8). Phosphorescence in frozen solutions that were cooled by liquid or air was observed in 1894 by James Dewar. Wiedemann published an important advance in the theory of phosphorescence in 1889. He postulated that a phosphor could exist in a stable form A and an unstable form B. The conversion from form A to form B was caused by the absorption of light. Form B then returned to form A with the emission of light. In 1935, A. Jablonski (9) proposed the scheme of electronic energy levels that has become the basis for the interpretation of luminescence phenomena. A very important paper in 1944 by G. N. Lewis and M. Kasha (10) established that the metastable state in a molecule from which phosphorescence could occur was a triplet state. Lewis and Kasha also suggested the use of phosphorescence as a qualitative tool for the identification of organic compounds. However, the first paper on quantitative phosphorimetry did not appear until 1957 and was published by Kiers, Britt, and Wentworth (11). But, as O'Haver (8) has emphasized, most of the quantitative applications of phosphorescence did not appear in the literature until after 1962. An important monograph that gives a historical perspective, as well as applications, on the fluorescence and phosphorescence of proteins and nucleic acids was written by Konev (12) and published in 1967.

The paper by Kiers et al. (11) was important primarily for two reasons. Phosphorescence analytical curves were reported for a number of organic compounds, and two- and three-component mixtures were analyzed quantitatively by phosphoroscopic resolution, selective excitation, and simultaneous equations. Their paper helped stimulate method development work in phosphorimetry.

Another important paper, published in 1962 by Parker and Hatchard (13), described a photoelectric phosphorescence spectrometer. They obtained phosphorescence spectra, lifetimes, and quantum efficiencies for several compounds. Their results helped to generate interest in phosphorimetry among chemists. Several applications began to appear in 1963. Amino acids, enzymes, and peptides were determined by Freed and Vise (14), and McGlynn et al. (15) discussed the analysis of mixtures of polycyclic aromatic hydrocarbons by room-temperature fluorescence, low-temperature fluorescence, and low-temperature phosphorescence. An important paper, published by Winefordner and Latz (16) in 1963, described a low-temperature phosphorimetric method for aspirin in blood. This was the first paper that described an application of phosphorescence to a complex sample. This work furthered the acceptance of phosphorimetry as a useful analytical method. It was also the first of numerous manuscripts to be published on low-temperature phosphorimetry by Professor J. D. Winefordner's research group at the University of Florida. This group had established themselves as leaders in the field of low-temperature phosphorescence analysis. After 1965, and until 1974, the field of analytical low-temperature phosphorimetry was developed mainly by Professor Winefordner's group.

In 1975, Aaron and Winefordner (17) reviewed primarily low-temperature phosphorimetry in terms of theory, instrumentation, and applications. In a 1981 review on phosphorimetry, Ward et al. (18) emphasized that very little work had been done in the field of low-temperature phosphorescence since 1975.

Roth (19) suggested the use of room-temperature phosphorescence for analysis from his observations that several organic compounds adsorbed on cellulose gave room-temperature phosphorescence. Lloyd and Miller (20) have discussed earlier observations of room-temperature phosphorescence. Schulman and Walling (21,22) independently measured room-temperature phosphorescence from several compounds adsorbed on solid surfaces. In 1974 Paynter et al. (23) published detailed results of the analytical development and application of the phenomenon of solid-surface room-temperature phosphorescence using filter paper as a solid surface. In other developments in phosphorescence, Cline Love and co-workers (4,24) reported the new analytical approach of solution micelle-stabilized room-temperature phosphorescence in 1980. Donkerbroek et al. (25), in 1981, introduced sensitized room-temperature solution phosphorescence as a viable analytical approach for certain classes of compounds. Hurtubise (26) has summarized many of the new analytical phosphorescence approaches.

1.3. Phosphorimetry as an Analytical Tool

Essentially all of the analytical applications for phosphorimetry have been for organic compounds. It has shown itself to be very sensitive and selective in organic trace analysis. As a technique that is complementary to fluorescence spectrometry, it permits both fluorescence and phosphorescence information to be obtained for many samples. In some analytical situations, it may be the only viable analytical approach to achieve the desired sensitivity and selectivity. Phosphorimetry has been employed to obtain structural information from proteins and polymers in addition to the routine characterization and determination of materials such as drugs and pollutants. With the advent of new room-temperature phosphorescence techniques, fresh analytical approaches are being developed in phosphorimetry. These developments have opened up new avenues in theory, instrumentation, and applications that will be applied to some of the many challenging analytical problems of organic trace analysis in the future.

References

1. Hurtubise, R.J. In *Trace Analysis: Spectroscopic Methods for Molecules*; Christian, G.D.; Callis, J.B., Eds.; Wiley: New York, 1986; Chapter 2.
2. Hurtubise, R.J. *Solid-Surface Luminescence Analysis: Theory, Instrumentation, Applications*; Marcel Dekker: New York, 1981.
3. Vo-Dinh, T. *Room-Temperature Phosphorimetry for Chemical Analysis*; Wiley: New York, 1984.
4. Cline Love, L.J.; Skrilec, M.; Habarta, J.G. *Anal. Chem.* 1980,

$\underline{52}$, 754.

5. Donkerbroek, J.J.; Gooijer, C.; Velthorst, N.H.; Frei, R.W. *Anal. Chem.* 1982, $\underline{54}$, 891.
6. Vanderkool, J.M.; Calhoun, D.B.; Englander, S.W. *Science* 1987, $\underline{236}$, 568.
7. Guillet, J. *Polymer Photophysics and Photochemistry*; Cambridge University Press: New York, 1985; Chapter 8.
8. O'Haver, T.C. *J. Chem. Educ.* 1978, $\underline{55}$, 423.
9. Jablonski, A. *Z. Physik* 1935, $\underline{94}$, 38.
10. Lewis, G.N.; Kasha, M. *J. Am. Chem. Soc.* 1944, $\underline{66}$, 2100.
11. Kiers, R.J.; Britt Jr., R.D.; Wentworth, W.E. *Anal. Chem.* 1957, $\underline{29}$, 202.
12. Konev, S.V. *Fluorescence and Phosphorescence of Proteins and Nucleic Acids*; Plenum Press: New York, 1967.
13. Parker, C.A.; Hatchard, C.G. *Analyst* 1962, $\underline{87}$, 664.
14. Freed, S.; Vise, M.H. *Anal. Biochem.* 1962, $\underline{5}$, 338.
15. McGlynn, S.P.; Neely, B.T.; Neely, W.C. *Anal. Chim. Acta* 1963, $\underline{28}$, 472.
16. Winefordner, J.D.; Latz, H.W. *Anal. Chem.* 1963, $\underline{35}$, 1517.
17. Aaron, J.J.; Winefordner, J.D. *Talanta* 1975, $\underline{22}$, 707.
18. Ward, J.L.; Walden, G.L.; Winefordner, J.D. *Talanta* 1981, $\underline{28}$, 201.
19. Roth, M. *J. Chromatogr.* 1967, $\underline{30}$, 276.
20. Lloyd, J.B.F.; Miller, J.N. *Talanta* 1979, $\underline{26}$, 180.
21. Schulman, E.M.; Walling, C. *Science* 1972, $\underline{178}$, 53.
22. Schulman, E.M.; Walling, C. *J. Phys. Chem.* 1973, $\underline{77}$, 902.
23. Paynter, R.A.; Wellons, S.L.; Winefordner, J.D. *Anal. Chem.* 1974, $\underline{46}$, 736.
24. Skrilec, M.; Cline Love, L.J. *Anal. Chem.* 1980, $\underline{52}$, 1559.
25. Donkerbroek, J.J.; Elzas, J.J.; Gooijer, C.; Frei, R.W.; Velthorst, N.H. *Talanta* 1981, $\underline{28}$, 717.
26. Hurtubise, R.J. *Anal. Chem.* 1983, $\underline{55}$, 669A.

CHAPTER 2

PHOTOPHYSICAL ASPECTS OF LUMINESCENCE

2.1.   General Considerations

The theory of molecular luminescence has been discussed extensively in the literature (1-8).   In this chapter only the basic theoretical principles of molecular luminescence will be considered. It is important to present the fundamental aspects of both fluorescence and phosphorescence so that a relatively comprehensive understanding of photoluminescence processes can be formulated.

Most of the organic compounds that fluoresce or phosphoresce are aromatic. Some highly unsaturated aliphatic compounds with π electronic systems also yield luminescence.   Each occupied orbital of a ground-state molecule has a pair of electrons.   The Pauli exclusion principle states that two electrons in an orbital must have opposing spins, and therefore the net spin for most ground-state molecules is zero. A molecule in which all the electron spins are paired is said to be in a singlet state.   If an electron is promoted in a molecule by absorption of ultraviolet or visible electromagnetic radiation to a higher energy level, its spin will either be in the same direction or opposed to the spin of another electron in the orbital. For the case where the spins are in the same direction, the resulting state is known as a triplet state.   The energy associated with electrons of opposing spins is larger than with electrons of parallel spins (Hund's rule).

2.2.   Processes in Photoluminescence

Several processes that are important in luminescence analysis are shown in Figure 2.1. If a molecule does not undergo a

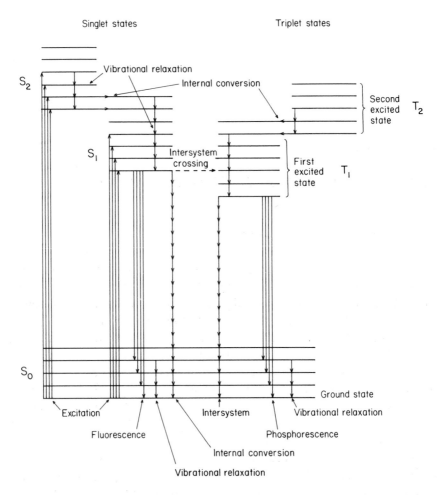

Figure 2.1. Partial energy level diagram for a photoluminescent molecule.

photochemical reaction after it absorbs electromagnetic radiation in the ultraviolet or visible region, it is normally promoted to a vibrational level in an excited electronic singlet state. Generally, the molar absorptivity ϵ can be used as a measure of the probability of absorption of electromagnetic radiation (3). For example, a molar absorptivity of 10^5 at a specific wavelength indicates a high probability that an electronic transition will occur. The time for a molecule to pass from the ground state to an electronically excited state is about 10^{-15} s. Because this time period is so short, the atomic nuclei in the organic molecule do not appreciably change their relative positions. Thus, electronic transitions essentially occur without change in the positions of the nuclei. This is a statement of the Franck-Condon principle.

Immediately after excitation by electromagnetic radiation, a molecule has the same geometry and is in the same environment as it was in the ground state. For a molecule in solution, the upper vibrational levels for an excited electronic state of the molecule relax vibrationally in about 10^{-12} s. Within this time period, the excited molecule relaxes vibrationally to the lowest vibrational level of the lowest excited singlet state. At room temperature in solution, the solvent molecules reorient themselves to a state of equilibrium that is compatible with the new molecular polarity of the excited solute molecule. The vibrational and solvent relaxation processes are accompanied by a loss in thermal energy. Once the molecule is in the lowest vibrational level of the lowest excited singlet state, a radiative transition can occur in which the solute molecule drops to one of a number of possible vibrational levels of the ground electronic state. The radiative transition is called fluorescence and almost always occurs from the lowest vibrational level of the lowest excited singlet state in a molecule. The decay time of fluorescence is about the same order of magnitude as the lifetime of an excited singlet state, namely, 10^{-9} to 10^{-7} s. Because of spectroscopic selection rules, the transition from an excited singlet state to a singlet ground state is spin-allowed and thus occurs with a high degree of probability. After the fluorescence transition occurs, vibrational relaxation and solvent reorientation take place, and the molecule then arrives in a ground-state equilibrium configuration.

As shown in Figure 2.1, other processes can compete with fluorescence. The excited molecule can lose energy by other means such as internal conversion and intersystem crossing. Internal conversion is a radiationless process whereby a molecule passes from a higher to a lower electronic state without the emission of a photon. After internal conversion takes place, vibrational relaxation immediately follows to the lowest vibrational level of the lower electronic state. However, a radiationless transition between the first excited singlet state to the lowest excited triplet state is called intersystem crossing. A change in spin occurs with intersystem crossing, and thus, spin selection rules are not obeyed rigorously. However, intersystem crossing can take place because of spin-orbit coupling between an excited singlet state and an excited triplet state. The net result of spin-orbit coupling is the mixing of singlet character into the triplet state and the mixing of triplet character into the singlet state. Thus, the mixing of states removes the spin-forbidden nature of the transitions between pure singlet and pure triplet states. If intersystem crossing can favorably compete with fluorescence or internal conversion to the ground state, the molecule can pass from the lowest excited singlet state to a triplet state. Then the molecule will undergo vibrational relaxation to arrive at the lowest vibrational level of the lowest excited triplet state. From this state, the molecule can undergo a radiative transition or undergo intersystem crossing to the ground state. The radiative transition results in phosphorescence. The intersystem crossing processes are depicted in more detail in Figure 2.2. It should be realized that the T_1-S_0 transition in Figure 2.2 involves a change in spin. The radiationless deactivation of T_1 results from intersystem crossing to some higher vibrational level of S_0 as shown in Figure 2.2. The numbers associated with the various energy states are the approximate range of rate constants for the respective energy states. It should be noted in Figure 2.2 that there is a great difference between the S_1-T_1 and T_1-S_0 rates. Generally, this is related to the fact that the energy gap between T_1 and S_0 is greater than the energy gap between S_1 and T_1 (9). In addition, T_n states may intervene between S_1 and T_1; however, none may intervene between T_1 and S_0.

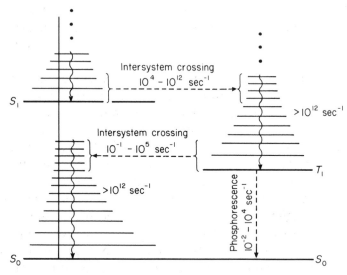

Figure 2.2. The intersystem-crossing processes $S_1 \rightarrow T_1$ and $T_1 \rightarrow S_0$. The ranges of rate constants which are shown cover many but not all situations. (McGlynn/Azumi/Kinoshita, *Molecular Spectroscopy of the Triplet State*, 1969, p.14. Reprinted by permission of Prentice Hall, Inc., Englewood Cliffs, New Jersey.)

Because phosphorescence originates from the lowest triplet state in a molecule, its decay time is similar to the lifetime of the triplet state, about 10^{-4} to 10 s. The radiative transition from a triplet excited state to a singlet ground state is a spin-forbidden transition because it involves a change in spin. Because phosphorescence is a spin-forbidden process, the rate of phosphorescence is relatively slow compared to other processes associated with excited molecules. This fact gives the electronic triplet state its long lifetime. For molecules in solution at room temperature, the long lifetime of the triplet state substantially increases the probability of collisional transfer of energy with solvent molecules or impurity molecules. This process is very efficient in solution at room temperature and is often the main pathway for loss of triplet-state excitation energy. Regular solution phosphorescence is rarely observed at room temperature unless special, and sometimes extreme, measures are taken to lower the rates of competing processes. Solution phosphorescence that is useful analytically can be observed for a phosphor by dissolving the solute in a solvent that freezes to form a rigid glass at the temperature of liquid nitrogen (77 K). Also, recent analytical advances have permitted phosphorescence to be obtained readily for a large number of compounds at room temperature at the nanogram and subnanogram levels (10-12).

2.3.   Luminescence Quantum Yields

The efficiency of luminescence can be discussed in terms of quantum yield. The fluorescence quantum yield is defined as follows:

$$\phi_f = \frac{\text{quanta emitted as fluorescence}}{\text{quanta absorbed to a singlet excited state}} \qquad (2.1)$$

Phosphorescence quantum yield can be defined similarly:

$$\phi_p = \frac{\text{quanta emitted as phosphorescence}}{\text{quanta absorbed to a singlet excited state}} \qquad (2.2)$$

It is important to consider the relationships of quantum yields to the rate constants of the different excited-state processes (1-3,8). Figure 2.3 gives a summary of the important rate constants for excited-state processes, and Equations (2.3) and (2.4) define the fluorescence and phosphorescence quantum yields in terms of rate constants.

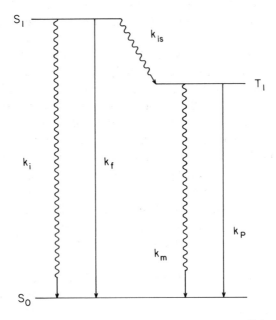

Figure 2.3. Summary of rate constants for exicted-state processes: (S_0) ground state; (S_1) lowest excited signlet state; (T_1) lowest triplet state; (k_i) internal conversion for lowest excited signlet state; (k_f) fluorescence; (k_m) intersystem crossing between lowest triplet state and ground state; (k_p) phosphorescence; (k_{is}) intersystem crossing. Wavy lines indicate nonradiative transitions.

$$\phi_f = \frac{k_f}{k_f + k_i + k_{is}} \qquad (2.3)$$

$$\phi_p = \left(\frac{k_p}{k_p + k_m}\right)\left(\frac{k_{is}}{k_f + k_i + k_{is}}\right) \qquad (2.4)$$

These equations indicate how excited-state processes compete with one another. For example, if $k_f \gg k_i$ and k_{is}, the fluorescence yield will be large and approach unity. Equation (2.4) shows that the definition of ϕ_p is more complicated than the definition of ϕ_f. The quantum yield of phosphorescence depends on the competition between intersystem crossing from T_1 to S_0 states and phosphorescence emission (Figure 2.3). It also depends on the rate of intersystem crossing from S_1 to T_1 relative to the rate of fluorescence and the rate of internal conversion from S_1 to S_0 (Figure 2.3). Usually, k_f and k_p are dependent on molecular structure and are affected only mildly by the environment. The term k_{is} is dependent on molecular structure, but can also be dependent on the environment. The k_i and k_m rate constants are highly dependent upon the molecular environment and to a lesser extent on the nature of the electronic states in the molecule. As Figure 2.3 indicates, fluorescence and phosphorescence are only two of the various processes that can occur after molecular excitation. It is important to realize that other phenomena could also compete with fluorescence and phosphorescence. For example, bimolecular quenching, energy transfer, and photochemical reactions are some of the other processes that could effectively diminish ϕ_f or ϕ_p (1,3,6). Demas and Crosby (13) have given an extensive review for the measurement of photoluminescence quantum yields. Bridges (14) has discussed some of the more recent developments for the determination of fluorescence quantum yields.

2.4. Luminescence Lifetimes

Fluorescence lifetime and phosphorescence lifetime are important parameters in describing luminescence phenomena. Demas (15) has given a comprehensive discussion of excited state lifetime measurements. For example, ideally, fluorescence intensity decays after the removal of the exciting source by a first-order equation

$$I = I_0 e^{-t/\tau_f} \qquad (2.5)$$

where I_0 is the initial intensity, I is the intensity at some time t, and τ_f is the observed fluorescence lifetime. If t is equal to τ_f, the fluorescence intensity then is equal to $1/e$ of its initial value. The term τ_f is also considered the mean decay time for fluorescence or the mean lifetime of the excited state. Experimental lifetimes, whether for fluorescence or phosphorescence, indicate the overall rate at which the excited state is deactivated, which consists of both radiative and nonradiative processes. The intrinsic or natural lifetime of fluorescence τ_f^o is obtained experimentally if fluorescence is the only mechanism by which the excited molecule returns to the ground state. The intrinsic lifetime of fluorescence is equal to $1/k_f$. The relationship between τ_f and τ_f^o is given by the following equation.

$$\phi_f \tau_f^0 = \tau_f \qquad (2.6)$$

Ideally, phosphorescence decay also follows equation (2.5). The intrinsic or natural lifetime of phosphorescence is defined in the same fashion as for fluorescence, namely, $\tau_p^0 = 1/k_p$. However, the relationship between the quantum yield and intrinsic and observed lifetimes is different for phosphorescence compared to fluorescence (16). This is shown by equation (2.7)

$$\frac{\phi_p \tau_p^0}{\phi_t} = \tau_p \qquad (2.7)$$

where ϕ_t is the triplet formation efficiency and τ_p is the observed phosphorescence lifetime.

It is important to consider τ_f and τ_p in terms of rate constants because this helps to emphasize some of the fundamental differences between fluorescence and phosphorescence lifetimes. Equation (2.8) defines τ_f as a function of rate constants. (See Figure 2.3.)

$$\tau_f = \frac{1}{k_f + k_i + k_{is}} \qquad (2.8)$$

If a bimolecular quenching process were competing for fluorescence, then the term $k_q[q]$ would be added to the denominator, where k_q is the rate constant for bimolecular quenching and [q] is the concentration of quencher. The experimental phosphorescence lifetime is given by the equation below. (See Figure 2.3.)

$$\tau_p = \frac{1}{k_p + k_m} \qquad (2.9)$$

As with fluorescence, if a quenching process were present, the term $k_q[q]$ would be added to the denominator of equation (2.9). Comparison of equations (2.8) and (2.9) shows that the experimental lifetimes of fluorescence and phosphorescence are a function of different rate processes. Also, if no photochemical reactions are occurring and no quenching agents are present, then the fluorescence lifetime is a function of three rate constants, whereas phosphorescence lifetime is a function of two rate constants. If other phenomena are operative such as delayed fluorescence or energy transfer processes, then the situation is more complex. However, normally in organic trace analysis these phenomena are not present or can be minimized experimentally.

2.5. Polarization

Emission polarization can be used to determine the orientation of the emission transition dipole relative to that of a particular absorption band. Polarization results because electronic transitions are characterized by transition moments that have unique orientations with respect to molecular structure. It is known for aromatic hydrocarbons that allowed transitions are in-place polarized (1). Thus, with this information, it is possible to obtain facts about the direction of the emission. An important parameter for the differentiation between the π,π^* and the n,π^* character of an emitting triplet state is phosphorescence polarization (1,17). Figure 2.4 shows a typical experimental setup used for the measurement of

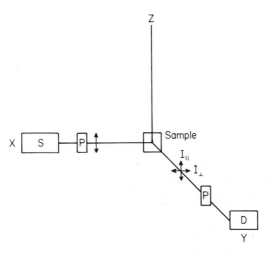

Figure 2.4. Block diagram of apparatus for fluorescence polarization measurements: (S) source; (P) polarizer; (D) detector; ($I_\parallel$) intensity of fluorescence polarized in same plane as excitation radiation; ($I_\perp$) intensity of fluorescence in plane perpendicular to excitation radiation. (Reprinted with permission for W.R. Seitz, "Luminescence Spectrometry," in P.J. Elving, E.J. Meehan, and I.M. Kolthoff, Eds., *Treatise on Analytical Chemistry*, 2nd Ed., Part I, Vol. 7, Sec. H, Wiley, New York, 1981. Chap.4.)

polarization. The exciting radiation is polarized in the xz-plane. The emission that occurs can be resolved into $I_{\parallel}$, the intensity of the emission in the yz-plane, and $I_{\perp}$, the intensity of emission in the xy-plane. The definition of polarization p is given by equation (2.10).

$$p = \frac{I_{\parallel}\text{-}I_{\perp}}{I_{\parallel}\text{-}I_{\perp}} \tag{2.10}$$

Additional discussions on polarization can be found in the literature (1-3,7).

2.6. Excitation and Emission Spectra

Both phosphorescence and fluorescence emission spectra are recorded by measuring the luminescence intensity as a function of wavelength at a fixed excitation wavelength. However, an excitation spectrum is recorded by measuring luminescence intensity at a fixed emission wavelength while varying the exciting wavelength. Because the energy transitions in fluorescence and phosphorescence are smaller in magnitude than those associated with excitation, the luminescence spectra will appear at longer wavelengths than the wavelengths of the excitation spectrum. Since the lowest triplet state is at lower energy than the first excited singlet state, phosphorescence will occur at longer wavelengths than fluorescence. Figure 2.5 gives examples of typical room-temperature excitation and fluorescence emission spectra and a low-emperature phosphorescence emission spectrum for benzo[f]-quinoline.

The spectra in Figure 2.5 are somewhat broad due to vibrational transitions associated with the electronic transitions (Figure 2.1) and perturbations caused by the solvent. In general, there is greater loss in vibrational detail with more polar solvents. For samples in solid solutions at low temperature, spectra are usually more structured than those from solution at room temperature.

2.7. Excited-State Acid-Base Chemistry and Other Aspects

The acid-base chemistry of the excited states for lumines-

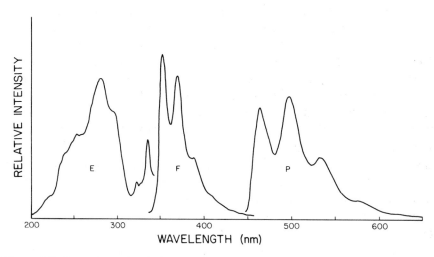

Figure 2.5. Excitation and emission spectra for benzo[f]quinoline in ethanol: (E) excitation; (F) fluorescence; (P) phosphorescence.

cent molecules is important for a variety of reasons. For example, in trace analysis, if a molecule loses a proton in the excited state before emission, then the emission properties of the deprotonated species could be quite different from the properties of the original molecule. The excitation of a molecule from the ground electronic state to its lowest excited singlet state involves a change in the electronic dipole moment, and the excited molecule normally has different properties compared to the molecule in the ground state. Thus, the chemical properties of the molecule in the excited state can be very different from the chemical properties of the molecule in the ground state. The triplet state, because of its relatively long lifetime, would permit photoexcited molecules adequate time to encounter a potential reactant. The general properties of the triplet state are the basis of a considerable amount of photochemistry. The lowest excited singlet state has a much shorter lifetime than the triplet state, which would lower the probability that a fluorescent molecule would encounter a reactant and thus alter its fluorescence properties. Chemical reactions such as proton exchange with acids and bases in solution can occur during the lifetime of the lowest excited singlet state and the lowest excited triplet state.

Schulman (18) has given a detailed treatment of the acid-base chemistry of excited singlet states. Three major conditions can arise: (a) excited-state proton exchange is considerably slower than fluorescence in an acid or conjugate base; (b) excited-state proton exchange is much faster than fluorescence in an acid or conjugate base; (c) excited-state proton exchange and fluorescence are of comparable rates. In the first situation, fluorescence will occur from the excited molecule without an excited-state proton transfer reaction. The quantum yield of fluorescence from the excited species should be pH independent. In addition, the fluorescence intensity will depend only upon the absorbance at the excitation wavelength and the concentration of the ground-state species. In the second situation discussed by Schulman (18), it is the dissociation constant of the excited-state proton exchange which essentially determines the fluorescence behavior of the molecule. The excited singlet-state pK_a is normally very different from the ground-state pK_a because the electronic distribution of an electronically excited molecule is different from the ground state. An example of the third case mentioned by Schulman is one in which the rates of proton transfer between an excited acid and solvent or between an excited base and solvent are comparable to the rates of deactivation of acid and conjugate base by fluorescence.

In general, there are fewer reports of triplet-state acidity constants compared to singlet-state acidity constants. Most earlier analytical phosphorescence work has been done in rigid solutions at liquid-nitrogen temperature. Under liquid-nitrogen conditions, acid-base equilibrium cannot be quickly established after excitation. Usually, the triplet-acidity constants are much closer to the ground-state acidity constants rather than excited singlet-state acidity constants. Jackson and Porter (19) obtained the acidity constants of the lowest triplet state for several aromatic compounds. They used two approaches. With the first method, the triplet state was populated by flash photolysis and the acid-base equilibrium was studied spectrophotometrically. In the other approach, the difference between the acidity constants in the triplet and ground states was calculated from the energy levels of the acid and base that were derived from phosphorescence spectra. Triplet-state acidity constants for hydroxy- and amino-substituted anthraquinones, and related compounds were reported by Richtol and Fitch (20). Also, Vander Donckt (21) reported several pK_a values for compounds in the ground state, the lowest excited singlet state, and lowest excited triplet state. For example, pK_a values for 2-naphthol are 9.49 for the ground state, 2.81 for the lowest excited singlet state, and 7.7 for the lowest excited triplet state (21).

2.8. Delayed Fluorescence

In trace analysis, it is important to be aware that delayed fluorescence can take place in solution and may also occur from compounds adsorbed on solid surfaces. Delayed fluorescence from solutions has been discussed by Parker (22); however, very few reports have appeared on delayed fluorescence from compounds adsorbed on solid surfaces. Generally, a delayed fluorescence spectrum is identical to the regular fluorescence spectrum of a molecule. In addition, delayed fluorescence has an unusually long lifetime and is sometimes mistaken for phosphorescence because of this characteristic. Three types of delayed fluorescence will be discussed. In E-type delayed fluorescence, a molecule is excited to the first excited singlet state followed by intersystem crossing to the triplet state. If the energy difference between the singlet and triplet state is small, then it is probable that the singlet state can be repopulated by thermal excitation. Fluorescence emission can then occur from the singlet state. This type of fluorescence is highly temperature-dependent and has a lifetime approximately equal to the phosphorescence lifetime. Anthraquinone and eosin are examples of compounds which show E-type delayed fluorescence (22-24).

Another type of delayed fluorescence has been designated P-type delayed fluorescence. The intensity of this type of fluorescence is proportional to the square of the intensity of the excitation source, which indicates two photons are needed to produce the fluorescence. Two molecules are excited to the singlet state, and each molecule then undergoes intersystem crossing to the triplet states. The triplet states interact with one another to give an excited singlet state and a ground state. The regenerated excited singlet state then gives the delayed fluorescence. This type of fluorescence has been obtained for several aromatic hydrocarbons. A third type of delayed fluorescence can occur when the excited singlet state undergoes photooxidation with the ejection of an electron followed by recombination to regenerate an excited singlet state that emits fluorescence. This type of fluorescence has been labeled recombination fluorescence. Acriflavine shows this type of fluorescence (22-24).

2.9. Heavy-Atom Effects

In molecules with light atoms, the spin angular momenta and

orbital angular momenta of the electrons are coupled and quantized separately (25). Primarily because of this condition, a transition between states of different spin multiplicity may not be observed. Any observable transition must obey the selection rule $\Delta S=0$, where S is the spin quantum number. Nevertheless, in light atoms, there is some interaction between the spin and orbital angular momenta of an electron. As a result, transitions for which $\Delta S \neq 0$ can occur, but with a small probability. This interaction is known as spin-orbit coupling. In molecules containing heavier atoms, strong interactions occur between the spin angular momentum and the orbital angular momentum of the same electron. This causes what is called the internal heavy-atom effect. An external heavy-atom effect can also occur with a heavy atom external to the organic molecule. For example, a weak, transient complex can be formed between 1-chloronaphthalene and ethyl iodide (25).

If the heavy-atom effect is operative, spin-orbit coupling is enhanced, a triplet state acquires some singlet character as a result of mixing with a singlet state, and a singlet state takes on some triplet character. This condition favors a higher probability of intersystem crossing from the lowest excited singlet state to the lowest excited triplet state in a molecule (26). Frequently, one sees a decrease in the fluorescence quantum yield and an increase in the phosphorescence quantum yield when the heavy-atom effect occurs. The higher phosphorescence quantum yield has been used many times in the analysis of organic compounds by phosphorescence to improve the sensitivity of phosphorimetry (27).

2.10. Excited Complexes and Other Phenomena

Solute-solute interactions may occur in the lowest excited singlet state at relatively high concentrations. The formation of a complex of an excited molecule with a ground-state molecule of the same type results in the formation of an excimer. The excimer dissociates to two ground-state singlet molecules accompanied by the emission of fluorescence. The excimer emission occurs at longer wavelengths than normal fluorescence because the excimer is lower in energy than the unassociated excited singlet state. In addition, the intensity of excimer emission is proportional to the square of the concentration of the fluorescent molecule. It is well known that pyrene gives excimer fluorescence. Parker (2) has discussed excimers in some detail.

When excited-state complex formation occurs between two different solute molecules, an exciplex is formed. Froehlich and Wehry (28) have given a detailed discussion of exciplexes. Exciplex formation is usually observed with fluorescence phenomena because diffusion of the excited species is needed for the excited-state complex to form. However, Froehlich and Wehry (28) have also considered triplet exciplexes. Generally, the subject of triplet exciplex formation is more complicated than that of singlet exciplexes, and the observation of emission from triplet exciplexes in fluid media is rare. The literature on the study of triplet exciplexes is rather limited and is usually discussed in terms of excitation of ground-state complexes. Nagakura (29) has reviewed the phosphorescence spectroscopy of excited states of complexes that are stable in their ground-states. Examples of such complexes are between tetrachlorophthalic anhydride (acceptor) and various donors such as toluene, p-xylene, and durene.

The possibility of photochemical reactions in luminescence analysis work is an important consideration. Relatively high energy sources are used in luminescence work, and thus, it is possible to have photochemical reactions by overexposing the sample to the source radiation. It is beyond the scope of this book to consider photochemistry. Turro (8) has given a detailed discussion of photochemistry. In analytical work, erratic results or changes in the excitation or emission spectra suggest the probability of a photochemical reaction.

Luminescence quenching is an important phenomenon that has been discussed extensively in the literature (30,31). Molecular oxygen is a well known quencher of fluorescence and phosphorescence. Quenching is a process by which an electronically excited molecule gives up its energy by interacting with another solute molecule. Bimolecular luminescence quenching is represented by the following kinetic scheme (32)

$$D + h\upsilon \longrightarrow D^* \tag{a}$$

$$D^* \xrightarrow{k_c} D + h\upsilon \tag{b}$$

$$D^* \xrightarrow{k_q} D + \Delta \tag{c}$$

$$D^* + Q \xrightarrow{k_2} D + Q + \Delta \tag{d}$$

Step (a) indicates the excitation of ground state D to form an excited species D^*; step (b) represents the radiative decay of D^* with the emission of either fluorescence or phosphorescence; step (c) shows the first-order and pseudo-first-order quenching of D^* by intra-molecular interactions or solvent and impurity quenching; step (d) shows the bimolecular quenching of D^* by quencher Q. Most of the discussion in the literature centers around fluorescence quenching because, historically, phosphorescence was measured at 77 K where diffusion controlled interactions by a quencher would be minimized.

Noncollisional energy transfer (sensitized luminescence) can take place over distances larger than the contact distances of molecular collision. Because noncollisional energy transfer is not spin-forbidden, singlet-triplet, triplet-singlet, singlet-singlet, and triplet-triplet transfer can occur. Noncollisional energy transfer is thought to arise from vibrational coupling interaction between the excited states of the donor and the acceptor. The triplet-triplet energy transfer between benzophenone as donor and naphthalene as acceptor is an example of noncollisional energy transfer. If benzophenone and naphthalene are in solution at 77 K and excited at 365 nm, benzophenone absorbs the exciting radiation, then undergoes intersystem crossing to its triplet state, transfers its excitation energy to naphthalene, and the phosphorescence of naphthalene is observed. The phosphorescence of naphthalene is not excited by 365-nm radiation (33).

2.11. Environmental Effects on Luminescence

Environmental conditions can alter one or more of the rate constants in excited state processes or alter the energy levels in a luminescent molecule. Becker (1) has discussed several aspects of the environmental effects on fluorescence and phosphorescence. Wehry (30) has summarized several environmental factors that influence fluorescence. Zander (26) has considered low-tempera-ture phosphorimetry and Hurtubise (10) and Vo-Dinh (11) have covered environmental considerations in room-temperature phos-phorescence.

Useful analytical luminescence signals have been reported for molecules in the gas phase, in solution, in the crystalline form, and in various rigid matrices. The gas phase is acceptable for luminescence measurements. Frequently, though, high pressure is needed so sufficient sample can be obtained for luminescence measurements. However, at low pressure in the gas phase, molecular interactions are minimized and unperturbed luminescence spectra can be approximated.

The most common condition for observing fluorescence is the liquid phase at room temperature. The environmental factors affecting fluorescence have been extensively covered in the literature (1-6,8,24,30,34,35). These references can be consulted for details on environmental factors that influence the fluorescence yield. Below, a general discussion is given on the factors that affect phosphorescence yield. Several of these parameters are discussed in detail in later chapters.

Phosphorescence intensities of solutes in liquid solution are generally too low to be useful in analytical work. Donkerbroek et al. (36,37) investigated the room-temperature phosphorescence of various phosphors in solution at room temperature. They concluded that direct solution room-temperature phosphorescence of phosphors with intensities high enough to be of analytical use is rather rare. This is mainly due to the long lifetime of the triplet state, which favors collisional quenching by oxygen and other quenchers in solution at room temperature. Parker and Joyce (38) reported the phosphorescence quantum yields and phosphorescence lifetimes of nine aromatic carbonyl compounds in very pure fluorocarbon solvents at 20°C with low dissolved oxygen. Also, Turro et al. (39) observed the room temperature solution phosphorescence of bromo- and dibromonaphthalene after nitrogen purging of acetonitrile solutions.

Turro (40) has made additional comments about phosphorescence in fluid solution at room temperature. He has commented that if phosphorescence is obtained at 77 K, it can also generally be observed in fluid solution at room temperature, if two conditions are fulfilled. These are (a) impurities capable of quenching triplet states are rigorously excluded; (b) the triplet state does not undergo an activated unimolecular deactivation which has a rate constant greater than or about equal to 10^4 times the rate constant for phosphorescence (k_p). To observe phosphorescence, the quantum yield should be at least 10^{-4}. Turro (40) has pointed out that k_p for a n,π^* triplet state is near 10^2 s^{-1}, and an approximate k_p for a π,π^* triplet state is 10^{-1} s^{-1}. With these k_p values and some additional assumptions, it is possible to calculate the maximum concentration of quencher for the observation of phosphorescence if the quencher is a diffusional quencher. For nonviscous organic solvents where the rate of diffusion is about 10^{10} M^{-1}s^{-1}, the maximum value of quencher that is tolerable is about 10^{-4} M for a n,π* triplet state and 10^{-7} for a π,π* triplet state (40). These results indicate that it should be easier to observe room-temperature phosphorescence from n,π* triplet states in fluid solutions compared to π,π* triplet states in fluid solutions. This has generally been the situation. It is very difficult to attain a concentration of 10^{-7} M quencher, and phosphorescence in fluid solutions from a π,π* triplet state is rarely observed. Even though the phosphorescence signal can be obtained at room temperature in fluid solutions, generally, the signals are not strong enough to be useful analytically except for relatively few compounds. Thus, room-temperature phosphores- cence from fluid solutions is not used very much in analytical chemistry. However, analytically useful room-temperature phosphorescence can be obtained from compounds adsorbed on surfaces, micelle-stabilized systems, and by solution-sensitized phosphorescence (12). These approaches are briefly discussed below and are more fully covered in later chapters.

Reviews and monographs have appeared on analytical solid-surface room-temperature phosphorescence (10-12,41). With this approach, a solution of the phosphor of interest is adsorbed on a solid matrix, the solvent is evaporated, and then the room-temperature phosphorescence of the adsorbed phosphor is measured. No general model has been developed to account for the interactions needed for the room temperature phosphorescence from compounds adsorbed on solid surfaces. For filter paper, it has been proposed that hydrogen bonding of ionic organic molecules to hydroxyl groups of filter paper is the main interaction providing a rigid matrix for solid-surface room-temperature phosphorescence (42). Apparently, the rigid matrix minimizes collisional deactivation of the phosphor molecules in the triplet state.

Cline Love and co-workers (43,44) developed the analytical approach of solution micelle-stabilized room-temperature phosphorescence. Phosphorescence can be observed from several aromatic compounds in fluid solution at room temperature by incorporating the phosphor into a micellar system. In general, a detergent concentration above the critical micelle concentration is used to ensure micelle formation, heavy atoms are used, and the system is deoxygenated with an inert gas. The micelle can organize reactants on a molecular scale and increase the interaction of the phosphor and heavy-metal counterion. Apparently, the net effect is to raise the effective concentration of the heavy metal and increase the probability of spin-orbit coupling (43).

Sensitized phosphorescence in solution at room temperature has been investigated by Donkerbroek et al. (36,37) as an alternative room-temperature phosphorescence technique. Sensitized phosphorescence can be described as follows. After excitation and before radiationless decay of the analyte, the analyte transfers its triplet-state energy to an acceptor molecule and then the acceptor molecule emits phosphorescence. 1,4-Dibromonaphthalene and biacetyl have been investigated as acceptor molecules (36,37).

An important environmental factor in phosphorescence is the heavy-atom effect. As considered in Section 2.9, heavy atoms tend to increase phosphorescence quantum yields. Heavy-atom solvents usually decrease fluorescence quantum yield and increase the efficiency of intersystem crossing. Normally, the external heavy-atom effect increases the rates of both intersystem crossing and phosphorescence. There is some uncertainty about the actual mechanism for external heavy-atom effects (30). For halogen-containing solvents or alkali halide salts, there is substantial evidence that a 1:1 complex is formed between an excited state of the solute and the species containing the heavy atom. The exciplexes formed appear to be charge-transfer in nature. Nevertheless, the previous model does not give an explanation of all the observations from heavy-atom perturbations. The general analytical result from the use of external heavy-atom solvents is usually the enhancement of sensitivity and selectivity in phosphorescence analysis.

There is a variety of other environmental factors that can influence the phosphorescence quantum yield such as temperature, viscosity, and hydrogen bonding effects. Several of these factors will be considered in later chapters.

2.12. n,π* and π,π* States

An important consideration in the luminescence behavior of a compound is the nature of the lowest lying electronic state (45,46). For aromatic molecules excited by ultraviolet or visible light, the molecules usually arrive at the lowest excited singlet state. Some of the molecules can undergo intersystem crossing and arrive at the lowest excited triplet state. These considerations mean that the behavior of a molecule after excitation will not depend primarily on the initial excited state, but rather on the nature of the lowest excited state. For example, if the lowest excited singlet state is a

Table 2.1. Comparison of n, π* and π, π* Singlet States

Property	n,π* State	π,π* States
ϵ_{max}	10-10^3	10^3-10^5
Lifetime	10^{-7} to 10^{-5}	10^{-9} to 10^{-7}
Singlet-triplet split	Small	Generally large
Rate of intersystem crossing	Greater than for fluorescence	Of the same order as fluorescence

Reprinted with permission from D. M. Hercules, Ed., *Fluorescence and Phosphorescence Analysis*, Wiley, New York, 1966.

π,π* state, then the molecule will show the characteristics of a π,π* state. Also, if the lowest excited singlet state is an n,π* state, it will show the characteristics of an n,π* state. Table 2.1 compares some of the properties of n,π* and π,π* singlet states (47). Usually n—>π* transitions are less intense than π—>π* transitions, and therefore n,π* excited singlet states have longer lifetimes. This aspect favors enhanced intersystem crossing from the n,π* state. Also, the smaller singlet-triplet energy gap for the n,π* excited electronic state tends to enhance intersystem crossing. Normally, if the lowest excited singlet state is a n,π* state, intersystem crossing is favored over fluorescence.

Several properties of the intersystem crossing in polynuclear aromatic compounds have been detailed by El-Sayed (48,49) and summarized by Vo-Dinh (50). The rate of intersystem crossing between electronic states with different characteristics such as n,π* and π,π* electronic states is about 1000 times faster than between electronic states with the same properties. These aspects can be summarized as follows (48-50):

Relatively Low Intersystem Crossing Rate
    Singlet(π,π*) ——> Triplet(π,π*)
    Singlet(n,π*) ——> Triplet(n,π*)
Relatively High Intersystem Crossing Rate
    Singlet(n,π*) ——> Triplet(π,π*)
    Singlet(π,π*) ——> Triplet(n,π*)

Aromatic carbonyl compounds frequently have high intersystem crossing rates. For example, benzophenone has an intersystem crossing rate of 10^{11}, and it undergoes $S_1(n,\pi^*)$——>$T_2(\pi,\pi^*)$ intersystem crossing. However, naphthalene has an intersystem crossing rate of 10^6, and it undergoes $S_1(\pi,\pi^*)$——>$T_1(\pi,\pi^*)$ intersystem crossing (51). Table 2.2 gives the criteria that are used for the assignment of triplet states as n,π^* or π,π^* types. It is emphasized that there are several exceptions to the generalizations in both Tables 2.1 and 2.2. Thus, for the identification of a state, it is important that it conform to several of the appropriate criteria for a singlet state or a triplet state given in Table 2.1 or Table 2.2 (52).

2.13. Effects of Molecular Structure on Luminescence Properties

Luminescence properties depend on both structural and environmental factors. The effects of structure on luminescence have been reviewed by several authors (1,6-8,11,24,30,34,35,45). The following discussion will emphasize the phosphorescence of important compound classes. Zander (45) has given considerable detail on the low-temperature phosphorescence properties of compound classes. In addition, it will be assumed that the phosphorescence is from a solution at 77 K. Discussion of room-temperature phosphorescence from various compound classes will be presented in Chapters 6 and 7.

Aromatic hydrocarbons normally yield phosphorescence, and there is a trend to longer wavelength phosphorescence emission as the number of rings increases. Zander (45) and Becker (1) have given a detailed discussion of the phosphorescence characteristics of numerous polycyclic aromatic hydrocarbons.

Generally the phosphorescence lifetimes and quantum yields of aromatic hydrocarbons are altered by halogen- substitution. For example, halogen-substituted products of naphthalene show significant changes in their phosphorescence properties, which are dominated primarily by the internal heavy- atom effect. The introduction of additional halogen atoms into aromatic hydrocarbons results in a red-shift in the phosphorescence (45). The introduction of -OH, -SH, or -NH$_2$ groups to an aromatic hydrocarbon also results in a red-shift of the phosphorescence spectrum.

Aromatic nitro compounds show little or no fluorescence but usually show n,π^* phosphorescence. Lewis and Kasha (46) have

Table 2.2. Criteria for Assignment of Triplet States as $T_{n,\pi*}$ or $T_{\pi,\pi*}$

Property	$T_{\pi,\pi*}$	$T_{n,\pi*}$
Lifetime (sec) of phosphorescence	>1	$<10^{-1}-10^{-2}$
Polarization 0-0 band of phosphorescence	Predominantly out-of-plane	Predominantly in-plane
Vibrational structure phosphorescence	Variable	Prominent CO, NO or C-N-C progressions
Triplet-singlet split	$>3000 \text{ cm}^{-1}$	$\lesssim 2500 \text{ cm}^{-1}$
Intensity $T \longleftarrow S$ absorption	$f \sim 1-30 \times 10^{-9}$	$f \sim 1-7 \times 10^{-7}$
External heavy atom (a) Intensity $T \longleftarrow S$ increase absorption (b) Lifetime of phosphorescence	(a) $\sim$twofold (b) Decrease	Very small Very small
EPR of triplet	ZSF parameter D^*, positive and small	Insufficient data

Reprinted with permission from R. S. Becker, *Theory and Interpretation of Fluorescence and Phosphorescence*, Wiley, New York, 1969.

published phosphorescence data on several aromatic nitro compounds. For nitroanilines and nitronaphthylamines, either fluorescence or phosphorescence is observed.

The majority of aromatic aldehydes and ketones that have been investigated give n,π* phosphorescence. The phosphorescence lifetimes are very short for these compounds, and usually they do not give fluorescence.

Phosphorescence spectra have been obtained for several

aromatic carboxylic acids and acid derivatives such as p-amino-benzoic acid. In addition, similar measurements have been reported for nitriles (45).

As with many compound types, if the lowest excited singlet state is n,π* in character, then fluorescence is absent or the fluorescence quantum yield is small. Quinones are no exception, and if the lowest excited singlet state is a n,π* type, then quinones usually show weak fluorescence. However, quinones that have the lowest triplet state as an n,π type show intense phosphorescence (45).

The phosphorescence of nitrogen heterocycles occurs from either the lowest excited n,π* triplet or the lowest excited π,π* triplet state. An interesting example of how experimental conditions determine the luminescence properties of a nitrogen heterocycle is given by quinoline. In nonpolar solvents, only phosphorescence appears, and no fluorescence can be detected. It has been concluded that the lowest excited singlet state in a nonpolar solvent is an n,π* state. In polar solvents, both fluorescence and phosphorescence are observed for quinoline. This has been attributed to the lowest excited singlet state being a π,π* state in polar solvents (45).

The simple five-membered-ring heterocycles, furan, pyrrole, and thiophene show neither fluorescence nor phosphorescence. Indole, carbazole, and mono- and dibenzocarbazoles give phosphorescence. The phosphorescence of several benzologs of thiophene such as thionaphthene have been studied. Also, the phosphorescence of diphenylene oxide and benzodiphenylene oxides has been measured (45).

A variety of aliphatic compounds has been investigated for phosphorescence over the years. For example, ethylene and its simple derivative do not phosphoresce. However, some aliphatic polyacetylenes yield phosphorescence. Also, various aliphatic aldehydes and ketones have been investigated. Formaldehyde shows no phosphorescence. A weak phosphorescence has been observed for glyoxal (CHOCHO). Biacetyl ($CH_3COCOCH_3$) has been studied extensively and gives intense phosphorescence. The phosphorescence can be observed in the solid, liquid and vapor states. Simple aliphatic ketones such as acetone, methyl ethyl ketone, diethyl ketone, and cyclopentanone phosphoresce from 440-450 nm, and their lifetimes are a few milliseconds or less.

In this section, a general survey of some compound types that give low-temperature phosphorescence has been given. However, numerous drugs, alkaloids, biochemical systems containing potential phosphors, and a variety of other compounds show phosphorescence. Many of these compounds are discussed throughout the book.

References

1. Becker, R.S. *Theory and Interpretation of Fluorescence and Phosphorescence*; Wiley-Interscience: New York, 1969.
2. Parker, C.A. *Photoluminescence of Solutions*; Elsevier: New York, 1968.
3. Hercules, D.M., Ed; *Fluorescence and Phosphorescence Analysis*; Wiley-Interscience: New York, 1966.
4. Wehry, E.L., Ed; *Modern Fluorescence Spectroscopy*; Vol. 1, 1976; Vol. 2, 1976; Vol. 3, 1981; Vol. 4, 1981; Plenum Press: New York.
5. Winefordner, J.D.; Schulman, S.G.; O'Haver, T.C. *Luminescence Spectrometry in Analytical Chemistry*; Wiley Interscience: New York, 1972.
6. Birks, J.B. *Photophysics of Aromatic Molecules*; Wiley Interscience: New York, 1970.
7. McGlynn, S.P.; Azumi, T.; Kinoshita, M. *Molecular Spectroscopy of the Triplet State*; Prentice Hall: Englewood Cliffs: NJ, 1969.
8. Turro, N.J. *Modern Molecular Photochemistry*; Benjamin/Cummings: Menlo Park, CA, 1978.
9. Reference 7, pp 13-17.
10. Hurtubise, R.J. *Solid-Surface Luminescence Analysis: Theory, Instrumentation, Applications*; Marcel Dekker: New York, 1981.
11. Vo-Dinh, T. *Room-Temperature Phosphorimetry for Chemical Analysis*; Wiley Interscience: New York, 1984.
12. Hurtubise, R.J. *Anal. Chem.* 1983, 55, 669A.
13. Demas, J.N.; Crosby, G.A. *J. Phys. Chem.* 1971, 75, 991.
14. Bridges, J.W. In *Standards in Fluorescence Spectrometry*; Miller, J.N., Ed.; Chapman and Hall: New York, 1981; Chapter 8.
15. Demas, J.N. *Excited-State Lifetime Measurements*; Academic Press: New York, 1983.
16. Reference 2, p 88
17. Reference 11, pp 22-24.
18. Schulman, S.G. *Modern Fluorescence Spectroscopy*; Wehry,

E.L., Ed.; Vol. 2; Plenum Press: New York, 1976; Chapter 6.

19. Jackson, G.; Porter, G. *Proc. Roy. Soc., Ser. A* 1961, 260, 13.
20. Richtol, H.H.; Fitch, B.R. *Anal. Chem.* 1974, 46, 1860.
21. Vander Donckt, E. *Prog. React. Kinet.* 1970, 5, 273.
22. Parker, C.A. *Advances in Photochemistry*; Noyes, W.A.; Hammond, G.S.; Pitts, J.N., Eds.; Wiley Interscience: New York, 1964; Vol. 2; pp 305-383.
23. McCarthy, W.J.; Winefordner, J.D. *J. Chem. Educ.* 1967, 44, 136.
24. Seitz, W.R. In *Treatise on Analytical Chemistry*; Elving, P.J.; Meehan, E.J.; Kolthoff, I.M., Eds.; 2nd ed., Part I, Vol. 7, Sec H; Wiley Interscience: New York, 1981; Chapter 4.
25. Reference 7, pp 40-43.
26. Zander, M. *Phosphorimetry*; Academic Press: New York, 1968; pp 20-27.
27. Reference 11, pp 49-58.
28. Froehlich, P.; Wehry, E.L. In *Modern Fluorescence Spectroscopy*; Wehry, E.L., Ed.; Plenum Press: New York, 1976; Vol. 2, Chapter 8.
29. Nagakura, S. In *Excited States*; Lim, E.C., Ed.; Academic Press: New York, 1975; Vol. 2, pp 321-383.
30. Wehry, E.L. In *Practical Fluorescence: Theory, Methods, and Techniques*; Guilbault, G.G., Ed.; Marcel Dekker: New York, 1973; Chapter 3.
31. Eftink, M.R.; Ghiron, C.A. *Anal. Biochem.* 1981, 114, 199.
32. Demas, J.N. *J. Chem. Educ.* 1976, 53, 657.
33. Reference 3, pp 33-36.
34. Hurtubise, R.J. In *Trace Analysis: Spectroscopic Methods for Molecules*, Christian, G.D.; Callis, J.B., Eds.; Wiley Interscience: New York, 1986; Chapter 2.
35. Lakowicz, J.R. *Principles of Fluorescence Spectroscopy*; Plenum Press: New York, 1983.
36. Donkerbroek, J.J.; Gooijer, C.; Velthorst, N.H.; Frei, R.W. *Anal. Chem.* 1982, 54, 891.
37. Donkerbroek, J.J.; Elzas, J.J.; Gooijer, C.; Frei, R.W.; Velthorst, N.H. *Talanta* 1981, 28, 717.
38. Parker, C.A.; Joyce, T.A. *Trans. Faraday Soc.* 1969, 65, 2823.
39. Turro, N.J.; Liu, K.C.; Chow, M.F.; Lee, P. *Photochem. Photobiol.* 1978, 27, 523.
40. Reference 8, pp 129-130.
41. Parker, R.T.; Freedlander, R.S.; Dunlap, R.B. *Anal. Chim.*

Acta 1980, <u>119</u>, 189; 1980, <u>120</u>, 1.

42.   Schulman, E.M.; Parker, R.J. *J. Phys. Chem.* 1977, <u>81</u>, 1932.

43.   Cline Love, L.J.; Skrilec, M.; Habarta, J.G. *Anal. Chem.* 1980, <u>52</u>, 754.

44.   Skrilec, M.; Cline Love, L.J. *Anal. Chem.* 1980, <u>52</u>, 1559.

45.   Reference 26, Chapter 2.

46.   Lewis, G.N.; Kasha, M. *J. Am. Chem. Soc.* 1944, <u>66</u>, 2100.

47.   Reference 3, p 24.

48.   El-Sayed, M.A. *J. Chem. Phys.* 1962, <u>36</u>, 573.

49.   El-Sayed, M.A. *J. Chem. Phys.* 1963, <u>38</u>, 2834.

50.   Reference 11, pp 26-27.

51.   Reference 8, p 186.

52.   Barltrop, J.A.; Coyle, J.D. *Principles of Photochemistry*; Wiley: New York, 1978.

CHAPTER 3

INSTRUMENTATION

3.1. General Aspects

Hurtubise (1) has reviewed the instrumentation for fluorescence and phosphorescence; Vo-Dinh (2) has surveyed the recent instrumentation for room-temperature phosphorimetry; and Hurtubise (3) has discussed the earlier instrumentation for solid-surface room-temperature phosphorescence. To acquire luminescence data, the following basic components are necessary: a source for the exciting radiation, a device for selecting the excitation wavelength, a sample cell, a device for selecting the emission wavelength, and a detector. For phosphorescence measurements, there is a need to distinguish between fluorescence and phosphorescence signals. This is usually accomplished with a phosphoroscope or a pulsed-source gated-detector system. Figure 3.1 shows a general schematic for luminescence instrumentation.

In many commercial and research luminescence instruments, the components employed consist of the same components that are used in ultraviolet-visible spectrophotometers. Thus, there are some similarities in the instrumental design of absorption spectrophotometers and luminescence spectrophotometers. However, there are also major differences in instrumental design of the two types of instruments. In comparing luminescence measurements with absorption measurements, it is important to realize that the intensity impinging on the detector normally is much lower for luminescence measurements for several reasons (4):

1. Analytical luminescence measurements are normally made at low absorbances (A < 0.05).
2. The fraction of the emitted luminescence impinging on the detector system is usually quite small.

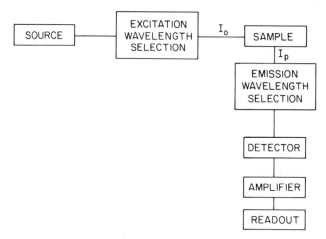

Figure 3.1. General schematic for phosphorescence instrumentation. Normally, a phosphoroscope or a pulsed-source gated-detector system is needed to measure phosphorescence in the absence of fluorescence. (See Sections 3.6 and 3.7)

3.   To minimize spurious optical effects, luminescence instruments are frequently designed to view a narrow interval of sample in the middle of the cell rather than viewing the entire length of the sample cell.

4.   Frequently the fluorescence or phosphorescence quantum efficiency is considerably less than one; thus, luminescence instrumentation is designed to maximize instrumental factors that affect the signal.   One of the main objectives in instrumental design is to obtain the highest possible signal-to-noise ratio.

An important consideration that affects luminescence instrumentation and measurements is that many of the instrumental components affecting the measured luminescence signals are wavelength dependent. To obtain the true excitation spectrum and emission spectrum of a luminescent component, it is necessary to correct for the instrumental wavelength dependence.

3.2.   Sources

3.2.1. Conventional Sources

The xenon arc lamp (gas discharge lamp) and high-pressure mercury lamp are still the most widely used sources in luminescence instrumentation. A high-pressure mercury lamp is important in situations where a luminescent species can be excited at one of the mercury emission lines. The xenon arc lamp is more frequently used because of the continuum output of this type of lamp (1).

Other types of lamps are used to excite luminescence. For example, incandescent lamps and low-pressure mercury-vapor lamps have been used (5). Tungsten lamps (incandescent) are inexpensive and stable and do not require elaborate power supplies. The primary disadvantage of incandescent lamps is the limited useful output range. A tungsten lamp covers the visible range and has no output below 300 nm. Therefore, a tungsten lamp is useful only for excitation in the visible range. Because many compounds are excited in the ultraviolet range, the tungsten lamp has found little use in luminescence work. The low-pressure mercury-vapor lamp is used to some extent in luminescence analysis. It does not require a sophisticated power supply and is very stable. The mercury-vapor lamp gives intense emission at 253.7, 296.5, 302.2, 312.2, 313.3, 365.0, 404.7, 435.8, 546.1, 577.0, and 579.0 nm (6). A low-pressure mercury-vapor lamp is used with filter fluorometers because specific lines can be used to efficiently excite samples. It is also used as a source to calibrate monochromators. Because of the line emission, the mercury-vapor lamp is usually not employed with scanning luminescence spectrometers.

Because fluorescence and phosphorescence intensity is proportional to source intensity, the stabilization of the excitation source is an important consideration. Source intensity fluctuations occur because of power supply variation, changes in arc or discharge conditions, and geometry or sputtering of the quartz envelope with electrode material (7). Techniques for source stabilization have been discussed by Hamilton et al. (7).

3.2.2. Lasers

The majority of the laser-excited luminescence spectrometry studies of analytical importance have dealt with fluorometry (8-18). However, excitation with lasers has also been applied to phosphorimetry (10). Vo-Dinh (19) has considered the general properties of various types of lasers. In addition, the investigations in which lasers have been used in fluorometry can be very helpful in accessing their usefulness in phosphorimetry.

The theoretical capability of a xenon arc lamp compared to a typical laser for excitation in fluorometry has been discussed by Wright (17). Lytle (16) has further expanded the comparison developed by Wright and various aspects of laser-excited molecular fluorescence of solutions such as limit of detection. A 150-W xenon lamp can deliver 10 µW at 500 ± 5 nm (2.5×10^{13} photons s^{-1}) to a monochromator slit. If the throughput efficiency is 30%, 7.5×10^{12} photon s^{-1} will be irradiating the sample (16). Lytle assumed that each of the photons absorbed would be reemitted at the specified fluorescence wavelength, the emission monochromator passed 10% of the signal, and there was a lens collection efficiency of 5%, a monochromator throughput of 30%, and a photomultiplier quantum yield of 20%. With these conditions, 2.3×10^8 counts s^{-1} would be produced at a ratemeter. For an absorptance (the ratio of radiant energy absorbed by a sample to that incident upon it) of 1.7×10^{-9}, 4 counts s^{-1} would be produced for a S/N of 2. If it is assumed that the path length is 1 cm and the molar absorptivity is 10^5, then this would correspond to a 7.4×10^{-15} M solution. In general, a tunable laser is more efficient than a xenon arc lamp because a laser linewidth is normally narrower than the absorption linewidth, and all the photons that are incident upon the sample have the potential of being absorbed. As an example, a continuous-wave laser can deliver 1 W of 555-nm ($\sim 2.5 \times 10^{18}$ photons s^{-1}) radiation to an absorbing sample in solution (16). Using the same conditions as discussed for a xenon lamp, the fluorescent solution would be 2.3×10^{-20} M. Such a great improvement in a laser compared to a xenon lamp is difficult to realize experimentally. In addition, the calculated concentrations for both the xenon arc lamp and laser source are rarely attained experimentally. This results mainly because of experimental blank limitations. Boutilier et al. (20) have derived steady-state molecular luminescence radiance expressions assuming narrowband excitation. Also, Omenetto and Winefordner (10) have considered several aspects of limits of detection with lasers for absorption, fluorescence, phosphorescence, and Raman spectrometry. In addition, Winefordner and Rutledge (21) have calculated detection limits for molecular luminescence with laser excitation.

Most likely, lasers will not displace arc lamps as sources for routine analysis in the near future (4). Nevertheless, the important advantages of lasers compared to mercury and xenon arc lamps are greater intensity, degree of monochromaticity, and spatial coherence. Pulsed lasers yield very short pulses and are good sources for lifetime measurements. The most widely used lasers for fluorescence excitation are the nitrogen laser (337.1 nm), the argon ion laser (either the 488.0- or 514.5-nm line), and tunable dye lasers.

Boutilier and Winefordner (22,23) studied the external heavy-atom effect using time-resolved laser-excited low-temperature phosphorimetry. The major conclusions they made concerning laser-excited time-resolved phosphorimetry are given below (22).

1. The pulsed nitrogen laser is an excellent excitation source for phosphors that have a molar absorptivity as small as 10 at 337 nm. This is so because of the high peak power and excellent pulse-to-pulse reproducibility of the pulsed nitrogen laser.
2. The flashlamp pumped dye laser requires a ratio system to compensate for pulse-to-pulse variation, dye decomposition, and frequency-doubling crystal drift for it to be of analytical use in phosphorimetry.
3. The major noise sources were associated with immersion cooling in liquid nitrogen. Thus, it is possible that the noise from these sources could be reduced by conduction cooling.

3.3. Wavelength Selection

The optical components used for wavelength selection in luminescence work are practically the same as those used in ultraviolet-visible absorption analysis. A substantial amount has been written about the properties and characteristics of the optical components used in both absorption and luminescence analysis (24-27). Thus, in the next two sections on filters and monochromators, only a general overview will be given on these wavelength selection devices.

3.3.1. Filters

The use of filters is the easiest means of selecting excitation and emission wavelengths. However, filters are much more widely used for fluorescence measurements than for phosphorescence measurements. Normally, fluorescence is measured in solution at room temperature and phosphorescence is usually very weak or nonexistent under these conditions. Thus, it is relatively straightforward to routinely measure fluorescence using filters without interference from phosphorescence. However, to measure phosphorescence without interference from fluorescence, a phosphoroscope or a pulsed-source gated-detector system is needed to obtain phosphorescence signals without interference from fluorescence. These approaches are needed for both room-temperature and low-temperature phosphorescence measurements. Filters can be used in the measurement of phosphorescence, but a device such as a phosphoroscope is needed in the instrumentation for the measurement step to minimize interference from any fluorescence signal.

Figure 1 shows where the wavelength selection components would appear in a luminescence instrument. Filters can select a very narrow ($\leq$ 1-nm) or very large ($\sim$ 100-nm) bandpass with peak transmissions up to 90%. Frequently, filters are used to minimize the stray background from grating monochromators. Both interference and absorption filters can be used in wavelength selection. Guilbault (27) has given a detailed listing of filters used in luminescence work. Generally absorption filters are less expensive than interference filters and are employed in many luminescence applications.

3.3.2. Monochromators

A monochromator produces a beam of radiation of narrow bandwidth while permitting the wavelength to be varied. The fundamental elements of a monochromator are an entrance slit, a dispersing element, and an exit slit. The dispersing element is either a grating or a prism. Prisms are not used very much in luminescence instruments because their greatest dispersion is in the ultraviolet. Most luminescence, and especially phosphorescence measurements, are made in the visible region. Also, grating mono-chromators give a higher light flux throughput than prism monochromators for a given resolution. With gratings, dispersion changes only slightly with wavelength, and they are less expensive than prisms. There is usually a tradeoff between light throughput and resolving power for any monochromator. For example, decreasing slit width enhances resolution, but at the expense of source intensity (1).

In the majority of commercial luminescence spectrophotom-eters, two monochromators are used, one for excitation and one for emission measurements. Because gratings are the most widely used dispersing elements in luminescence instruments, the properties, advantages, and disadvantages will be considered briefly.

There are two commonly used monochromator mountings in commercial luminescence monochromators. These are the Ebert mounting and the Czerny-Turner mounting (1). With these mountings, the grating is rotated to change wavelengths while the angle between the incident and diffracted rays remains constant.

Blazed reflection-type gratings are usually employed in luminescence instruments. The grooves in the grating are aluminized and have a precise shape so that the maximum amount of radiation is diffracted at a particular angle. For example, if the grating is blazed at 450 nm, its maximum output is at 450 nm. A grating's efficiency will diminish as the wavelength changes from the wavelength at which it has been blazed.

The advantages of gratings can be summarized as follows: (a) gratings have uniform resolution and linear dispersion at all wavelengths; (b) up to 50% of the incident radiation can be diffracted into the first order with a blazed grating; (c) gratings are less expensive than prisms. The major disadvantage of a grating is that several orders of spectra are obtained. One way of minimizing this problem is to use filters in the optical path. As an example, to observe the 600-nm spectral line without interference from the second-order 300-nm spectral line, a filter cutting off radiation below 400 nm should be used (5).

3.4. Slits and Cell Position

The type of slits and their widths are important parameters in determining the resolution of a luminescence instrument. If the entrance and exit slits are of equal width, the distribution of energy as a function of wavelength for radiation passing through the exit slit of a monochromator can be represented as an isosceles triangle. The wavelength at peak transmittance is defined as the nominal wavelength and is the value read on the dial or digital readout of a luminescence spectrometer. The bandpass in terms of wavelengths is the bandwidth at one-half the peak transmittance and is considered equilvalent to the width of the exit slit. With grating instruments, the bandpass for a particular slit width is constant throughout the spectrum and depends on the number of rulings in the grating (28). The relationship between resolution and slit width has been discussed in the literature (5,24,28). As general guidelines, the smaller the slit width, the better the resolution, and conversely, the larger the slit width, the greater the sensitivity.

The most common cell position is one in which there is a 90° angle between the source and detector with the cell at the 90° position. This arrangement provides for minimal background interference from source radiation.

3.5.   Detectors and Detector Systems

3.5.1. Photomultiplier Tubes

Photomultiplier tubes are extensively used for the detection of both fluorescence and phosphorescence. One of the most important characteristics of a photomultiplier tube is its relative sensitivity as a function of wavelength. The sensitivity depends mainly on the composition of the photocathode. Materials such as Cs-I, Cs-Te, Sb-Cs, and multialkali have been used in preparing photocathodes. The large change in efficiency for some photomultiplier tubes over certain wavelength regions is an important factor causing recorded luminescence spectra to differ from true spectra and to vary from instrument to instrument.

The time response of detectors is important in measuring lifetimes, time-dependent spectra, or in pulse counting. The response time of a photomultiplier is about 10^{-8} s which is sufficiently fast for the measurement of all but the fastest luminescence decay times (24). In complete darkness, there will still be some anode current in a photomultiplier tube. This is called dark current and arises for several reasons. Some of these are field emission from sharp points on electrodes, radioactivity, cosmic rays, but most importantly from thermionic emission from the photocathode. Thermionic emission has an approximately exponential dependence on cooling. Thus, cooling can minimize this contribution greatly. Two important sources of noise in a photomultiplier tube are the statistical probability of emission of a photoelectron from a photocathode and the fluctuation of secondary emission at the dynodes. There is a further noise component due to the random arrival of photons at the photocathode. Hamilton et al. (29) and Winefordner et al. (24) have considered these various noise components in some detail.

3.5.2. Photon Counting

Photon counting is used for the measurement of low light

intensities. In photon counting, the photoelectron pulses at the anode of a photomultiplier tube are counted with a high speed electronic counter. On the average, each photoelectron pulse contains L electrons for a total charge of eL coulombs. Assuming L is approximately 10^6, the resulting charge is large enough to cause an observable voltage pulse across the phototube load resistor R_L. One pulse occurs for each photoelectron ejected at the cathode. Therefore, the average pulse-count rate is proportional to the light level. This assumes that the cathode sensitivity is constant. The photoelectron pulses are usually around several millivolts and can be amplified by a wide band AC amplifier and counted by electronic counter circuits. Several of the advantages of photon counting are (24): (a) digital systems generally drift less than analog systems; (b) at very low light levels, long counting times may be used to accumulate the total count; (c) several types of noise are discriminated against; (d) digital readouts are easy to read and can be interfaced directly to computers.

One disadvantage of photon-counting detection is that the gain of the photomultiplier tube cannot be varied by changes in applied voltage. Also, photon counting has a somewhat limited range of intensity over which the count rate is linear. Generally, photon-counting detection is usually inconvenient with high signal levels. For example, to stay within the linear range, slit widths must be adjusted or fluorescence intensity adjusted using neutral density filters. Also, the signal-to-noise ratio becomes unsatisfactory at count rates below 10,000 photons per second (24).

Several of the photon-counting systems that have been developed for fluorescence use can be adapted to phosphorescence analysis. Franklin et al. (30) have described a relatively simple, inexpensive photon-counting system that can be connected to photomultiplier tubes. Several fundamental instrumental aspects are considered. Jameson et al. (31) discussed the construction and performance of a scanning photon-counting spectrofluorometer. Koester and Dowben (32) considered a subnanosecond single photon-counting spectrofluorometer with synchronously-pumped tunable dye laser excitation. Cova et al. (33) have given a careful analysis of the requirements for microspectrofluorometric measurements when measuring the fluorescence emitted by single cells. A fully digital instrument using single-photon detection and a multichannel analyzer was discussed. Darland et al. (34) presented results for the optimum measurement system parameters for several photon-counting systems. They concluded that two relatively simple measurements, namely, pulse height distributions and linearity studies, could provide important information about the optimum operating parameters for pulse-counting measurement systems. Later, Darland et al. (35) described a method for calculating the relative efficiency of pulse-counting experiments. Meade (36) reviewed the practical aspects of photon counting. Problems associated with optimizing the detector, amplifier, and discriminator were considered. Also, a number of counting configurations found in commercially available equipment were discussed.

3.5.3. Optoelectric Imaging Detectors

The use of image or array detectors in luminescence spectrometry is a somewhat recent development. Most of these types of detectors have been used in fluorescence work. Several different types of image detectors are commercially available (37-41). Christian et al. (42) have given a comprehensive discussion of the properties of array detectors. The major advantage of the imaging detectors is the potentially large multichannel advantage they present. For example, if equal total observation times are used, the signal-to-noise ratio obtained by using N detectors to simultaneously observe N channels is $N^{\frac{1}{2}}$ greater than that obtained by using a single detector to observe the N channels sequentially. In addition, the time required for N detectors simultaneously observing N channels to obtain a given signal-to-noise ratio is a factor of N less than that required for a single detector observing N channels sequentially. Normally the full multichannel advantage is not achieved because of nonideal conditions (42).

The most widely used low-light-level multichannel detector is the Silicon Intensified Target (SIT) vidicon. The main part of this detector is a 16-mm-diameter silicon target. The target consists of a 1000 x 1000 array of islands of p-type silicon deposited on an n-type silicon wafer. Each of the islands forms a diode whose cathode is addressed by a scanning electron beam. The anode is connected in common with all of the other diodes. In operation, the target is continuously scanned so that each diode is sequentially back-biased, and its capacitance is charged.

The spectral response of the most common SIT vidicon (RCA 4804) is defined in the visible and near-IR regions by the composition of the photocathode. Extension into the ultraviolet region is limited by the transmission of the fiber-optic faceplate. This faceplate has a sharp cutoff at 380 nm. The response may be extended into the ultraviolet region by special techniques (42). The photometric properties of the SIT vidicon are very good when operated in the continuous scanning mode. Normally, the linearity is comparable to a photomultiplier tube. The dynamic range is limited at the high end by the capacitance of the photodiodes and at the low end by preamplifier noise. The SIT vidicon is an energy detector rather than a power detector like a photomultiplier tube. Because of this property, it can be operated in an integration mode for enhancing the signal-to-noise ratio.

Vo-Dinh et al. (43) replaced a photomultiplier detection system with a commercial SIT-optical multichannel analyzer (OMA) in a commercial spectrophotofluorometer. They considered the SIT-OMA response versus measurement time, resolving power of the SIT system, analytical calibration curves, and limits of detection. It was concluded that a SIT image detector tube with a commercial spectrofluorometer was very suitable to analytical applications in molecular spectrometry.

Goeringer and Pardue (44) described the development of a SIT vidicon camera system for phosphorescence studies and its application to room-temperature phosphorescence of salts of organic acids deposited on filter paper. The instrument permitted time resolved spectra to be recorded with a minimum scan time of 8 ms. (See Section 3.6.)

Cooney et al. (45) made a comprehensive study of image devices versus photomultiplier detectors in atomic and molecular luminescence spectrometry by signal-to-noise ratio calculations. They compared the signal-to-noise ratios of several image devices (Si-vidicon (V), SIT, intensified SIT (ISIT), secondary electron conduction vidicon (SEC), and image dissector (ID)) with each other and with several sensitive photomultiplier tubes commonly used in optical spectroscopy. Information was given for several hypothetical experimental situations in both atomic and molecular luminescence spectrometry. For the detectors considered, the parameters of internal gain, efficiency in the visible and ultraviolet regions, dark count rate, and area per channel (image detectors) were discussed.

Some of the conclusions regarding molecular luminescence spectrometry are given below.

1. Because the signal-to-noise ratios of image devices (ISIT,

SIT, and SEC) are close to the signal-to-noise ratios of photomultiplier tubes, for the case of photon noise limitation, the image devices possess a time advantage because they record many spectral components simultaneously.

2.   For quantitative analysis, similar detection limits should be obtained with SIT, ISIT, SEC, or photomultiplier tube.

3.   The integrating image devices (ISIT, SIT, SEC, and V) have substantial analytical potential.

Christian et al. (42) have considered one of the conclusions made by Cooney et al. (46) concerning the lower detection limit for a photomultiplier tube as a single-channel detector with sequential linear scan compared to the SIT as a parallel detector. Cooney et al. (46) apparently did not discuss the sensitivity variation from channel to channel, which is a factor to be considered in determining limit of detection.

Talmi (39) has discussed, in detail, the state-of-the-art of self-scanned photodiode arrays used as a spectrometric multi-channel detector for molecular absorption and molecular fluorescence. He compared the fluorescence emission spectra of fluorene measured with a 1024-element silicon photodiode (SPD) linear array, SIT, ISPD, and a photomultiplier tube (R375). He concluded the ISPD was the multichannel detector of choice because of its photomultiplier tube-like superior gain and its adequate UV spectral response. The ISPD is comprised of a SPD mechanically attached by an optical-fiber coupler to a microchannel plate image intensifier. Unfortunately the spectral resolution of the ISPD is substantially worse than the SPD. Recently, Ingle and Ryan (47) have discussed luminescence measurements with an intensified diode array.

In summary, image detectors are now widely used in luminescence analysis; however, the photomultiplier tube will continue to be used extensively. Because of the wide variety of image detectors available, it is essential for researchers not familiar with the properties of these detectors to consult the literature on image detectors about specific properties and possible applications.

3.6.   Rotating-Can Phosphoroscope

Phosphorescence can be measured in the absence of fluorescence with a minimum of scattered radiation by using a phosphoroscope. A phosphoroscope is a device which achieves this by allowing the measurement of phosphorescence while the sample is being excited out of phase with the measurement step. Several years ago the effects of phosphoroscope design in the measurement of phosphorescence intensity were discussed by O'Haver and Winefordner (48). They derived mathematical expressions which related the measured and instantaneous phosphorescence intensities to the decay time of the phosphor and also to the characteristic parameters of the phosphoroscope used in the phosphorescence measurement system. Two types of phosphoroscopes were considered: (a) the Becquerel phosphoroscope, in which the sample cell is placed between two circular rotating disks in a straight through arrangement; (b) the Aminco-Keirs phosphoroscope, which is widely used, in which the sample cell is placed in the middle of a rotating cylinder with two diametrically opposite apertures. With this arrangement, the emitted phosphorescence is collected at a right angle to the excitation radiation (49). In this section, only the Aminco-Keirs phosphoroscope will be considered because it is more extensively used than the Becquerel phosphoroscope. Other phosphoroscopes that have been developed for solid-surface room-temperature phosphorescence will be considered in Chapter 6.

In the rotating-can phosphoroscope shown in Figure 3.2, radiation from the excitation monochromator passes through the excitation shutter aperture and impinges onto the rotating can. When the phosphoroscope is positioned so that one of the openings is aligned with the excitation shutter aperture, the excitation radiation excites the phosphor in the sample cell. At the same instant, the emission shutter aperture is blocked by the phosphoroscope can, and thus, prompt fluorescence and scattered excitation radiation will not be detected. As the phosphoroscope can continues to rotate, the excitation radiation is blocked by the rotating can. Fluorescence and scattered excitation radiation then decay very rapidly, and only phosphorescence remains. It is assumed that the sample does not give any delayed fluorescence. As the phosphoroscope can continues to rotate, its other opening becomes aligned with the emission shutter aperture, and then the phosphorescence radiation goes to the emission monochromator and photodetector. The general dimensions of the phosphoroscope can are shown in the insert in Figure 3.2. The arc lengths D and L refer to the closed and open portions of the phosphoroscope can, respectively. The excitation and emission shutter apertures are assumed to have the same width, given by W in the insert in Figure 3.2.

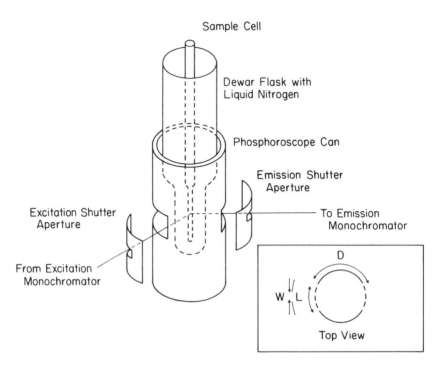

Figure 3.2. Schematic diagram of a rotating-can phosphoroscope. (Reprinted with permission from T.C. O'Haver and J.D. Winefordner, *Anal. Chem.*, <u>38</u>, 602. Copyright 1966 American Chemical Society.)

Figure 3.3 shows the operation of a phosphoroscope in considerable detail (48). In considering Figure 3.3 several assumptions are made: (a) the excitation radiation intensity is assumed to be proportional to the extent of opening of the excitation shutter aperture; (b) the fraction of the instantaneous phosphorescence radiation striking the emission shutter aperture and received by the photodetector is assumed to be proportional to the extent of the opening of the emission shutter aperture; (c) the magnitude of the openings of both the excitation and emission shutter apertures is assumed to vary linearly with time during opening and closing.

The events that occur during one cycle of excitation and observation are discussed below. Considering Figure 3.3A, from left to right, as the excitation shutter aperture starts to open, the excitation intensity at the sample begins to increase. It climbs to a maximum fractional value of 1 by the end of the period t_t, the shutter transit time, which is the time required for the edge of the phosphoroscope window to move past the shutter aperture of width, W. The phosphorescence intensity increases to some extent during

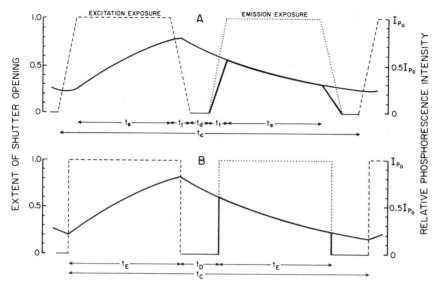

Figure 3.3. Graphical representation of operation of a phosphoroscope.

    A. True representation.

    B. Approximate representation.

Key to Symbols and Lines on Diagram. (See text for discussion of symbols.)

    t_e = exposure time

    t_d = shutter delay time

    t_t = shutter transit time

    t_C = time for one cycle of excitation and observation

    $t_E = t_e + t_t$

    $t_D = t_d + t_t$

Line - - - represents extent of opening of excitation shutter apertures according to left-hand vertical coordinate. The value zero means completely closed; 1/2 means half-way open; 1.0 means completely open, etc.

Line . . . . represents the extent of opening of the emission shutter aperture according to left-hand vertical coordinate.

Line —— represents the instantaneous phosphorescence intensity of the sample of any time according to the right-hand coordinate.

I_{po} = phosphorescence intensity which would be measured if the excitation and emission shutter apertures were opened for a length of time long enough to allow the phosphorescence intensity to reach a steady state value.

(Reprinted with permission from T.C. O'Haver and J.D. Winefordner, *Anal. Chem.*, **38**, 602. Copyright 1966 American Chemical Society.)

this time. The shutter aperture remains completely open for time t_e, the exposure time. During this time period, the phosphorescence intensity grows exponentially toward the value I_{po}. At the end of the exposure time, the shutter aperture starts to close, the excitation intensity impinging on the sample decreases, and the phosphorescence signal begins to decay. After the transit time, t_t, the shutter aperture is closed, and the phosphorescence decay becomes exponential. During the delay time, t_d, both shutter apertures are closed to reduce interference from fluorescence and scattered incident radiation. That is, t_d is the time required for the trailing edge of the phosphoroscope window to travel from the point where excitation is just ended to the point where the leading edge of the phosphoroscope just reaches the front edge of the emission shutter aperture. The phosphorescence continues to decay through the delay time and on until the beginning of the next excitation period. After the shutter delay time, t_d, the emission shutter aperture begins to open. This allows the phosphorescence radiation to pass onto the photodetector. The heavy solid line in Figure 3.3 represents the phosphorescence intensity, which passes through the emission shutter aperture and is detected by the photodetector. The phosphorescence intensity increases during the opening transit time, decays exponentially during t_e, and decreases to zero during the closing transit time. The emission shutter is completely closed before the next excitation begins. As indicated in Figure 3.3, the time for one complete cycle of excitation and observation is t_c. The solid line (phosphorescence intensity) in Figure 3.3A is the steady-state intensity, which results when an adequate number of cycles has elapsed so that the net phosphorescence signal does not change from cycle to cycle. The observed signal will be proportional to the area within the heavy solid line in Figure 3.3A. This is the integrated phosphorescence intensity observed in one cycle, P_p. This is only a fraction of what could be observed if both shutter apertures were continuously open. Under this condition, the phosphorescence intensity would reach a steady value of I_{po}, and the integrated phosphorescence intensity detected per cycle would be P_c. Normally it is not possible to have both shutter apertures open at the same time because of interference from fluorescence and scattered source radiation. Thus, when phosphorescence is measured and when a phosphoroscope is used in the measurement step, there will be a loss in the observed phosphorescence signal. The ratio of P_p and P_c is dependent upon the values of the times τ, t_e, t_t, t_d, and t_c. The term τ is phosphorescence lifetime. O'Haver and Winefordner (48) have derived equations relating the previous terms and have discussed the equations in detail.

Equation 3.1 gives the ratio P_p/P_c (48,50).

$$P_p/P_c = \alpha = \frac{\tau\exp(-t_D/\tau)[1 - \exp(-t_E/\tau)]^2}{t_c[1 - \exp(-t_c/\tau)]} \qquad (3.1)$$

The terms in Equation 3.1 are defined in Figure 3.3 or were discussed in the text. Equation 3.1 can be simplified for two limiting conditions.

Condition 1. If τ/t_c and τ/t_E are much less than one, then Equation 3.1 simplifies to

$$\alpha \cong \tau/t_c\exp(-t_D/\tau) \qquad (3.2)$$

Condition 2. If τ/t_c and if τ/t_E are greater than 100, then Equation 3.1 becomes

$$\alpha \cong (t_E/t_c)^2 \qquad (3.3)$$

For condition 2, t_E must be less than one-half of t_c, and thus, α must be less than 0.25. For shorter decay times, or slower phosphoroscope speeds, when $\tau < t_c$, α begins to decrease. For an Aminco phosphoroscope, this occurs at about $\tau = 10^{-3}$ s, operating at a maximum speed of 7000 rpm. Normally, the phosphoroscope is operated fast enough so that $\tau \gg t_c$ so undesirable changes in the phosphoroscope speed or alterations in the sample phosphorescence lifetime will have little effect on α.

Langouet (51) discussed the use of disk phosphoroscopes using operational calculus, and Hollifield and Winefordner (52) described a modular phosphorimeter with a single-disk phosphoroscope. Yen et al. (53) constructed an analog switch phosphoroscope. They used it for the measurement of phosphorescence from several organic compounds at 77 K. The system employed a quad analog switch to gate the input of the detector system as in a boxcar integrator. The device was relatively easy to construct and inexpensive. The analog switch phosphoroscope was comparable to the rotating can phosphoroscope in rejecting scattered or stray radiation and fluorescence. Also, it may be used at higher chopping frequencies than the frequencies used with commercial phosphoroscopes. In the next section, the more advanced pulsed-source time-resolved phosphorimeter is considered.

3.7.   Time-Resolved Phosphorimeters and Other Phosphorimeters

Pulsed-source time-resolved phosphorimetry is a useful approach for analyzing mixtures of fast-decaying phosphors. Some of the fundamental aspects of pulsed-source phosphorimetry were established theoretically by O'Haver and Winefordner (48), and Winefordner (54) discussed the experimental usefulness of pulsed-source gated-detector instrumentation. A relatively sophisticated pulsed-source phosphorimeter was reported by Fisher and Winefordner (55).

The fundamental equations derived by O'Haver and Winefordner (48) for mechanical choppers were expanded by O'Haver and Winefordner (56) to apply to systems that used pulsed flash tubes for excitation and pulsed photomultiplier tubes for the

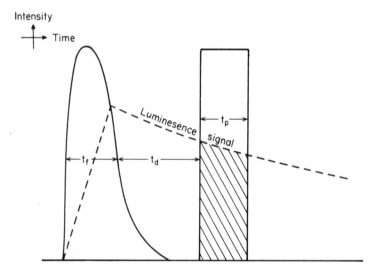

Figure 3.4. Schematic diagram of events occurring during one cycle of sample excitation and observation in a pulsed-source phosphorimeter system: (t_f) half-intensity width of the source flash; (t_d) delay time after the end of the excitation pulse (assuming excitation pulse to be retangular) to the beginning of observation of the phosphorescence signal "on" time of the detector or the read-out system; (t_p) "on" time of detector or read-out system; dashed line represents buildup and decay of phosphorescence; solid white curve represents flash intensity temporal distribution; rectangular pulse represents "on" time of detector or read-out system; yellow area represents measured integrated luminescence signal per source pulse. (Reprinted with permission from R.P. Fisher and J.D. Winefordner, *Anal. Chem.*, 44, 948. Copyright 1972 American Chemical Society.)

observation of phosphorescence.   The equations are useful for evaluating pulsed systems for analytical applications, for predicting optimum instrumental conditions, and for comparing pulsed systems with mechanically chopped systems.

Figure 3.4 illustrates one cycle of sample excitation and observation for a pulsed-source gated-detector phosphorimeter system (55). After an initial pulse of source energy, with a duration t_f, the phosphorescence intensity climbs to a maximum value and then decays exponentially. At a delay time, t_d, after the source flash has decayed substantially, the photomultiplier detector is turned on, and the phosphorescence signal is monitored. The photomultiplier detector is then turned off, and the sequence is repeated.   The integrated luminescence intensity is measured during the "on" time, t_p, of the phototube. The expression for integrated phosphorescence intensity I is given in Table 3.1. The term I is directly proportional to f, the source-pulsed repetition frequency, in the case of a d.c.

Table 3.1. Theoretical Expressions for Pulsed-Source Phosphorimetry

Integrated phosphorescence intensity	$I \ =$	$\dfrac{I_0 f\, t_f\, [\exp(-t_d/\tau)]\,[1 - \exp(t_p/\tau)]}{[1 - \exp(-1/f\tau)]}$
Binary system	$I_{tA} \ =$	$I_{0A}\exp(-t_d/\tau_A)$
	$I_{tB} \ =$	$I_{0B}\exp(-t_d/\tau_{B)}$
	$I_{tT} \ =$	$I_{tA} + I_{tB}$
Multicomponent system	$I_{tT} \ =$	$\displaystyle\sum_i I_{ti}\exp(-t_d/\tau_i)$

I = integrated phosphorescence intensity, arbitrary units; I_0 = steady-state integrated intensity, arbitrary units; f = source-pulse repetition frequency, Hz; t_f = flash duration halfwidth, sec; t_d = delay time, sec; t_p = "on" time of detector or read-out system, sec; τ = phosphorescence decay time, sec; I_{tA} and I_{tB} = phosphorescence intensity of molecules A and B, respectively, at a delay time t, arbitrary units; I_{0A} and I_{0B} = phos-phorescence intensities of molecules A and B, respectively, at time t = 0, arbitrary units; I_{tT} = total phosphorescence intensity of a mixture of i components, at a delay time t, arbitrary units. (Reprinted with permission from J. J. Aaron and J. D. Winefordner, *Talanta* 1975, 22, 707.)

read-out system (49,54,56). General expressions for binary and multicomponent systems of phosphors are also given in Table 3.1. In general, with the equations in Table 3.1, three methods may be used to evaluate the phosphor concentrations by pulsed-source phosphorimetry (49,55): (a) Multiple analytical curve method. With this approach, the dependence of the slopes of the analytical curves on delay time t_d is employed to determine the analytical concentrations in binary phosphor mixtures. (b) Exponential method. The decay times τ_a, τ_b...τ_i of the phosphors of a mixture are measured by obtaining the phosphorescence signal I_a, I_b...I_i of each component as a function of the delay time t_d. The main advantage of this method is that it may be readily applied to systems with more than two components. (c) Logarithmic decay time method. In this method, a semilogarithmic plot of phosphorescence signal versus time is employed to determine the phosphor concentration in a multicomponent mixture. The calculations are simpler with this approach than in the other two methods, but the accuracy for the analytical concentration of the phosphor is not as good as the other methods. Harbaugh et al. (57) used a xenon flashtube source in the application of pulsed-source phosphorimetry to the quantitative and qualitative analysis of drugs.

Figure 3.5 gives a block diagram of the first pulsed-source phosphorimeter designed by Fisher and Winefordner (55). The source they used was a short-arc high-pressure xenon flashtube pulsed by a trigger module. Read-out devices for pulsed-source systems are a little more complicated than the d.c. systems employed in commercial phosphorimeters. There are primarily three measurement systems (49): (a) pulsing of the photomultiplier tube, which is turned off and on at different moments of the cycle, and an integrating d.c. measurement system, which measures the signal during the on-time of the photomultiplier tube; (b) the photomultiplier tube is operated continuously, with an electronic gate (boxcar integrator) that detects the phosphorescence signal; (c) the photomultiplier tube is operated continuously as in method (b), but with a fast multichannel read-out device for scanning the entire decay curve.

In later work, Boutilier and Winefordner (22) described the instrumentation they used for time-resolved laser-excited phosphorimetry. Figure 3.6 shows a block diagram of an updated system (58) that has essentially the same design as the original one they used (22). The main reasons for using a laser source were the fundamental limitations with continuum sources. For example, with

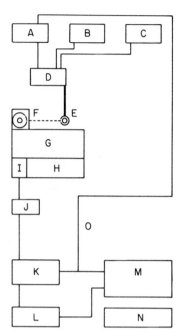

Figure 3.5. Block diagram of time-resolution pulsed-source phosphorimeter: (A) pulse generator; (B) power supply - low voltage; (C) power supply - high voltage; (D) trigger circuit for source; (E) xenon flash-tube source; (F) sample-cell compartment; (G) emission monochromator; (H) photomultiplier power supply; (I) photomultiplier; (J) variable-load resistor; (K) oscilloscope; (L) preamplifier; (M) signal averager; (N) potentiometric recorder; (O) synchronization for oscilloscope and signal averager. (Reprinted with permission from R.P. Fisher and J.D. Winefordner, *Anal. Chem.*, <u>44</u>, 948. Copyright 1972 American Chemical Society.)

a continuum source, whether continuous wave (cw) or pulsed, only a small fraction of spectral output is useful for excitation of phosphorescence. Even if one assumes fast collection optics and wide-band interference filters, the useful radiant flux transferred to the sample is only a small fraction of the total spectral output. The ideal source would be one of high intensity, tunable, and with monochromatic radiation. Dye lasers have all of these properties. In their work, Boutilier and Winefordner (22,23) used both a pulsed N_2 laser or a flashtube-pumped dye laser in place of the pulsed xenon source. They concluded that the pulsed nitrogen laser was an excellent excitation source for phosphors even with phosphors having a molar absorptivity as small as 10 at 337 nm. This resulted because of its high peak power and excellent pulse-to-pulse reproducibility. The flashlamp pumped dye laser required the use of a ratio system to compensate for pulse-to-pulse variation, dye

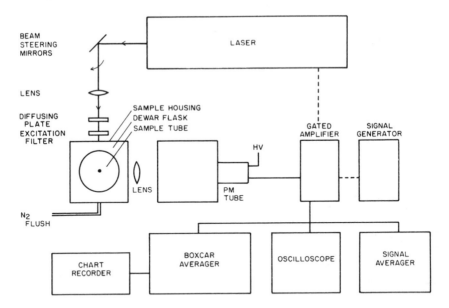

Figure 3.6. Block diagram of experimental system used for time-reolved phosphorimetry. (Reprinted with permission from C.G. Barnes and J.D. Winefordner, *Appl. Spectrosc.*, 1984 <u>38</u>, 214.)

decomposition, and frequency-doubling crystal drift for it to be of importance in phosphorimetry. The major sources of noise they found were associated with immersion cooling in liquid nitrogen. Thus, it would be possible that the noise from these sources could be reduced by conduction cooling (22). Barnes and Winefordner (58) derived expressions for the optimization of time-resolved phosphorimetry. Several of these expressions are discussed in Chapter 4.

Wilson and Miller (59) developed a computer-controlled laser phosphorimeter in which the phosphorescence spectra were recorded on magnetic tape as signal-averaged families of decay curves. With the phosphorescence data in this form, time-resolved phosphorimetric investigations were possible without the aid of mechanical or electronic chopping devices. Normally the data were obtained once and then could be displayed as desired using essentially any time window of interest. This permitted one to conduct detailed kinetic analyses of the decay processes influencing emission.

Goeringer and Pardue (44) discussed the design and performance characteristics of a time-resolved phosphorescence spectrometer with a silicon intensified target (SIT) vidicon camera system and a pulsed source. The instrument allowed time-resolved spectra to be recorded with a minimum scan time of 8 ms. Spectral decay data were processed by a variety of rate constants, initial intensities, and lifetimes. Figure 3.7 shows a block diagram of the system. Both temporal and spectral data were collected from the room-temperature phosphorescence of mixtures of phosphors adsorbed on filter paper.

Nithipatikom and Pollard (60) designed a room-temperature phosphorescence lifetime spectrometer that was capable of resolving the phosphorescence intensity from multicomponent decaying species into separate lifetimes. The main features of the phosphorescence lifetime spectrometer were careful blank subtraction, which was facilitated by a reproducible chopped-xenon arc source and data reduction with the use of non-linear least-squares reconvolution. Accuracy was assured by careful point-by-point subtraction of a proper blank. The chopped continuous excitation worked well for that purpose, but with electronics limitations, the

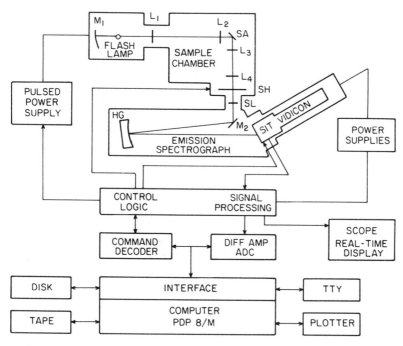

Figure 3.7. Block diagram of SIT based time-resolved phosphorimeter. (Reprinted with permission from D.E. Goeringer and H. C. Pardue, *Anal. Chem.*, <u>51</u>, 1054. Copyright 1979 American Chemical Society.)

lifetime measurements were from 0.5 ms or longer. Nithipatikom and Pollard indicated that a well-shielded tunable-dye laser or flash lamp and faster electronics would facilitate the measurement of shorter lifetimes with their system.

The primary advantages of pulsed-source time-resolved phosphorimeters compared to conventional phosphorimeters with mechanical modulation are the following (49): (a) the possibility of obtaining higher source peak intensities, and proportionately less noise, during the measurement period, which would permit lower detection limits; (b) the capability of obtaining greater selectivity for a short-lived phosphor compared to a long-lived phosphor; (c) the possibility of measuring phosphors with lifetimes as short as 10 μs for a flash-tube pulse, compared to about 30 μs for a disk chopper and approximately 100 μs for a typical rotating can phosphoroscope; (d) an improvement in signal-to-noise ratio and precision by gating the detector for a predetermined time; (e) the ability to scan the phosphorescence decay curve easily and to examine the linearity of the log I vs. t_d plot. This provides a quick check of the purity of the phosphor standards.

In related work, Ho and Warner (61) discussed an approach to phosphorimetry for multicomponent samples by the rapid acquisition of multiparametric data. The phosphorescence data are obtained in the form of emission-excitation matrices. In relationship to the work by Ho and Warner, researchers at the University of Washington (62) developed a fluorescence instrument, called a video fluorometer. The video fluorometer used a novel illumination method to obtain simultaneous multiwavelength excitation of the fluorescent sample. Mathematical expressions were used in representing the fluorescence emission-excitation matrix, and linear algebra and computer algorithms were used for both quantitative and qualitative analysis of fluorescence data (63,64). Equivalent mathematical techniques were employed by Ho and Warner (61) for the phosphorescence emission-excitation matrix. The video fluorometer system was modified slightly to obtain the phosphorescence emission-excitation matrix. With fluorescence measurements, the excitation source was continuous, and the observed fluorescence signal reached steady-state conditions before the data were obtained. For the phosphorescence experiments, the excitation source was cut off completely, and thus, only the phosphorescence signal was observed. The multidimensional phosphorimetric approach discussed by Ho and Warner (61) shows the need to couple instrumental methods with mathematical algorithms for simultaneous multiparametric luminescence measurements.

Pace et al. (65) described an inexpensive and highly automated instrument for obtaining high-resolution fluorescence and phosphorescence spectra of polycyclic aromatic hydrocarbons. Fluorescence and phosphorescence signals were processed by gated integration and photon counting, respectively. The instrument could be employed in both the frequency and time domains. Also, a method for measuring single and double exponential phosphorescence decay times was presented. Decay times ranging from 0.3 ms to 300 ms were measured for several naphthalene derivatives, with a precision of ~10%.

3.8. Phase-Resolved Phosphorimetry

Mousa and Winefordner (66) proposed the use of phase-resolved phosphorimetry. This approach involves the phase resolution of the phosphorescence signal from phosphors with different lifetimes. Because of the different amplitude and phase relationships of their luminescence signals, mixtures of phosphors have the potential of being quantitatively analyzed by this technique. Table 3.2 gives the equations derived by Winefordner and co-workers (49,66) describing the phase and frequency characteristics of luminescence. The expression for total luminescence intensity contains a constant intensity-term, and a sinusoidally varying intensity-term. The expression may be separated into individual intensities of fluorescence, I_f, and phosphorescence, I_p. The scattered light intensity, I_s, is assumed to be negligible. If a frequency-and phase-selective detection system is used in the luminescence system, only the a.c. terms of the luminescence intensity are observed.

With the correct choice of excitation and emission wavelengths and if strongly phosphorescent compounds are used, the fluorescence term can be neglected, and only the a.c. phosphorescence term is important experimentally. Nevertheless, one of the limitations of phase-resolved phosphorimetry is that frequently this selectivity cannot be obtained. This means that the a.c. fluorescence term in the theoretical expression in Table 3.2 must also be considered.

Based on the work of Mousa and Winefordner (66), Aaron and Winefordner (49) have summarized the main conclusions for phase-resolved phosphorimetry: (a) The phosphorescence param-

Table 3.2. **Theoretical   Expressions   for   Phase-Resolved Phosphorimetry**

Total luminescence intensity	d.c. term	a.c. term
	$I_L = k_L \cdot I'_o + m_L \cdot k_L \cdot I''_o \cdot cos(\omega t - \theta_L)$	

INDIVIDUAL INTENSITIES

Fluorescence $\quad\quad\quad\quad I_F = k_F \cdot I'_o + m_F \cdot k_F \cdot I''_o \cdot cos(\omega t - \theta_F)$

Phosphorescence $\quad\quad I_P = k_P \cdot I'_o + m_P \cdot k_P \cdot I''_o \cdot cos(\omega t - \theta_P)$

FREQUENCY FUNCTIONS

of the a.c. $\quad\quad\quad$ Degree of modulation   Phase-shift angle

term of $\quad\quad\quad\quad\quad m_P = (1 + 4\pi^2 f^2 \tau_P^2)^{-1/2};$

phosphorescence $\quad\quad\quad\quad\quad\quad\quad\quad\quad \theta_P = tan^{-1}(2\pi f \tau_P)$

---

k_L, k_F and k_P = factors taking into account the quantum efficiency and concentration factors for total luminescence, fluorescence, and phosphorescence respectively; I'_o and I''_o = constant-intensity term and sinusoidally-varying intensity term of the exciting-light function $I_0(t)$, respectively; m_L, m_F, and m_P = degree of modulation of total luminescence, fluorescence, and phosphorescence, respectively; θ_L, θ_F, and θ_P = phase-shift angle of total luminescence, fluorescence, and phosphorescence, respectively, degrees; τ_P = phosphorescence lifetime, sec; ω = angular frequency in Hz; f = linear frequency, Hz. (Reprinted with permission from J.J. Aaron and J.D. Winefordner, *Talanta* 1975, 22, 707.)

eters, m_p (degree of modulation) and θ_p (phase-shift angle), are a function of the frequency, f, of modulation of the source, and the lifetime, τ_p, of the phosphorescence.   An initial study of the variation of these parameters with frequency of the source must be made for a potential application so the optimal analytical conditions can be chosen for a given species in a mixture; (b) At a fixed frequency, the phase-shift angle, θ_p, (and the phase-angle difference $(\theta_p - \theta_R)$ between the analyte and reference signal) is a cosine function of the phosphorescence signal intensity. By adjusting the magnitude of the reference phase angle to a certain value, it is possible to maximize or minimize the measured analytical signal of the phosphor of concern; (c) For a mixture of two phosphors, the phase-resolution of the two components can be achieved either by changing the phase angle of the reference signal (phase method), or by varying the frequency of modulation (frequency method) to eliminate the signal from one of the two phosphorescent

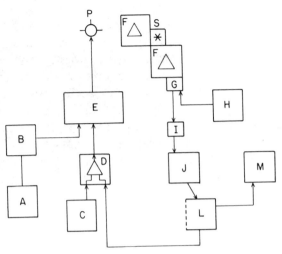

Figure 3.8. Block diagram of instrumental system for phase-resolved phosphorimetry: (A) 0-40V, 0-30A DC power supply; (B) starter circuit; (C) 0-100V, 0-0.2A DC power supply; (D) summing operational amplifier and current booster; (E) modulation circuit; (F) excitation and emission monochromators; (G) photomultiplier tube and housing; (H) photomultiplier tube and housing; (I) load resistors; (J) differential amplifier; (L) lock-in amplifier; (M) strip-chart recorder (optional); (P) xenon arc lamp; (S) sample compartment.
(Reprinted with permission from J.J. Mousa and J.D. Winefordner, *Anal. Chem.* 46, 1195. Copyright 1974 American Chemical Society.

components and to maximize the signal of the other phosphorescent component. The noise, however, is not eliminated. Figure 3.8 gives a block diagram of a phase-resolved phosphorimeter. The fundamental components are similar to the instrumentation used in phase fluorometry and modulation fluorometry.

Lue-Yen and Winefordner (67) considered the analytical possibilities of phase-resolved phosphorimetry and concluded that the technique does not have the superior noise-rejection capability of time-resolved phosphorimetry. In phase-resolved phosphorimetry, phosphors with short phosphorescence lifetimes have a larger response compared to phosphors with long phosphorescence lifetimes. This condition is advantageous when the background phosphorescence from the solvent is long-lived. However, an inherent disadvantage occurs because only phosphors with lifetimes between 1-50 ms can be easily measured using phase-resolved phosphorimetry (67). If the lifetime is greater than 50 ms, measurements must be made at a low frequency. At low frequencies, bubbling noise associated with liquid nitrogen and other low frequency noise make it difficult, and sometimes impossible, to obtain accurate and precise data. If the lifetime is less than 1 ms, the modulation frequency at which the most accurate phase measurements can be made must be high. This increases the probability for fluorescence and scattered radiation to interfere. These aspects further restrict the applicability of phase-resolved phosphorimetry. Lue-Yen and Winefordner (67) concluded that the phase-resolved phosphorescence technique was deemed undesirable as a routine general purpose analytical technique for phosphors. However, the phase resolution technique apparently works for strong phosphors or concentrated solutions of weak phosphors if no fluorescence or stray light interference is present.

3.9. Commercial Instruments

Commercial instruments are readily available for phosphorescence measurements. Generally, they are fluorescence instruments that contain a rotating-can phosphoroscope accessory or instruments with a pulsed source and gated detector. In addition, some instruments use a rotating chopper and pulsed-detector system. There are many commercial instruments that can be used to measure phosphorescence. Hurtubise (1) has considered several of these instruments.

3.10. Instrumentation for Lifetime Measurements

Luminescence lifetime measurements are important in many basic studies such as quenching and energy transfer. In analytical work, luminescence lifetimes give an additional parameter for characterizing compounds and offer the possibility of differentiating various components in mixtures based on lifetime differences.

Phosphorescence lifetimes are longer than prompt fluorescence lifetimes and are easier to measure. The simplest approach is to measure the phosphorescence signal as a function of time after termination of the exciting radiation. For lifetimes longer than approximately 1.0 s a strip-chart recorder or x-y recorder may be used. For more rapid phosphorescence decay times, a wide-band oscilloscope may be used. The general concepts and the important instrumentation for phosphorescence lifetime measurements were considered in Sections 3.6-3.8. Several of these instrumental systems can be used for phosphorescence lifetime measurements and are briefly discussed below.

Winefordner and co-workers (55,56) have developed pulsed-source phosphorescence instrumentation for time resolution phosphorimetry. Fisher and Winefordner (55) have provided an extensive comparison of pulsed-source and continuous operating mechanical phosphorescence systems. Boutilier and Winefordner (22) have described a pulsed laser (N_2 laser or flash-lamp pumped-dye laser) time resolved phosphorimeter for phosphorescence lifetime measurements. Also, Mousa and Winefordner (66) described the technique of phase-resolved phosphorimetry which can be used for phosphorescence lifetime measurements. Goeringer and Pardue (44) discussed a time-resolved phosphorescence spectrometer with a silicon intensified target vidicon camera system and a pulsed source. The instrument permitted time-resolved spectra to be recorded with a minimum scan time of 8 ms. Spectral decay data were processed by a variety of regression methods. Wilson and Miller (59) developed a computer-controlled laser phosphorimeter. The phosphorimeter design permitted time-resolved phosphorimetric studies without the aid of mechanical or electronic chopping devices. Charlton and Henry (68) constructed a simple apparatus for the determination of relatively long phosphorescence decay times. The main components of the system consisted of a flash gun, a shutter mechanism, and a simple photomultiplier tube. Dyke and Muenter (69) described the details of an inexpensive system for phosphorescent lifetime measurements constructed of a strobe lamp, optical components, a photomultiplier tube, and an oscilloscope. The system was capable of measuring phosphorescence lifetimes in the millisecond region. In the area of data manipulation, Cline Love and Skrilec (70) have discussed data reduction methods for first-order kinetics of phosphorescence decay. In a recent monograph, Demas (71) has given a comprehensive treatment of excited-state lifetime measurements.

3.11. Corrected Excitation and Emission Spectra

To obtain the luminescence properties of a compound free from instrumental artifacts, it is required to correct the luminescence spectra. The instrumental aspects that need to be corrected are wavelength-dependent efficiency of the excitation source and detector system, plus any wavelength dependency in the monochromators and other optics. Many commercial instruments have accessories available for correcting excitation and emission spectra. Corrected spectra are essential if spectra from different instruments are to be compared. In addition, corrected luminescence spectra are needed in calculating quantum yields and other fundamental luminescence parameters. Parker (72) and Demas and Crosby (73) have reviewed procedures for obtaining corrected spectra. Other authors have discussed several aspects of corrected spectra (74-81).

Excitation spectra are distorted primarily by the wavelength-dependent intensity of the exciting radiation. The intensity can be converted to a signal proportional to the number of incident photons by use of a quantum counter. A convenient quantum counter is Rhodamine B in ethylene glycol (3g/L). This solution practically absorbs all incident radiation from 220 to 600 nm. Generally, the quantum yield and emission maximum ($\sim$ 630 nm) are independent of the excitation wavelength from 200 to 600 nm. Therefore, the Rhodamine B solution gives a signal of constant wavelength and the signal is proportional to the photon flux of the exciting radiation (82). Quantum counters have been widely used to obtain corrected excitation spectra and the references cited can be consulted for detailed procedures (74-81).

To obtain corrected emission spectra, it is necessary to determine the efficiency of the detector system with respect to wavelength. One way of obtaining correction factors is to compare the emission spectrum of a standard material with the corrected spectrum for that substance. Another way to obtain correction factors is by determining the wavelength-dependent output from a calibrated light source. The wavelength distribution of radiation from a tungsten filament lamp can be approximated as a black body of equivalent temperature. Standard lamps are available from the National Bureau of Standards. In addition, one may obtain a corrected emission spectrum by using a magnesium oxide or barium sulfate scatterer and a quantum counter. The spectral output of the source, usually a Xe lamp, is determined and then the source is used as a calibrated light source. The relative photon output of the source can be obtained with a quantum counter in the sample compartment. Once the intensity distribution of the source is determined, the source output is directed onto the detector with a magnesium oxide scatterer. Magnesium oxide, or barium sulfate, is assumed to scatter all the wavelengths with equal efficiency. This procedure is summarized as follows:

1. The excitation wavelengths are scanned with the quantum counter in the sample compartment. The results give the lamp output as a function of wavelength.
2. The scatterer is then placed in the sample compartment, and the excitation and emission monochromators are scanned simultaneously. This step gives the product of lamp output times the sensitivity of the detector system.
3. Finally, the product in step 2 is divided by the function for lamp output, and the sensitivity factor for the detector system is acquired (82). The time in obtaining data for corrected excitation and emission spectra is minimized with modern computerized instrumentation.

Roberts (83) has considered several factors for the correction of excitation and emission spectra. Some of his recommendations follow:

1. The excitation source should be corrected by means of a quantum counter.
2. Two compounds that are widely accepted as quantum counters are Rhodamine B and 1-dimethylaminonaphthalene-5-sulfonate.
3. The primary calibration of the detector system is very conveniently done with the calibrated excitation system and a scatterer such as $BaSO_4$.
4. There is a need for secondary emission standards.
5. The usefulness of standard compounds would be improved if they were available as solid blocks. Both heavy-metal doped inorganic glasses and aromatic fluorophores in plastic deserve further attention.

References

1. Hurtubise, R.J. In *Trace Analysis: Spectroscopic Methods for Molecules*; Christian, G.D.; Callis, J.B., Eds.; Wiley-Interscience: New York, 1986; Chapter 2.
2. Vo-Dinh, T. *Room-Temperature Phosphorimetry for Chemical Analysis*; Wiley-Interscience: New York, 1984; Chapter 3.
3. Hurtubise, R.J. *Solid-Surface Luminescence Analysis*; Marcel Dekker: New York, 1981; Chapters 2 and 3.

4. Seitz, W.R. In *Treatise on Analytical Chemistry*, 2nd ed.; Elving, P.J.; Meehan, E.J.; Kolthoff, I.M., Eds.; Wiley: New York, 1981; Part I, Vol. 7, Sec. H, Chapter 4.

5. Guilbault, G.G. *Practical Fluorescence: Theory, Methods, and Techniques*; Marcel Dekker: New York, 1973; Chapter 2.

6. Schulman, S.G. *Fluorescence and Phosphorescence Spectroscopy: Physiochemical Principles and Practice*; Pergamon Press: Elmsford, NY, 1977; p 137.

7. Hamilton, T.D.S.; Munro, I.H.; Walker, G. In *Luminescence Spectroscopy*; Lumb, M.D., Ed.; Academic Press: New York, 1978; pp 149-238.

8. Hieftje, G.M.; Travis, J.C.; Lytle, F.E. *Lasers in Chemical Analysis*; Humana Press: Clifton, NJ, 1981.

9. Winefordner, J.D. In *New Applications of Lasers to Chemistry*; Hieftje, G.M., Ed.; ACS Symposium Series 85; American Chemical Society: Washington, DC, 1978; pp 50-79.

10. Omenetto, N.; Winefordner, J.D. *Crit. Rev. Anal. Chem.* 1981, <u>13</u>, 59.

11. Richardson, J.H. In *Modern Fluorescence Spectroscopy*; Wehry, E.L., Ed.; Plenum Press: New York, 1981; Vol. 4, Chapter 1.

12. Demtroder, W. In *Analytical Laser Spectroscopy*; Omenetto, N., Ed.; Wiley: New York, 1979; Chapter 5.

13. Latz, H.W. In *Modern Fluorescence Spectroscopy*; Wehry, E.L., Ed.; Plenum Press: New York, 1976; Vol. 1, Chapter 4.

14. Green, R.B. *J. Chem. Educ.* 1977, <u>54</u>, A365, A407.

15. Dovichi, N.J.; Martin, J.C.; Jett, J.H.; Keller, R.A. *Science* 1983, <u>219</u>, 845.

16. Lytle, F.E. *J. Chem. Educ.* 1982, <u>59</u>, 915.

17. Wright, J.C. In *Lasers in Chemical Analysis*; Hieftje, G.M.; Travis, J.C.; Lytle, F.E., Eds.; Humana Press: Clifton, NJ, 1981; Chapter 9.

18. *Analytical Applications of Lasers*; Piepmeier, E.G., Ed.; Wiley-Interscience: New York, 1986.

19. Reference 2, pp 92-96.

20. Boutilier, G.D.; Winefordner, J.D.; Omenette, N. *Appl. Optics* 1978, <u>17</u>, 3482.

21. Winefordner, J.D.; Rutledge, M. *Appl. Spectrosc.* 1985, <u>39</u>, 377.

22. Boutilier, G.D.; Winefordner, J.D. *Anal. Chem.* 1979, <u>51</u>, 1384.

23. Boutilier, G.D.; Winefordner, J.D. *Anal. Chem.* 1979, <u>51</u>, 1391.
24. Winefordner, J.D.; Schulman, S.G.; O'Haver, T.C. *Luminescence Spectrometry in Analytical Chemistry*; Wiley-Interscience: New York, 1972; Chapter III.
25. Meehan, E.J. In *Treatise on Analytical Chemistry*; Elving, P.J.; Meehan, E.J.; Kolthoff, I.M., Eds.; Wiley: New York, 1981; Part I, Vol. 7, Sec. H, Chapter 3.
26. Ingle, J.D.; Crouch, S.R. *Spectrochemical Analysis*; Prentice Hall: Englewood Cliffs, NJ, 1988; Chapter 3.
27. Reference 5, pp. 38-47.
28. Olsen, E.D. *Modern Optical Methods of Analysis*; McGraw-Hill: New York, 1975; Chapters 1 and 8.
29. Reference 7, pp 199-205.
30. Franklin, M.L.; Horlick, G.; Malmstadt, H.V. *Anal. Chem.* 1969, <u>41</u>, 2.
31. Jameson, D.M.; Spencer, R.D.; Weber, G. *Rev. Sci. Instrum.* 1976, <u>47</u>, 1034.
32. Koester, V.J.; Dowben, R.M. *Rev. Sci. Instrum.* 1978, <u>49</u>, 1186.
33. Cova, S.; Prenna, G.; Mazzini, G. *Histochem. J.* 1974, <u>6</u>, 279.
34. Darland, E.J.; Leroi, G.E.; Enke, C.G. *Anal. Chem.* 1979, <u>51</u>, 240.
35. Darland, E.J.; Leroi, G.E.; Enke, C.G. *Anal. Chem.* 1980, <u>52</u>, 714.
36. Meade, M.L. *J. Phys. E.* 1981, <u>14</u>, 909.
37. Talmi, Y. *Anal. Chem.* 1975, <u>47</u>, 658A, 697A.
38. Talmi, Y.; Baker, D.C.; Jadamec, J.R.; Saner, W.A. *Anal. Chem.* 1978, <u>50</u>, 936A.
39. Talmi, Y. *Appl. Spectrosc.* 1982, <u>36</u>, 1.
40. Talmi, Y., Ed. *Multichannel Image Detectors*; ACS Symposium Series; American Chemical Society: Washington, DC, 1979; Vol. 102.
41. Talmi, Y., Ed. *Multichannel Image Detectors, Volume 2*; ACS Symposium Series; American Chemical Society: Washington, DC, 1983; Vol. 236.
42. Christian, G.D.; Callis, J.B.; Davidson, E.R. In *Modern Fluorescence Spectroscopy*; Wehry, E.L., Ed.; Plenum Press: New York, 1981; Vol. 4, Chapter 4.
43. Vo-Dinh, T.; Johnson, D.J.; Winefordner, J.D. *Spectrochim. Acta* 1977, <u>33A</u>, 341.
44. Goeringer, D.E.; Pardue, H.C. *Anal. Chem.* 1979, <u>51</u>, 1054.

45. Cooney, R.P.; Boutilier, G.D.; Winefordner, J.D. *Anal. Chem.* 1977, 49, 1048.
46. Cooney, R.P.; Vo-Dinh, T.; Walden, G.; Winefordner, J.D. *Anal. Chem.* 1977, 49, 939.
47. Ingle, J.D.; Ryan, M.A. In *Multichannel Image Detectors, Volume 2*; Talmi, Y., Ed.; ACS Symposium Series; American Chemical Society: Washington, DC, 1983; Vol. 236, Chapter 7.
48. O'Haver, T.C.; Winefordner, J.D. *Anal. Chem.* 1966, 38, 602.
49. Aaron, J.J.; Winefordner, J.D. *Talanta* 1975, 22, 707.
50. Reference, 24, pp 203-206.
51. Langouet, L.L. *Appl. Opt.* 1972, 11, 2358.
52. Hollifield, H.C.; Winefordner, J.D. *Chem. Instru.* 1969, 1, 341.
53. Yen, E.L.; Boutilier, G.D.; Winefordner, J.D. *Can. J. Spectrosc.* 1977, 22, 120.
54. Winefordner, J.D. *Accounts Chem. Res.* 1969, 2, 361.
55. Fisher, R.P.; Winefordner, J.D. *Anal. Chem.* 1972, 44, 948.
56. O'Haver, T.C.; Winefordner, J.D. *Anal. Chem.* 1966, 38, 1258.
57. Harbaugh, K.F.; O'Donnell, C.M.; Winefordner, J.D. *Anal. Chem.* 1974, 46, 1206.
58. Barnes, C.G.; Winefordner, J.D. *Appl. Spectrosc.* 1984, 38, 214.
59. Wilson, R.M.; Miller, T.L. *Anal. Chem.* 1975, 47, 256.
60. Nithipatikom, K.; Pollard, B.D. *Appl. Spectrosc.* 1985, 39, 109.
61. Ho, C.-N.; Warner, I.M. *Trends Anal. Chem.* 1982, 1, 159.
62. Johnson, D.W.; Callis, J.B.; Christian, G.D. *Anal. Chem.* 1977, 49, 747A.
63. Warner, I.M.; Christian, G.C.; Davidson, E.R.; Callis, J.B. *Anal. Chem.* 1977, 49, 564.
64. Fogarty, M.P.; Warner, I.M. *Anal. Chem.* 1981, 53, 259.
65. Pace, C.F.; Thornberg, S.M.; Maple, J.R. *Appl. Spectrosc.* 1988, 42, 891.
66. Mousa, J.J.; Winefordner, J.D. *Anal. Chem.* 1974, 46, 1195.
67. Lue Yen, E.; Winefordner, J.D. *Anal. Chem.* 1977, 49, 1262.
68. Charlton, J.L.; Henry, B.R. *J. Chem. Educ.* 1974, 51, 753.
69. Dyke, T.R.; Muenter, J.S. *J. Chem. Educ.* 1975, 52, 251.
70. Cline Love, L.J.; Skrilec, M. *Anal. Chem.* 1981, 53, 2103.
71. Demas, J.N. *Excited-State Lifetime Measurements*; Academic Press: New York, 1983.

72. Parker, C.A. *Photoluminescence of Solutions*; Elsevier: New York, 1968.

73. Demas, J.N.; Crosby, G.A. *J. Phys. Chem.* 1971, 75, 991.

74. Parker, C.A.; Rees, W.T. *Analyst* 1960, 85, 587.

75. Argauer, R.J.; White, C.E. *Anal. Chem.* 1964, 36, 368.

76. Melhuish, W.J. *J. Opt. Soc. Am.* 1962, 52, 1265.

77. Drushel, H.V.; Sommers, A.L.; Cox, R.C. *Anal. Chem.* 1963, 35, 2166.

78. Landag, I.; Kremen, J.C. *Anal. Chem.* 1974, 46, 1694.

79. Chen, R.F. *Anal. Biochem.* 1967, 20, 339.

80. Corliss, D.A.; West, S.S.; Golden, J.F. *Appl. Opt.* 1980, 19, 3290.

81. Mielenz, K.D., Ed. *Optical Radiation Measurements: Measurement of Photoluminescence*; Academic Press: New York, 1982; Vol. 3,

82. Lakowicz, J.R. *Principles of Fluorescence Spectroscopy*; Plenum Press: New York, 1983.

83. Roberts, G.C.K. In *Standards in Fluorescence Spectrometry*; Miller, J.N., Ed.; Chapman and Hall: New York, 1981, Chapter 7.

CHAPTER 4

IMPORTANT ANALYTICAL CONSIDERATIONS

The relationship between phosphorescence intensity and concentration is important both in quantitative work and in fundamental studies. Because phosphorescence signals can be obtained under a variety of conditions, a knowledge of the expressions that relate phosphorescence intensity to concentration becomes even more important. As will be discussed in this chapter, theoretical expressions that relate phosphorescence intensity and concentration under all experimental conditions have not been fully developed.

Other important analytical considerations are factors that affect signal-to-noise ratios in phosphorimetry. In this area, theoretical expressions have been derived for solution low-temperature phosphorimetry. Other significant aspects that will be discussed in this chapter are limits of detection and selectivity.

4.1. Relationships between Phosphorescence Intensity and Concentration in Low-Temperature Solution Phosphorimetry

The phosphorescence intensity P of a solution is proportional to the intensity of source radiation absorbed by the phosphor and the phosphorescence quantum yield, ϕ_p, of the compound. If the intensity of the radiation absorbed is written as the difference between the incident intensity, I_0, and transmitted intensity, I_t, then the intensity of phosphorescence is given by the following equation.

$$P = \phi_p(I_0 - I_t) \tag{4.1}$$

The Beer-Lambert law can be written in the form, $I_t = I_0 \cdot 10^{\epsilon bc}$, where I_0 is incident source intensity, ϵ is molar absorptivity, b is pathlength, and c is the concentration in moles/liter. Substitution of I_t into Equation (4.1) using the Beer-Lambert law gives Equation (4.2).

$$P = \phi_p I_0 (1 - 10^{-\epsilon bc}) \tag{4.2}$$

Expanding the exponential term as a series gives

$$P = 2.3\phi_p I_0 \epsilon bc \left[\frac{2.3\epsilon bc}{2!} + \frac{(2.3\epsilon bc)^2}{3!} - \cdots \frac{(2.3\epsilon bc)^n}{(n+1)!} \right] \tag{4.3}$$

If the product, ϵbc, is less than 0.01, all of the ratios with a factorial term become small and can be neglected in Equation (4.3). Thus, Equation (4.3) becomes

$$P = 2.3\phi_p I_0 \epsilon bc \tag{4.4}$$

Equation (4.4) is of fundamental importance and shows that in dilute solutions a linear relationship exists between phosphorescence intensity and the concentration of the phosphor [1,2]. Equation (4.4) also shows that P can be increased by increasing the terms ϕ_p, I_0, ϵ, and b. Which of the terms would be increased in an experimental situation would depend mainly on the type of instrumental system being employed and the chemical and physical properties of the phosphor. For example, increasing the slit width on the excitation side of the phosphorescence spectrometer would result in I_0 increasing and thus P would also increase. With the phosphor, it may be possible to change solvents and thus increase both ϕ_p and ϵ.

Equation (4.3) shows that at some point phosphorescence intensity becomes a nonlinear function of concentration. Experimentally, the most important phenomenon responsible for a nonlinear calibration curve is the "inner-filter effect". The solution of the phosphor becomes so concentrated that the front part of the solution in the cell acts as an inner-filter by absorbing a relatively large fraction of source radiation. This prevents the solution in the back part of the sample cell from being excited or not being excited as effectively as the solution in the front part of the cell. This problem can be eliminated or minimized by diluting the sample.

The range of concentrations over which an analytical curve is linear is somewhat larger in low-temperature solution phosphorimetry than in fluorimetry (3). There are several possible reasons for this. The phosphorescence emission spectrum is farther removed from the fluorescence emission spectrum; thus there will be less self-absorption. Many times wider slits can be used in exciting the sample because a phosphoroscope is employed, which would minimize scattered radiation. In addition, the low temperature would minimize collisional deactivation. Finally, the smaller diameter cells used in low-temperature phosphorimetry would permit more uniform excitation of the sample at higher concentration relative to fluorimetry.

4.2. Effects of Instrumental and Spectral Parameters on the Photodetector Response

The discussion of the relationship between phosphorescence intensity and concentration given in Section 4.1 was very general and did not take into consideration the effects of instrumental and spectral parameters in affecting the actual instrumental response of a photodetector in a phosphorimeter. St. John et al. (4) have derived expressions that related the detector response in a phosphorimeter to a variety of important instrumental parameters. Figure 4.1 shows a typical arrangement for a sample cell in luminescence measurements along with some important definitions. In discussing the signal expressions, St. John et al. (4) made several assumptions. A summary of the assumptions is given below.

1. The entrance and exit slit widths and heights are equal for the excitation and emission monochromators.
2. The spectral bandwidth of the excitation monochromator, s, is appreciably less than the half-intensity width of the excitation spectral band, and the spectral bandwidth of the emission monochromator, s', is appreciably less than the half-intensity width of the emission spectral band.
3. The luminescence emission band shape is given by a Gaussian function of wavelength.
4. The intensity of absorption and emission are approximately

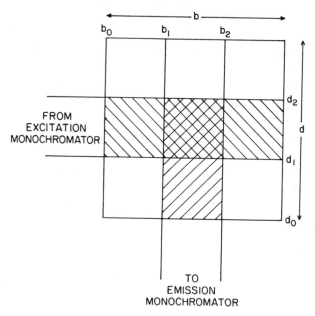

Figure 4.1. General arrangement for sample cell used in luminescence measurements: (b) width of cell parallel to excitation beam; (d) length of cell parallel to emission beam, ($b_2 \rightarrow b_1$) region over which emission is measured; ($d_2 \rightarrow d_1$) region over which excitation occurs. (Reprinted with permission from P.A. St. John, W.J. McCarthy, and J.D. Winefordner, *Anal. Chem.*, <u>38</u>, 1828. Copyright 1966 American Chemical Society.)

constant over the spectral bandwidths **s** and **s'**.

5. The measured emission band results from only a single electronic transition.

Normally, a dynamic shutter device such as a rotating can is used to modulate out-of-phase the exciting radiation and the phosphorescence radiation to minimize the signals due to fluorescence and incident-scattered radiation. Thus, the measured photodetector signal depends on the time between initiation of excitation and measurement of the phosphorescence signal. O'Haver and Winefordner (5) have reported an expression which accounts for the loss of signal due to intermittent excitation and observation. They defined a parameter α' which is the ratio of the observed power emitted when using the intermittent excitation and observation to the power emitted if using continuous excitation and observation. The α' term would be unity if continuous excitation and observation were employed.

The photoanodic current, i, from the detector due to a

phosphorescence signal is given by Equation (4.5).

$$i = \frac{P_{abs}\phi_s f_3 \gamma \alpha'}{4\pi A_s \Delta\lambda'} \tag{4.5}$$

For Equation (4.5), P_{abs} is the radiant power absorbed by the sample from b_1 to b_2 (Figure 4.1); ϕ_s is the energy yield of the luminescent process for the sample (ergs emitted s^{-1} per ergs adsorbed s^{-1}); f_3 is the fraction of emitted radiation reaching d_o (Figure 4.1); γ is the sensitivity factor and has units of amperes at the photodetector anode per watt of radiant power incident on the photocathode. A_s is the area of emitting sample surface (cm^2); and $\Delta\lambda'$ is the half-intensity width of the emission band (nm). St. John et al. (4) discussed two limiting cases for Equation (4.5) which describe most experimental situations and can be employed to investigate various possibilities of optimizing experimental conditions.

The first limiting case represents the phosphorescence of a compound in a dilute solution of a solvent which does not absorb much at the wavelengths of interest. This situation generally describes most quantitative phosphorescence measurements from solution. Using a number of assumptions which approximately obey the situation for quantitative measurements, St. John et al. (4) arrived at Equation (4.6).

$$i = 2.3I^0 k_a^2 a_s bC_s \phi_s \left[\frac{\gamma \alpha'}{4\pi A_s \Delta\lambda'} \right] \tag{4.6}$$

In Equation (4.6), I^0 is the intensity of source radiation (watts cm^{-2} $ster^{-1}$ nm^{-1}); k_a is a function of slit height, reciprocal linear dispersion, transmission factor of the monochromator, area of collimating mirror or lens, focal length of collimating mirror or lens, and slit width of the monochromator; a_s is the molar absorptivity (L $mole^{-1}$ cm^{-1}) of the phosphorescent sample at the excitation wavelength; b is the cell length (cm); C_s is the concentration of the phosphorescent sample (moles L^{-1}); ϕ_s is the energy yield of the phosphorescence process for the sample (ergs emitted per s per erg absorbed per s). The most fundamental conclusion from Equation (4.6) is that i is proportional to C_s. This is essentially the same conclusion reached with Equation (4.4);

however, Equation (4.6) shows more fully the effects of instrumental and spectral parameters on i. Of course, i is proportional to the phosphorescence intensity impinging on the detector.

In the other limiting case, the sample is highly concentrated. With this condition, the fraction of incident radiant power absorbed by the sample in region b_1 to b_2 in Figure 4.1 is one, and the fraction of emitted radiation reaching d_0 in Figure 4.1 is also one. The photoanodic current, i, is then given by Equation (4.7).

$$i = I^0 k_a^2 \phi_S \left(\frac{\gamma \alpha'}{4\pi A_S \Delta \lambda'} \right) \tag{4.7}$$

Equation (4.7) shows that i is independent of the sample concentration. The phosphor solution now behaves like a quantum counter. The photoanodic current is proportional to the energy yield of the phosphorescence process.

The shapes of the analytical curves of i vs. sample concentration, C_S, can be predicted from Equations (4.6) and (4.7). At low phosphor concentrations, Equation (4.6) shows that the slope of a log i vs. log C_S plot would be unity. At high phosphor concentration, Equation (4.7) indicates that the slope of the calibration curve would be zero. In practice, at very high concentrations of phosphor, the signal, i, would likely achieve a negative slope due to a decrease of ϕ_S with increasing concentration. The ϕ_S term would decrease because of self-quenching.

Van Geel et al. (6) have derived general intensity expressions that may be used for explaining the shapes of luminescence analytical calibration curves in fluorimetry and phosphorimetry. They took into consideration pre-filter (inner filter) effect due to absorption of exciting radiation by analyte and impurities in a region not viewed by the detector between the front edge of the cell and the measured luminescence region, primary absorption of exciting light by analyte and impurities within the viewed region, re-absorption of luminescence by analyte and impurities within the illuminated and viewed region, and post-filter effect due to absorption of luminescence by analyte and impurities in a non-illuminated region. The previous region is the cell region between the excitation region and detection. Their treatment is the first one in which all of the previous processes were considered. The expression they derived can be applied to solution phosphorimetry.

4.3. Relationships between Phosphorescence Intensity and Concentration in Low-Temperature Optically Inhomogeneous Matrices

Zweidinger and Winefordner (7) considered the experimental situation of nonreproducible phosphorescence intensities as a result of cracks or crystals created in the solvents employed in solution low-temperature phosphorimetry. In general, these systems were defined as optically inhomogeneous media. Based on the work of Hollifield and Winefordner (8), Zweidinger and Winefordner (7) investigated the quantitative aspects of optically inhomogeneous matrices for quantitative low-temperature phosphorimetry by using a rotating sample cell to minimize sample inhomogeneities. Intensity expressions for the phosphorescence of species in these matrices were derived by using the classical equations of Kubelka and Munk (9). The derived equations reliably predicted the shapes of the analytical curves in phosphorimetry for either clear rigid glasses or inhomogeneous matrices such as snows.

The sample (analyte plus solvent) was assumed to be an ideal diffuse, that is, the sample scattered incident-source radiation and phosphorescence in all directions. Using this primary assumption, other assumptions, and concepts defined by the Kubelka and Munk theory, Equation (4.8) was derived (7).

$$I_p = 2\phi_p \beta I_o \; x \left[\frac{(1 + \beta)\exp(k\bar{b}) + (1 - \beta)\exp(-k\bar{b}) - 2}{(1 + \beta)^2\exp(k\bar{b}) - (1 - \beta)^2\exp(-k\bar{b})} \right] \qquad (4.8)$$

The terms k and β are defined by Equations (4.9) and (4.10), respectively, and b is the average cell path length for diffuse reflectance ($\bar{b} = 2b$ for diffuse reflectance in snows and $\bar{b} = b$ for clear glasses). I_p is phosphorescence intensity, ϕ_p is phosphorescence quantum yield, and I_0 is exciting source intensity.

$$k = \sqrt{k(k+2s)} \qquad (4.9)$$

$$\beta = \sqrt{\frac{k}{k+2s}} \qquad (4.10)$$

In Equations (4.9) and (4.10), k is the fraction of radiation absorbed per average path length and is given by $k = 2.303 \, \epsilon C$, where ϵ is the molar absorptivity of the phosphor, C is the concentration of the phosphor, and s is the fraction of radiation scattered per average path length. Also, the term s is considered the scattering coefficient of the sample system and is primarily determined by the matrix. It is independent of analyte concentration for all "analytical" concentrations of the analyte.

For a clear glass or liquid solution, s is zero, and Equation (4.8) yields the well-known phosphorescence intensity equation.

$$I_p = \phi_p I_0 [1 - \exp(-kb)] \qquad (4.11)$$

At low concentrations (c) of phosphor where kb<<1:

$$I_p = \phi_p I_0 kb \qquad (4.12)$$

and at high concentrations of analyte where kb>>1:

$$I_p = \phi_p I_0 \qquad (4.13)$$

Thus, for the situation where s = o, all analytical curves (log I_p vs. log c) have a slope of unity at low concentration and a slope of zero at high concentrations of analyte.

For optically inhomogeneous matrices, for example, a snow or densely cracked glass where s>o, Equation (4.8) gives I_p. However, for low concentrations of analyte, where kb<<1, Equation (4.8) reduces to

$$I_p = 2\phi_p I_0 kb \left[\frac{1}{1+2sb} \right] \qquad (4.14)$$

and if 2sb<<1, then

$$I_p = 2\phi_p I_0 kb \qquad (4.15)$$

Comparison of Equation (4.12) with Equation (4.15) shows that a factor of 2 appears in Equation (4.15). If 2sb≥1, then I_p for a snow will be of the same order or smaller than I_p for a clear medium assuming the same kb (compare Equations (4.12) and 4.14)). For a high concentration of analyte where kb>>1, Equation (4.8) gives:

$$I_p = 2\phi_p I_0 \left[\frac{\sqrt{k}}{\sqrt{k+2s} + \sqrt{k}} \right] \qquad (4.16)$$

If k>>s, the Equation (4.16) reduces to Equation (4.13). If s>>k, then Equation (4.16) becomes

$$I_p = 2\phi_p I_0 \sqrt{\frac{k}{2s}} \qquad (4.17)$$

In summary, analytical curves (log I_p vs. log c) in inhomogeneous diffusely scattering media should have a slope of 0.5 at intermediate concentrations, and a slope of zero at high analyte concentrations. Zweidinger and Winefordner (7) showed that by comparing experimental and theoretical analytical curves, the equations they derived predicted quite accurately the general shapes of the analytical curves for optically inhomogeneous media. Also, Zweidinger and Winefordner showed that snows are essentially as analytically useful as clear rigid glasses.

4.4. Relationships between Phosphorescence Intensity and Amount of Adsorbed Phosphor in Solid-Surface Room-Temperature Phosphorescence

When considering the luminescence of organic compounds adsorbed on solid surfaces, one has to primarily consider absorption, reflection, and transmission from media that diffusely scatter radiation. In addition, under some experimental conditions, specular reflection has to be considered in the interpretation of the data. Several ways have been developed for describing the optical behavior of diffusely scattering media. However, more research remains to be done to experimentally verify theoretical expressions. In particular, there has been no detailed study in the correlation of experimental solid-surface RTP calibration curves with theoretical

RTP calibration curves. Work by Burrell and Hurtubise (10) has shown that calibration curves extended well beyond the normal linear range for solid-surface fluorescence, and solid-surface phosphorescence showed unique characteristics for benzo[f]quinoline adsorbed on silica gel chromatoplates. At high amounts of adsorbed benzo[f]quinoline, the solid-surface fluorescence intensity became constant and independent of the amount of the adsorbed fluorophor. However, the solid-surface phosphorescence intensity passed through a maximum and then decreased with increasing amounts of adsorbed phosphor. It was shown that fluorescence could occur from multilayers of fluorophor molecules, which would account for the fluorescence shape of the calibration curve at high amounts of fluorophor (10). The general shape of the fluorescence calibration curves were the same as that predicted by theoretical fluorescence equations derived for scattering media by Goldman (11). It was also shown that phosphorescence only occurred from molecules adsorbed on the surface of the adsorbent and not in multilayers. This would account for the general shape of the RTP calibration curve. Vo-Dinh (12) postulated that after all the available adsorption sites are occupied by the phosphor molecules, the molecules in higher molecular layers will not produce RTP. This postulation was supported by the results of Burrell and Hurtubise (10). Figure 4.2 shows typical RTF and RTP calibration curves obtained for the protonated form of benzo[f]quinoline adsorbed on a silica gel chromatoplate. A similar RTP calibration curve was obtained previously for p-aminobenzoic acid adsorbed on sodium acetate (13). Because there have been no detailed reports on theoretical calibration curves for solid-surface RTP, the remaining part of this section will consider, in general, the theoretical models that have been applied mainly to fluorescence from compounds adsorbed on scattering media. Several of these models should prove to be useful in solid-surface RTP. However, they should be considered with some caution because of the different shapes of calibration curves that have been obtained for RTF and RTP data (10).

As considered in Section 4.2, Zweidinger and Winefordner (7) derived theoretical intensity expressions for low-temperature phosphorescence for clear rigid solvents, cracked glasses, and snowed matrices. Also, experimental verification was given for the equations. Their intensity expressions have not been applied directly to solid-surface luminescence analysis. However, with some modifications and experimental corroboration, the intensity

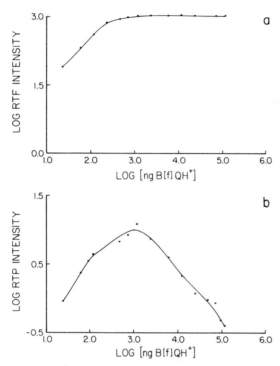

Figure 4.2. (A) log (B[f]QH$^+$ RTF intensity) vs. log (ng of B[f]QH$^+$) adsorbed on an EM chromatoplate. Each data point represents the average of three to five runs. (B) log (B[f]QH$^+$ RTP intensity) vs. log (ng of B[f]QH$^+$) adsorbed on an EM chromatoplate. Each data point represents the average of three to five runs. (Reprinted with permission from G.J. Burrell and R.J. Hurtubise, *Anal. Chem.*, _59_, 965. Copyright 1987 American Chemical Society.)

equations should prove useful in solid-surface phosphorescence work.

A general summary of theories and equations that have been considered for diffusely scattering media related to solid-surface luminescence analysis has been reported (12). Kortüm (9) has given a detailed discussion of theories and equations applicable to diffuse reflectance spectroscopy. Several of the equations have their origins from developments in the foggy atmosphere of radiation from stars. For example, Schuster (14) derived the fundamental differential equations for diffusely scattering materials to describe the behavior in a foggy atmosphere of radiation from a star. Kubelka and Munk (15) derived differential equations similar to those reported by Schuster. Later, Kubelka (16) published a system of differential equations that has been used in reflectance spectroscopy and luminescence spectroscopy (12,17).

There have been at least two different models reported using statistical approaches in diffuse reflectance spectroscopy (12,18). One model considers a powdered sample as a collection of plane-parallel layers. The thickness is equal to the average diameter of the sample particles. The other model assumes a collection of randomly shaped and rough-surfaced particles which have random orientations compared to the medium surface (12). These particular models have not been applied in a practical way to solid-surface luminescence analysis.

Most adsorbates used in solid-surface luminescence analysis scatter both exciting radiation and luminescence radiation. Also, the compound adsorbed on the solid matrix can absorb a fraction of the exciting radiation, and it is possible for the solid matrix to absorb exciting radiation. Several reports have been published on the relationship of the Kubelka-Munk theory to thin-layer densitometry, as applied in the transmission-absorption and reflection-absorption modes (19-25). The general approaches and equations in these reports are important because they detail various aspects of the Kubelka-Munk theory in densitometry and several of these aspects are applicable to solid-surface luminescence analysis.

Goldman and Goodall (21), Goldman (11), and Pollak and Boulton (19) discussed the assumptions of the Kubelka-Munk theory related to solid-surface absorption and fluorescence analysis. An extensive discussion of several of these assumptions can be found in the textbook by Kortüm (9). Several of the assumptions of the Kubelka-Munk theory which apply to solid surfaces are given below.

1. Source or luminescence radiation within the medium propagates only in the forward and background directions perpendicular to the plane-parallel boundary surfaces of the medium (19).

2. Once the radiation reaches the surface, the radiation is scattered in all directions (19).

3. The factors that define the optical response of the medium are k, the absorption coefficient of exciting radiation; s, the scattering coefficient of fluorescent radiation; and x, the thickness of the scattering medium (11,16,19).

4. The direction of the exciting radiation is perpendicular to the surface (19).

5. The medium is assumed to be homogeneous between the

boundary surfaces.   Theoretically, this means that the absorption coefficient and scattering coefficients are independent of the thickness of the medium (19).

It is important to briefly consider some of the work that has been done in developing equations that describe the relationship between fluorescence intensity and amount of fluorescent compound adsorbed on thin-layer chromatoplates and other surfaces. The general approaches used to derive the equations, with some alterations, should find use in solid-surface room-temperature phosphorescence (17). In fluorescence densitometry, Goldman (11) discussed the theoretical aspects of fluorescence reflected and transmitted from thin-layer chromatoplates. He considered both excitation radiation and fluorescent radiation in scattering media and obtained two pairs of differential equations.   One pair corresponded to fluorescent radiation emitted in the transmitted and reflected directions, and the other pair corresponded to absorption of exciting radiation in transmitted and reflected directions.   After further mathematical manipulations of the two pairs of differential equations, Goldman arrived at two general equations that were very complex.   One equation defined fluorescence reflected from the surface, and the other equation defined fluorescence transmitted through a surface such as a thin-layer chromatoplate.   Goldman's complex differential equations can be simplified considerably, and it can be shown theoretically that for small amounts of fluorophor, both the reflected fluorescence and the transmitted fluorescence is proportional to the amount of adsorbed fluorescent component (11).

Experimental data were acquired by Hurtubise (26) for fluoranthene as a model compound from thin-layer chromatoplates that gave experimental support for Goldman's simplified equations. A spectrodensitometer and aluminum oxide and silica gel glass-backed chromatoplates with an n-hexane mobile phase were employed in the experimental work. Prosek et al. (27) considered the advantages of the combination of fluorescence scanning from the far side and near side of chromatoplates for simultaneous measurement of fluorescence in quantitative analysis. They used the complex Goldman equations (11) and showed that experimental calibration curves gave very good fits to the appropriate theoretical calibration curves for transmitted fluorescence, reflected fluorescence, and simultaneous measurement of transmitted and reflected fluorescence.   No detailed comparison of experimental and theoretical results were reported by them. Nevertheless, the results from Prosek et al. (27) are in overall agreement with Goldman's theory.

A general theory for fluorescence from thin-layer chromatoplates based on the Kubelka-Munk theory and an electrical transmission-line model was developed by Pollak and Boulton (19) and Pollak (28,29).   Pollak and Boulton (19) employed the assumptions of the Kubelka-Munk theory and considered both excitation and fluorescence radiation in scattering media.   They reported a differential equation that was basically identical to an equation describing an electrical transmission line with purely resistive parameters.   The electrical transmission line model was used to simulate the optical behavior of turbid media in both the absorption and fluorescence modes.   Pollak (30) has discussed the transmission-line model in detail.   Hurtubise (17) has presented a general comparison of Goldman's equations, and Pollak and Boulton's equations as applied to solid-surface fluorescence analysis.   It should be emphasized again that these equations have not been applied in solid-surface room-temperature phosphorescence analysis.   Thus, it would be necessary to correlate experimental phosphorescence data with the theoretical equations to decide if the equations would be widely applicable to solid-surface phosphorescence.

In related work, Seely (31) derived equations for fluorescence generated within highly scattering media using the Kubelka and Munk treatment. An "apparent" quantum yield, calculated from observed intensities of fluorescence and of back-scattered light from the front surface of the solid matrix, can be corrected by Seely's expressions to give the true quantum yield of fluorescence. He also discussed some instrumental considerations that are needed to conform to Kulbeka and Munk boundary conditions.

4.5.   Signal-to-Noise Ratio Aspects

The ratio of signal-to-noise in luminescence spectroscopy is very important in luminescence measurements (4,32-37). There are many factors which influence the measured signal and noise in fluorescence and phosphorescence experiments. In this section, some of the signal-to-noise (S/N) ratio equations for photodetector signals will be considered. In Section 4.1 the important equations for the photodetector signal were discussed. The total root-mean square (rms) noise current Δi_T in amperes, at the photoanode, is given by Equation (4.18)

$$\overline{\Delta i_T} = (\overline{\Delta i_p}^2 + \overline{\Delta i_s}^2 + \overline{\Delta i_c}^2)^{1/2} \tag{4.18}$$

where Δi_p is the rms phototube noise due to the shot effect, Δi_s the rms effective luminescence flicker noise, and Δi_c the rms luminescence convection noise (4). St. John et al. (4) have discussed the noise level in detail and have derived Equation (4.19),

$$\overline{\Delta i_T} = [\Delta f k_D(i_d + i_B + i) + \xi^2[\sqrt{\Delta f_s} i + \sqrt{\Delta f_B} i_B]^2 + \epsilon^2 \Delta f[i + i_B]^2]^{\frac{1}{2}} \tag{4.19}$$

where Δf is the frequency response bandwidth of electrometer-read-out system (s^{-1}); k_D a function of the overall gain of the phototube and charge on an electron (coulombs); i_d the photoanodic dark current (A); i_B the background luminescence current (A); ξ the effective source flicker factor $(s^{1/2})$; Δf_s the effective frequency response bandwidth of electrometer readout system for sample (s^{-1}); Δf_B the effective frequency response backwidth of electrometer readout system for background (s^{-1}) and ϵ the convection flicker factor $(s^{1/2})$. St. John et al. (4) combined Equations (4.6) and (4.19) to obtain a relatively complex signal-to-noise equation. The equation was used for predicting the variation of $i/\Delta i_T$ as a function of a variety of experimental parameters. By plotting the signal-to-noise ratio versus the value of an experimental parameter, it was possible to decide if there was an optimum value of the parameter. If an optimum value were not acquired, it was possible to predict the best experimental conditions for an analysis. In general, the monochromator slit width, W, is an easily varied experimental parameter. In Figure 4.3 plots are presented of log(S/N) versus log(slit width) for several sample concentrations of analyte (4). It can be shown that the signal-to-noise ratio varies approximately with W^4 at small values of W (Figure 4.3) (4). This means that the total noise is determined primarily by the shot noise which is due essentially to dark current, i_d. For small values of W, $i/\Delta i_T$ is given by Equation (4.20),

$$\left(\frac{i}{\Delta i_T}\right)_{small\ W} = \frac{K_s W^4}{[\Delta f k_D i_D]^{\frac{1}{2}}} \tag{4.20}$$

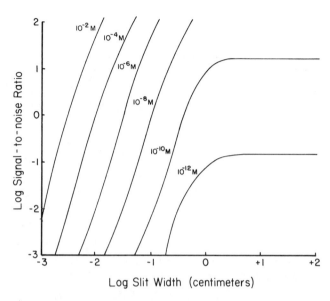

Figure 4.3. Calculated plots of signal-to-noise ratio vs. monochromator slit width for several sample concentrations of a hypothetical molecule. (Reprinted with permission from P.A. St. John, W.J. McCarthy, and J.D. Winefordner, *Anal. Chem.*, 38, 1828. Copyright 1966 American Chemical Society.)

where K_S is equal to the terms on the right side of Equation (4.6). For large values of W, $i/\overline{\Delta i}_T$ is given by Equation (4.21).

$$\left(\frac{i}{\overline{\Delta i_T}}\right)_{\text{large W}} = \frac{K_S}{(\xi^2[\sqrt{\Delta f_s}K_s+\sqrt{\Delta f_B}K_B]^2+\epsilon^2\Delta f[K_s+K_B]^2)^{\frac{1}{2}}} \qquad (4.21)$$

Equation (4.21) and Figure 4.3 show that the signal-to-noise ratio is independent of slit width. St. John et al. (4) emphasized that there is no strictly optimum slit width where the signal-to-noise ratio reaches a maximum value and decreases at both larger and smaller values of W (See Figure 4.3). Nevertheless, there is an analytically useful slit width, W_u, for measurement of luminescence. This is the slit width which would give the greatest spectral resolution with near maximum signal-to-noise ratio. The slit width, W_u, can be obtained equating Equations (4.20) and (4.21) and solving for the slit width. This would be the value of W that results at the intersection point when the curve with a low value of W is extrapolated to meet a curve with a high value of W (Figure 4.3). In a practical experimental situation, it may not be possible to attain, W_u. This could result from mechanical limitations of the slit mechanism and because the spectral band width corresponding to W_u may be appreciable compared to the half-intensity width of the excitation or emission spectral bands (4).

Signal-to-noise ratio considerations have to be treated somewhat differently for luminescent components adsorbed on solid surfaces (17). There has been no detailed development of signal-to-noise ratio equations for solid-surface room-temperature phosphorescence. However, several aspects related to signal-to-noise ratio for fluorescent components adsorbed on solid-surfaces have been discussed in the literature (29,38-40). A summary of these important concepts will be given because several of the concepts would be applicable to solid-surface phosphorescence.

For the measurement of direct fluorescence from scattering media, the fluorescent material stands out as a bright zone on a dark background. Theoretically, the dark background yields almost an ideally flat baseline. Nevertheless, electrical noise and residual fluorescence from impurities in the scattering medium can result in baseline problems. Generally, by integrating the output signal over a period of time, the electrical noise can be minimized (29,38). Because the residual fluorescence from impurities is essentially invariant, it would not be diminished by integration.

Optical noise can be a problem in direct fluorescence measurements from solid surfaces. Optical noise is caused by random fluctuations of the optical transfer in the scattering medium. Generally, optical noise can be a significant factor in limiting the performance in solid-surface fluorescence analysis. At very low light levels, electrical noise is important, and it is generated in the photodetector and preamplifier stages. Pollak (29,38) considered several sources of optical noise and these are summarized below.

1. Optical noise can result from source radiation instability.
2. Treatment of the medium with chemical reagents can cause nonrandom changes in optical parameters.
3. Density fluctuations of the medium and nonuniform particle size can cause optical noise.
4. Local variation in the thickness of the medium can cause

optical noise.

5.	When measurements are made in the reflection mode, it is possible that a fraction of a specularly reflected luminescence would reach the photodetector and cause optical noise. This problem can be reduced by careful optical design.

6.	Any specular component of the scattered radiation at the surface of the matrix varies randomly, and the intensity of the light entering the medium shows random fluctuations from point to point. The fluctuations appear as optical noise in both the transmission and reflection modes.

It can be shown theoretically that the signal-to-noise ratio in solid-surface fluorescence has a definite value which does not depend on the fluorescence or the amount of adsorbed fluorescent component (29). Pollak (29) estimated the signal-to-noise ratio for transmitted fluorescence and reflected fluorescence. The relationships are given in Equation (4.22) and (4.23), respectively.

$$\left(\frac{S}{N}\right)_{FT} \simeq \frac{1}{\delta_\gamma} \qquad (4.22)$$

$$\left(\frac{S}{N}\right)_{FR} \simeq \frac{\gamma}{\delta_\gamma} \qquad (4.23)$$

The optical noise fluctuations are incorporated in the coefficient γ. The γ term includes contributions from excitation radiation and fluorescence radiation. The term δ_γ represents the rms value of the fluctuation of the γ term. Equations (4.22) and (4.23) indicate that the signal-to-noise ratios are approximately constant and independent of the amount of fluorescer. However, the photodetector output signal is still dependent on the stability of the excitation radiation.

Generally, by increasing the intensity of the excitation radiation, the fluorescence intensity or phosphorescence intensity of an adsorbed compound will increase. However, this does not necessarily mean that the sensitivity will increase, because there may be an increase in impurity luminescence in the scattering media. However, the increased output from fluorescent or phosphorescent radiation may mask electrical noise and improve sensitivity and accuracy (29).

Pollak (39) has considered the nonuniform concentration distribution with depth in a solid surface.   He came to the theoretical conclusion that fluorescence determinations in the transmission mode would yield results that are almost independent of the distribution of the analyzed material with depth.   For fluorescence measurements in the reflection mode, he concluded that the measurements are strongly dependent on the distribution of concentration in the solid matrix. Pollak (40) also discussed the relationship between the dimensions of the scattering medium and sensitivity. No experimental data were offered for the theoretical conclusions (39,40). However, several of Pollak's conclusions for solid-surface fluorescence should be helpful in considering the fundamental advantages and limitations in solid-surface phosphorescence analysis.

4.6.   Limits of Detection

Limits of detection in molecular luminescence spectrometry have been reported for a large number of compounds. Long and Winefordner (41) considered limits of detection and the problems associated with various definitions of limit of detection. A general definition of the limit of detection is a concentration or an amount which describes the lowest concentration or amount of a component that is statistically different from an analytical blank.   The International Union of Pure and Applied Chemistry (IUPAC) adopted a model for the limit of detection calculation.   The ACS Subcommittee on Environmental Analytical Chemistry has supported this model (42). The limit of detection is based on the relationship between the gross analyte signal S_t, the blank S_b, and the variability of the blank σ_b. The limit of detection is defined by the magnitude to which the gross signal exceeds a defined multiple K_d of S_b (42).

$$S_t - S_b \geq K_d \sigma_b \tag{4.24}$$

If a single sample is being analyzed for which there is no blank data or if blanks are not available, then the limit of detection can be based on the peak-to-peak noise measured on the baseline close to the actual analyte peak. Generally, it has been recommended that detection should be based on a minimal K_d value of 3 (42). Therefore, the limit of detection is located at $3\sigma_b$ above the blank signal S_b. Long and Winefordner (41) recommended that limits of detection be reported using the IUPAC approach because errors in the analyte measurements can be incorporated into the detection limit.

Limit of detection can be viewed somewhat differently by using signal-to-noise ratio theory. For example, the smallest analytical signal that can be detected with a given level of confidence has been derived from small-sample statistical theory. Equation (4.25) defines the limiting signal-to-noise ratio at the limiting detectable sample concentration C_m (43,44).

$$\left(\frac{S}{N}\right)_{C_m} = \frac{t\sqrt{2}}{\sqrt{n}} \qquad (4.25)$$

In Equation (4.25), t is the student "t" and n is the number of combined sample and blank (or background) readings. The number of degrees of freedom is 2n-2. The noise N is the total root-mean square noise due to all sources and is approximately equal to the peak-to-peak noise divided by five (33). Equations have been described in the literature for the evaluation of C_m based on $(S/N)_{C_m}$ for molecular absorption spectrometry, molecular fluorescence spectrometry, and molecular phosphorescence spectrometry (4,44,45). The equations are somewhat complex and contain several instrumental factors that are important in defining C_m. However, the equations are important in relating the S/N ratio to important instrumental parameters, and various aspects of the equations are considered in Section 4.5.

4.7. Comparison of Absorption, Fluorescence, and Low-Temperature Solution Phosphorescence Spectrometry

Winefordner and co-workers (44,45) have considered expressions for the limiting detectable sample concentration in molecular absorption and in molecular fluorescence and low-temperature solution phosphorescence spectrometry. As mentioned in Section 4.5, some of the equations discussed contain several instrumental parameters that determine the limiting detectable sample concentration. With the equations, it is possible to compare the calculated limiting detectable sample solution concentrations for the three spectral techniques. It is assumed that the limits of detection are determined by the limiting signal-to-noise ratio expression given by Equation (4.25). In order to use Equation (4.25) to estimate C_m values, a number of conditions have to be met (44). All variance components can be described by the total rms noise expression which depends upon random noises such as phototube shot noise, source flicker noise, and amplifier noise. It is assumed that there is no contribution to the total statistical variance by the random and systematic sources of error, namely, nonreproducibility in cell positioning; variation in sample cell transmission; errors in reading signal output; dc drift in the light source-detector-amplifier-read-out system; and sampling errors (44). When employing the signal-to-noise ratio approach, the lowest value of C_m will be defined for given instrumental and experimental conditions. Measured values of C_m which are larger than the value calculated with Equation (4.25) implies the presence of other errors such as the random and systematic errors mentioned above. Cetorelli et al. (44) calculated values of C_m for several compounds using the expressions for absorption, fluorescence, and phosphorescence spectrometry and assuming the same spectrometric system. For example, the C_m values in mole/liter for biphenyl were 2.6 x 10^{-8}, 1.0 x 10^{-10}, and 3.5 x 10^{-10} for absorptiometry, fluorometry, and phosphorimetry (rotating-can), respectively. Thus, under conditions where various systematic errors and random errors are minimized, very low C_m values can be achieved by the three spectral techniques. The calculated C_m results for fluorometry were usually lower than the calculated results for phosphorimetry.

The magnitude of the source intensity has a large effect on the value of C_m in fluorescence and phosphorescence spectrometry. Figure 4.4 shows the variation of the minimum detectable sample concentration, C_m, with source intensity, I^o, for absorption, fluorescence, and phosphorescence spectrometry for biphenyl. Source intensity, I^o, had essentially no influence on the value of C_m in absorptiometry for the molecules Cetorelli et al. (44) investigated. This was expected because source flicker noise is normally dominant in absorption measurements. For fluorescence and phosphorescence, the value of C_m varies linearly with I^{o-1} because source flicker noise is insignificant compared to the phototube shot noise. It should be pointed out that I^o cannot be increased indefinitely because of the possibility of photochemical reactions being induced. Also, in fluorescence and phosphorescence, source flicker only starts to cause a slight effect at large light

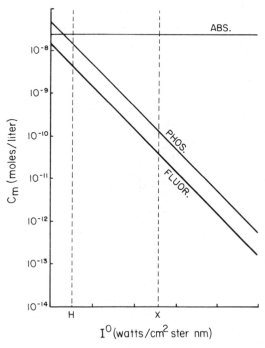

Figure 4.4. Variation of minimum detectable sample concentration, C_m, with source intensity, I^0, in absorption, fluorescence, and phosphorescence spectrometry for biphenyl. Lines H and X represent typical source intensities of a hydrogen arc lamp such as is used in absorption spectrometry and 150-w xenon arc lamp such as is used in luminescence spectrometry, respectively. (Reprinted with permission from J.J. Cetorelli, W.J. McCarthy, and J.D. Winefordner, *J. Chem. Educ.* 1968, 45, 98.)

levels, namely, at large values of I^0 and slit width.

For low-temperature solution phosphorescence there is a contribution to total noise due to convection of the thermostating medium. That is, low temperature solution phosphorescence measurements are made with the sample cell immersed in liquid nitrogen. Nevertheless, the value of C_m is not influenced significantly for values of the convection and source flicker factor encountered in low-temperature solution phosphorimetry (44).

Slit width (W) is most conveniently varied in luminescence and absorption measurements, and Figure 4.5 shows the changes in C_m as a function of slit width for these measurements. For absorption measurements, slit width does not influence the value of C_m appreciably when the noise is determined primarily by source flicker. However, at low source intensities, phototube shot noise becomes significant at small values of W (Figure 4.5). The

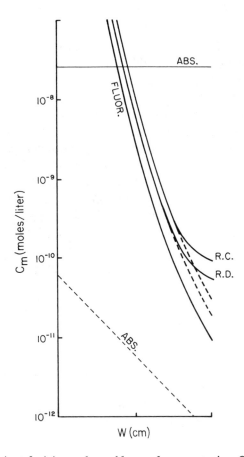

Figure 4.5. Variation of minimum detectable sample concentration, C_m, with monochromator slit width, W, for absorption, fluorescence, and phosphorescence rotating-can, (RC), and rotating-disc, (RD), spectrometry for biphenyl. The solid lines represent the actual curves for conditions given in Reference 44. The dashed lines indicate the C_m versus W curves if the only noise were shot noise (in the case of fluorescence the dashed and solid lines are indistinguishable). (Reprinted with permission from J.J. Cetorelli, W.J. McCarthy, and J.D. Winefordner, *J. Chem. Educ.* 1968, <u>45</u>, 98.)

influence of monochromator slit width on C_m for fluorescence and phosphorescence measurements is also given in Figure 4.5. As indicated in Figure 4.5, curvature only occurs at large values of slit width. As W increases, source flicker noise and convection flicker noise become significant. At sufficiently large values of W, the slit width will have no effect, and C_m will be constant as W increases.

The value of C_m is also dependent upon the photomultiplier spectral response curve when making fluorescence and phosphorescence measurements. Under ideal conditions, to obtain the smallest value of C_m for a luminescent molecule, the photodetector response curve should peak at the maximum luminescence intensity. This can be approximately achieved in many experimental situations with present-day photomultiplier tubes which have relatively constant response curves over a wide wavelength range. For absorption measurements, the value of C_m is less dependent on the photomultiplier spectral response. However, at long wavelengths, phototube shot noise becomes the major contribution to the total noise (44).

From the signal-to-noise ratio expressions, an indication of the limiting detectable sample concentration and the precision of measurement at any concentration is obtainable. However, the signal-to-noise ratio gives no information about the accuracy or selectivity of the analysis (33,44).

For most luminescent molecules studied, fluorescence spectrometry has been employed. Also, signal-to-noise ratio measurements favor fluorescence spectrometry. However, there are some severe spectral limitations in fluorescence spectrometry which may prevent the experimental attainment of calculated values of C_m. The spectral limitations include first and second order Rayleigh and Raman scattering of incident radiation. These are potentially significant spectral interferences in fluorescence studies. These spectral interferences present essentially no problems in phosphorescence spectrometry when a phosphoroscope or pulsed-source gated-detector system is used during the measurement step. In addition, for absorption spectrometry, Rayleigh and Raman scattering are not problems with the use of a double monochromator with both monochromators adjusted to the same wavelengths (44).

Because of the small influence of spectral interference in phosphorescence spectrometry, frequently greater selectivity is achieved in phosphorimetry than in fluorimetry or in absorption investigations. For example, in all three approaches, a molecule can be selectively excited. However, only with the luminescence approach can the emission be selectively measured by appropriate selection of wavelength position and spectral bandwidth of the emission monochromator. In addition, in phosphorimetric studies, even greater selectivity can be obtained by the use of time-resolved techniques (46,47). In general, the order of selectivity of the three spectral techniques are: phosphorescence > fluorescence > absorption.

There are some practical limitations to low-temperature

solution phosphorimetry. Several of these are: difficulty in aligning the sample cell reproducibly from measurement to measurement; sample cell size limitations (sample cells are usually cylindrical and about 1 mm in diameter); few solvents that form a clear, rigid glass at liquid nitrogen temperature. Most of these problems are eliminated by using room-temperature phosphorescence techniques. (See Chapter 6-8.)

Phosphorimetry and fluorimetry should be considered as complimentary as well as competing techniques. For many luminescent compounds, smaller values of C_m result by phosphorimetry compared to fluorimetry. For many other luminescent compounds, the opposite is true, and this leads to the complimentary nature of the two methods. The combination of fluorimetry and phosphorimetry will normally result in lower limits of detection and greater selectivity compared to absorption spectrometry.

4.8.   Time-Resolved Phosphorimetry

Instrumentation for time-resolved phosphorescence measurements was considered in Chapter 3. The resolution of spectrally similar phosphors based on differences in lifetime is known as time-resolved phosphorimetry (46,47). Generally, time-resolved phosphorimetry is important when instrumental or spectroscopic methods for resolution of components fail, when it may be difficult to separate the components in the sample, or when chemical treatment of the sample may increase analysis time or introduce contamination. Barnes and Winefordner (47) have considered optimization of time-resolved phosphorimetry for single channel detection instrumentation, which is less expensive and simpler to employ than multichannel detection. They discussed conditions for a pulsed-source time-resolved phosphorimetry system using a conventional single channel ("boxcar" averager or integrator) measurement system. (See Fig. 3.4 in Chapter 3 for pulsed-source gated-detector schematic.) They derived equations that could be employed to ease the trial-and-error optimization approach for a pulsed-source gated-detector phosphorimeter. The expressions were verified with experimental results obtained for several phosphors with widely different lifetimes, employing a pulsed-laser boxcar-averager phosphorimeter. Equation (4.26) can be used to determine the relative effects of various parameters

$$I = I_0 \left(\frac{t_p}{t_g} \right) \left(\frac{[1 - \exp(- t_g/\tau)]\exp(- t_d/\tau)}{[1 - \exp(- 1/f\tau)]} \right) \qquad (4.26)$$

where I is the observed phosphorescence intensity; I_0 is the CW phosphorescence intensity; t_p is the excitation pulse width, s; t_g is gate width, s; τ is the phosphorescence decay lifetime, s; t_d is the delay time, s; f is the pulse repetition rate, Hz. Equation (4.26) is similar to equations derived by O'Haver and Winefordner for modulated and pulsed sources (48,49). Barnes and Winefordner (47) obtained experimental data for several compounds to validate their theoretical expressions. Experimentally, the relative changes in phosphorescence intensity (I/I_0) with t_d, t_g, and f for several phosphors of different lifetimes were measured and compared with the relative changes predicted by theory. Table 4.1 gives the average results for intensity measurements for three phosphors. A theory/experiment ratio of unity indicates perfect correlation. Generally, the correlation of the theory with experimental results is

Table 4.1. Average Results of Intensity Measurements for Three Phosphors

	Average theory/experiment ratio		
Effect Studied	Thioridazine HCl	Vanillin	Nupercaine HCl
t_d	0.952	0.993	1.07
t_g	1.02	1.07	0.986
f	0.988	0.928	1.20
t_g, t_d	0.950	1.12	1.04
f, t_d	0.972	0.913	1.25
f, t_g	0.974	0.970	1.24

Reprinted with permission from C.G. Barnes and J.D. Winefordner, *Appl. Spectrosc.* 1984, 38, 214.

very good for all the parameters and all three phosphors. The expressions and results reported by Barnes and Winefordner (47) can be used to correctly predict the direction and magnitude of changes in phosphorescence intensity with gate width, repetition rate, delay time, and lifetime. This will permit easier and more rapid optimization of experimental conditions for time-resolved phosphorimetry experiments.

Laserna et al. (50) combined constant-energy synchronous scanning with time resolution to increase spectral selectivity in low-temperature phosphorimetry. The technique is called time-resolved constant-energy synchronous phosphorimetry (TRCESP). Inman et al. (51) have considered the theoretical optimization of parameter selection in constant energy synchronous luminescence spectrometry. Laserna et al. (50) concluded that TRCESP was a more complicated technique than time-resolved phosphorimetry or constant energy synchronous phosphorimetry alone. They recommended that one should first try constant energy synchronous phosphorimetry, then time-resolved phosphorimetry, and then TRCESP. However, this assumes that the analyst has attempted simpler approaches such as room-temperature phosphorimetry to see if the desired selectivity and/or sensitivity could be achieved.

References

1. Zander, M. *Phosphorimetry*; Academic Press: New York, 1968; pp 132-137.
2. Keirs, R.J.; Britt, R.D.; Wentworth, W.E. *Anal. Chem.* 1957, 29, 202.
3. McGlynn, S.P.; Neely, B.T.; Neely, C. *Anal. Chim. Acta* 1963, 28, 472.
4. St. John, P.A.; McCarthy, W.J.; Winefordner, J.D. *Anal. Chem.* 1966, 38, 1828.
5. O'Haver, T.C.; Winefordner, J.D. *Anal. Chem.* 1966, 38, 602.
6. Van Geel, F.; Voightman, E.; Winefordner, J.D. *Appl. Spectrosc.* 1984, 38, 228.
7. Zweidinger, R.; Winefordner, J.D. *Anal. Chem.* 1970, 42, 639.
8. Hollifield, H.C.; Winefordner, J.D. *Anal. Chem.* 1968, 40, 1759.
9. Kortüm, G. *Reflectance Spectroscopy*; Springer-Verlag: New York, 1969.

10. Burrell, G.J.; Hurtubise, R.J. *Anal. Chem.* 1987, 59, 965.
11. Goldman, J. *J. Chromatogr.* 1973, 78, 7.
12. Vo-Dinh, T. *Room Temperature Phosphorimetry for Chemical Analysis*; Wiley-Interscience: New York, 1984; pp 193-219.
13. von Wandruszka, R.M.A.; Hurtubise, R.J. *Anal. Chem.* 1977, 49, 2164.
14. Schuster, A. *Astrophys.* 1905, 21, 1.
15. Kubelka, P.; Munk, F. *Zeits Tech. Physik* 1931, 12, 593.
16. Kubelka, P. *J. Opt. Soc. Am.* 1948, 38, 448.
17. Hurtubise, R.J. In *Molecular Luminescence Spectroscopy: Methods and Applications - Part II*; Schulman, S.G., Ed.; Wiley: New York, 1988; Chapter 1.
18. Simmons, E.L. *Appl. Optics* 1975, 14, 1380.
19. Pollak, V.; Boulton, A.A. *J. Chromatogr.* 1972, 72, 231.
20. Pollak, V. *J. Chromatogr.* 1975, 105, 279.
21. Goldman, J.; Goodall, R.R. *J. Chromatogr.* 1968, 32, 24.
22. Hecht, H.G. *Anal. Chem.* 1976, 48, 1775.
23. Huf, F.A.; DeJong, H.J.; Schute, J.B. *Anal. Chim. Acta* 1976, 85, 341.
24. Huf, F.A. *Anal. Chim. Acta* 1977, 90, 143.
25. Huf, F.A. In *Quantitative Thin-Layer Chromatography and Its Industrial Applications*; Treiber, L.R., Ed.; Marcel Dekker: New York, 1987; Chapter 2.
26. Hurtubise, R.J. *Anal. Chem.* 1977, 49, 2160.
27. Prosek, M.; Kucan, E.; Katic, M.; Bano, M.; Medja, A. *Chromatographia* 1978, 11, 578.
28. Pollak, V. *Opt. Acta* 1974, 21, 51.
29. Pollak, V. *J. Chromatogr.* 1977, 133, 49.
30. Pollak, V. *IEEE Trans. Biomed. Eng.* 1970, 17, 287.
31. Seely, G.R. *Biophys. J.* 1987, 52, 311.
32. Winefornder, J.D.; Schulman, S.G.; O'Haver, T.C. *Luminescence Spectrometry in Analytical Chemistry*; Wiley-Interscience: New York, 1972.
33. Winefordner, J.D.; McCarthy, W.J.; St. John, P.A. *J. Chem. Educ.* 1967, 44, 80.
34. Enke, C.G.; Nieman, T.A. *Anal. Chem.* 1976, 48, 705A.
35. Cooney, R.P.; Vo-Dinh, T.; Walden, G.; Winefordner, J.D. *Anal. Chem.* 1977, 49, 939.
36. Cooney, R.P.; Boutilier, G.D.; Winefordner, J.D. *Anal. Chem.* 1977, 49, 1048.
37. Winefordner, J.D.; Rutledge, M. *Appl. Spectrosc.* 1985, 39,

377.
38. Pollak, V. *J. Chromatogr.* 1976, <u>123</u>, 11.
39. Pollak, V. *J. Chromatogr.* 1977, <u>133</u>, 195.
40. Pollak, V. *J. Chromatogr.* 1977, <u>133</u>, 199.
41. Long, G.L.; Winefordner, J.D. *Anal. Chem.* 1983, <u>55</u>, 712A.
42. *Guidelines for Data Acquisition and Data Quality Evaluation in Environmental Chemistry, Anal. Chem.* 1980, <u>52</u>, 2242.
43. St. John, P.A.; McCarthy, W.J.; Winefordner, J.D. *Anal. Chem.* 1967, <u>39</u>, 1495.
44. Cetorelli, J.J.; McCarthy, W.J.; Winefordner, J.D. *J. Chem. Educ.* 1968, <u>45</u>, 98.
45. Cetorelli, J.J.; Winefordner, J.D. *Talanta* 1967, <u>14</u>, 705.
46. Winefordner, J.D. *Acc. Chem. Res.* 1969, <u>2</u>, 361.
47. Barnes, C.G.; Winefordner, J.D. *Appl. Spectrosc.* 1984, <u>38</u>, 214.
48. O'Haver, T.C.; Winefordner, J.D. *Anal. Chem.* 1966, <u>38</u>, 602.
49. O'Haver, T.C.; Winefordner, J.D. *Anal. Chem.* 1966, <u>38</u>, 1258.
50. Laserna, J.J.; Mignardi, M.A.; von Wandruszka, R.; Winefordner, J.D. *Appl. Spectrosc.* 1988, <u>42</u>, 1112.
51. Inman, E.L.; Files, L.A.; Winefordner, J.D. *Anal. Chem.* 1986, <u>58</u>, 2156.

CHAPTER 5

LOW-TEMPERATURE PHOSPHORIMETRY

5.1. General Theoretical Considerations

Most of the past work in phosphorimetry has been performed with solutions at low temperature. In Chapter 1, a general historical survey was given on the major developments in low-temperature phosphorimetry. Also, some of the photophysical aspects of phosphorescence were considered. In this section, many of the reasons for using low temperature in solution phosphorescence will be discussed. Sensitized and quenched phosphorescence in solutions at room temperature is considered in Chapter 9, and micelle-stabilized room-temperature phosphorescence, cyclodextrin solution phosphorescence, and phosphorescence in colloidal suspensions are discussed in Chapter 10.

Tsai and Robinson (1) observed phosphorescence from a highly purified fluid solution of naphthalene in 3-methylpentane. They concluded that quenching by impurities and bimolecular depopulation were responsible for earlier failures to observe long-lived phosphorescence in liquid solutions. Parker and Joyce (2) reported the phosphorescence efficiencies and lifetimes of nine aromatic carbonyl compounds in pure perfluorocarbon solvents at 20°C. Also, Clark et al. (3) and Saltiel et al. (4) have reported the room-temperature phosphorescence of aromatic ketones in organic solvents. In addition, phosphorescence was observed at room temperature from oxygen-free aqueous solutions of benzophenone and acetophenone, and Vander Donckt et al. (5) reported the room-temperature phosphorescence of various aromatic compounds in fluid dimethylmercury. Turro et al. (6) published several room-temperature phosphorescence spectra. For example, nitrogen purged, acetonitrile solutions of bromo- and dibromonaphthalene gave phosphorescence spectra as a result of the internal heavy-atom effect.   Also, the external enhancement of aromatic hydrocarbons permitted the observation of room-temperature solution phosphorescence for these compounds in N_2 purged fluid solutions (6). Donkerbroek et al. (7) discussed several analytical aspects of room-temperature phosphorescence in liquid solutions. All of the above reports are characterized by the use of very pure solvents that are essentially oxygen free. In addition, the heavy-atom effect or aromatic carbonyl compounds were employed in the experiments.

Donkerbroek et al. (7) investigated several fluoro-, chloro-, and bromo-derivatives for room-temperature phosphorescence in liquid solutions to evaluate the analytical potential of phosphorescence in liquid solutions.   They concluded that direct room-temperature phosphorescence in liquid solutions with intensities strong enough to be of analytical interest was a somewhat rare phenomenon. It is important to keep in mind that their conclusion does not apply directly to sensitized solution phosphorescence, micelle-stabilized phosphorescence, and cyclo- dextrin solution phosphorescence.

As mentioned in Section 2.11, Turro (8) has discussed phosphorescence in fluid solution at room temperature. He states that phosphorescence can "generally" be observed at room temperature in fluid solution if two conditions are fulfilled:

1.   Impurities that can quench triplet states are rigorously excluded.
2.   The triplet state does not undergo an activated unimolecular deactivation which possesses a rate of $\geq 10^4 k_p$ at room temperature.

The measurement of experimental phosphorescence requires a phosphorescence quantum yield (ϕ_p) of about 10^{-4}. The value of ϕ_p may be expressed approximately by Equation (5.1) (8) where k_p is

$$\phi_p \sim \frac{k_p}{k_m + k_q[q]} \tag{5.1}$$

the rate constant for phosphorescence, k_m is the rate constant for unimolecular radiationless deactivation, and $k_q[q]$ represents the bimolecular deactivation of the triplet state.

A typical value of k_p for a molecule in a n,π^* triplet state (T_1) is 10^2 s, and a typical value of k_p for a molecule in a π,π^* triplet state (T_1) is 10^{-1} s. Thus, for $\phi_p \sim 10^{-4}$ the following relationships are valid:

$$k_m + k_q[q] \sim 10^6 \text{ s}^{-1} \text{ for } T_1(n,\pi^*) \tag{5.2}$$

$$k_m + k_q[q] \sim 10^3 \text{ s}^{-1} \text{ for } T_1(\pi,\pi^*) \tag{5.3}$$

Turro (8) calculated the maximum value of [q] tolerable for the observation of phosphorescence if [q] is considered as a diffusional quencher. For a nonviscous organic solvent, the rate constant for diffusion (k_{dif}) is about 10^{10} M^{-1} s^{-1}. If it is assumed that $k_{dif} = k_q$, then,

$$\text{if } k_{dif}[q] < 10^6 \text{ s, then } [q] < 10^{-4} \text{ M} \tag{5.4}$$

and

$$\text{if } k_{dif}[q] < 10^3 \text{ s, then } [q] < 10^{-7} \text{ M} \tag{5.5}$$

A concentration of 10^{-4} M for [q] can be readily obtained experimentally, but a concentration of 10^{-7} M is more difficult to obtain. This qualitative discussion permits one to understand why compounds with phosphorescence from T_1 (n,π^*) states can be observed at room temperature in fluid solutions but phosphorescence from $T_1(\pi,\pi^*)$ states are rarely observed unless extraordinary care is taken to eliminate bimolecular quenching.

In summary, based on the results that have been reported, analytically useful phosphorescence signals from fluid solutions at room temperature from a large number of compounds have not been achieved. Even if a phosphorescence signal could be obtained in fluid solution, the fundamental question to be asked is whether the phosphorescence signal is strong enough to be exploited analytically. This has not been the case normally, and thus, the sample has been cooled to liquid nitrogen temperature primarily to minimize bimolecular quenching. However, other factors can also be involved in increasing phosphorescence yield at low temperature (6).

5.2. Instrumental Aspects and Sample Preparation

The major instrumentation required for the measurement of low-temperature phosphorescence was considered in Chapter 3. In this section, the means employed for cooling the sample and general sample preparation techniques will be considered. For several years

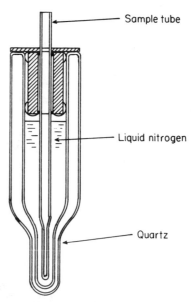

Figure 5.1. Dewar flask used for immersion cooling in phosphorimetry. (Reprinted with permission from J.D. Winefordner, P. A. St. John, and W. J. McCarthy, in *Fluorescence Assay in Biology and Medicine;* S. Udenfriend, Ed., Vol. II, Academic Press, New York, 1969.)

and up to the present time, the immersion method has been widely used for cooling samples to liquid nitrogen temperature (Figure 5.1). With this method, the sample tube is lowered into a Dewar that contains liquid nitrogen. Ward et al. (9) have discussed the disadvantages of this approach. There is some inconvenience involved in introducing the sample into the Dewar with liquid nitrogen. Normally, a long quartz capillary cell is slowly lowered into the quartz Dewar flask which contains liquid nitrogen. The rate of sample cooling can cause variation in the resulting sample matrix. For example, a clear glass, a cracked glass, or a snow can result. Analytical data from clear glasses are very reproducible if the sample cell is positioned in the instrument in the same way for each measurement. Once the capillary tube comes in contact with liquid nitrogen, the sample cools very quickly, and cracked glasses or snows result for the majority of solvents. The cracked glasses or snows can be avoided if the capillary tube is allowed to cool gradually over the liquid nitrogen before contact is made. This can involve a rather time-consuming procedure. However, there are several solvents available that form clear glasses at 77 K, most notably, pure ethanol and EPA (EPA is a mixture of ethanol, isopentane, and ether, 2:5:5 v/v/v). Winefordner et al. (10) gathered together an extensive list of solvents from the literature and their own work that may be employed for low-temperature phosphorimetry. There are relatively few single solvents which can be employed at low temperature. However, several solvents which are not useful alone do form good mixed solvents (10). Table 5.1 gives a partial listing of solvents that can be used in low-temperature phosphorimetry work.

A rotating sample-cell assembly was developed by Hollifield and Winefordner (11), and later modified by Zweidinger and Winefordner (12). It consisted of a Varian A60-A High Resolution Nuclear Magnetic Resonance Spinner Assembly mounted on a sample-compartment light-cover. The rotating sample-cell assembly was employed to minimize fluctuations of phosphorescence signals resulting from inhomogeneities in snowed media and variations in the diameter of the sample tube (11,12). Other advantages of the rotating sample-cell are (13): (a) the inner-filter effect is reduced; (b) sampling is simpler and more rapid; (c) precision of the measurement of the phosphorescence signal is improved.

Winefordner and co-workers (14,15) developed an open-ended quartz capillary tube which was suitable for aqueous solvents. The inherent photoluminescence of the capillary tube was minimized by using film polarizers. The following advantages were attained with the open-ended quartz capillary tube: (a) shattering of the sample cell by the strain caused by expansion of water at 77 K was prevented by the thick walls of the capillary tube; (b) the sampling procedure was relatively simple; (c) the size of the sample was reduced to about 20 µL.

Most commercial phosphorimeters employ immersion cooling of the sample. However, a more ideal cooling system is based upon conduction cooling of a short capillary sample cell in a copper block cooled by liquid nitrogen (9,16,17). Use of these systems reduces sample turn-over time to approximately one sample/min. Compared to the immersion technique a 3-5-fold reduction in time is realized. Also, there is a reduction in the amount of liquid nitrogen handling needed and the expense is reduced because of the elimination of optical-grade quartz Dewar flasks.

Jones and Winefordner (18) developed a device for rapidly cooling small amounts of sample to 15 K for low-temperature luminescence spectrometry. The sample cooling device was

Table 5.1. Solvents for Phosphorimetry

Type	Composition, v/v	Temperature K
Hydrocarbon	Methane	4.2
	Isopentane	77
	3-Methyl pentane	77
	Pentane (tech. grade 1:1 n-pentane, isopentane)	77
	n-Pentane:n-heptane 1:1	77
	Methyl cyclohexane:n-pentane 4:1-3:2	77
	Methyl cyclohexane:isopentane 4:1-1:5	77
	Methyl cyclohexane:methyl cyclopentane 1:1	77
	Pentene-2(cis)-pentene-2(trans)(mixed isomers)	77
	Paraffin oil (Nujol)	183-195
	Cyclohexane:decalin 1:3	20-77
Alcoholic	Ethanol, <5% H_2O	77
	n-Propanol	77
	Ethanol:methanol 4:1-5:2	77
	Isopropanol:isopentane 3:7	77
	Ethanol:isopentane:diethyl ether 2:5:5(EPA)	77
	Isopropanol:isopentane:diethyl ether 2:5:5	77
	n-Butanol:isopentane 3:7	77
	n- or Isopropanol:isopentane 2:8	77
	Ethanol (96%) and diethyl ether 2:1	77
	Ethanol:glycerol 11:1	77
	Ethanol:methanol:diethyl ether 8:2:1	77
	n-Propanol:diethyl ether 2:5	77
	n-Butanol:diethyl ether 2:5	77
	Diethyl ether:isooctane:isopropanol 3:3:1	77
	Diethyl ether:isooctane:ethanol 3:3:1	77
	Diethyl ether:isopropanol 3:1	77
	Isopropanol:methyl cyclohexane, isooctane 1:3:3	77
	Glycerol	183-195
	Propylene glycol	183
	Ethyl cellosolve:n-butanol:n-pentane 1:2:10	77

Reprinted with permission from J.D. Winefordner, P.A. St. John, and W.J. McCarthy, *Fluorescence Assay in Biology and Medicine*; S. Udenfriend, Ed., Vol. II, Academic Press, New York, 1969.

commercially available. It was a closed-cycle helium refrigerator that cools a copper cold tip to 11 K in 45 min, without the need for liquid helium. Jones and Winefordner designed the cold tip, and it was made of oxygen-free high-conductivity copper (Figure 5.2). A brass belt was threaded around the copper spool, and the other end of the belt was threaded around a plastic spool attached to a rotary motion vacuum feedthrough. The two spools were spaced so the brass belt was taut (Figure 5.3A). The vacuum chamber was also laboratory designed (Figure 5.3B). The spectrometric system used by Jones and Winefordner (18) was constructed from commercially available parts (Figure 5.4). The exciting radiation leaving the monochromator was focused onto the brass belt in contact with the cold tip. With the system, both fluorescence and phosphorescence were measured. Emitted radiation was detected with an optical spectrometric multichannel analyzer, which is labeled as OSMA in Figure 5.4. The OSMA was a linear photodiode array 25.6 mm long with 1024 elements. The cooling device developed by Jones and Winefordner (18) has several advantages: (a) the belt is self-cleaning; (b) samples can be run continuously; (c) there are very few moving parts; (d) no liquid coolant is needed. The disadvantages of the system are the following: (a) the refrigerator is large, heavy, and expensive; (b) obtaining reproducible injected samples is difficult. This problem would be partially solved if an automated injection device were employed. Overall, the cooling device was simple, fast, and easy to use.

Scharf et al. (19) described a novel sample compartment that could be employed for room-temperature and low-temperature phosphorimetry (-65°C) of samples held in solid substrates. The

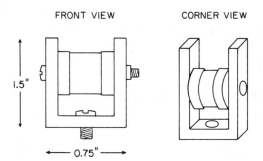

FRONT VIEW   CORNER VIEW

1.5"

0.75"

Figure 5.2. Detailed drawing of the copper cold tip. (Reprinted with permission from B.T. Jones and J.D. Winefordner, *Anal. Chem.*, 60, 412. Copyright 1988 American Chemical Society.)

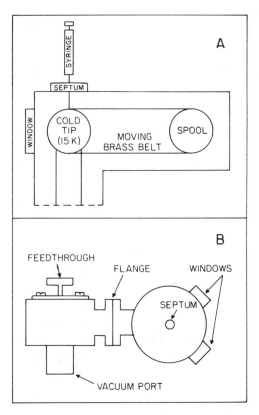

Figure 5.3. Simplified cross-sectional side view of the upper portion of the vacuum chamber (A); Top view of the vacuum chamber (B). (Reprinted with permission from B.T. Jones and J.D. Winefordner, *Anal. Chem.*, 60, 412. Copyright 1988 American Chemical Society.)

sample compartment assembly was designed for the Perkin-Elmer LS-5 fluorometer and contained four positions for samples and blank and permitted convenient subtraction of the blank from the sample spectra.

Lehotay et al. (20) evaluated a cryogenic miniature refrigeration system for low-temperature luminescence measurements. With the system, it was not necessary to handle liquid nitrogen and Dewar flasks. In addition, the refrigeration system was easier to handle than a Joule-Thompson refrigerator. Joule-Thompson refrigerators are usually large and bulky, often requiring elaborate vacuum systems and expensive gas compressors. With their system, a small glass cold finger is cooled by the expansion of

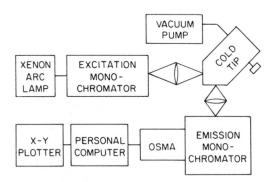

Figure 5.4. Block diagram of the spectrometric system. (Reprinted with permission from B.T. Jones and J.D. Winefordner, *Anal. Chem.*, 60, 412. Copyright 1988 American Chemical Society.)

nitrogen gas at 1800 psi input pressure from a standard gas cylinder. In a moderate vacuum, the cold finger reached a stable temperature of 80 K in approximately 15 min.

Gifford et al. (21) constructed a thin-layer, single-disk, multislot phosphorimeter for the direct measurement of phosphorescence from separated components on thin-layer chromatograms near liquid nitrogen temperature. Their phosphorimeter could also be used to measure room-temperature phosphorescence. Later, Miller et al. (22) described the construction of an improved thin-layer phosphorimeter. The thin-layer phosphorimeter was designed to fit the sample compartment of a spectrofluorometer. An aluminum-backed thin-layer chromatoplate was affixed with elastic bands to the outside of a hollow copper sample drum which could be filled with liquid nitrogen. The bottom of the drum was lipped to permit positioning on a turntable in the holder compartment. The rate of rotation of the turntable was controlled by a variable-output transformer and a scanning rate of 3-40 cm min^{-1} was provided.

Leigh and Leaback (23) designed an apparatus for the photographic recording of phosphorescent spots on thin-layer chromatoplates at 77 K. The apparatus is shown schematically in Figure 5.5. The system included a thermally insulated tray (D) which contained the developed thin-layer chromatoplate. In some experiments, the temperature of the thin-layer chromatoplate was estimated by using a digital thermometer with a thermocouple probe. In a typical experiment, the cooled thin-layer chromatoplate was exposed to a uniform area (10-cm diameter) of ultraviolet

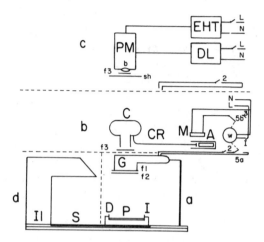

Figure 5.5. Schematic diagrams of apparatus for recording phosphorescence on thin-layer chromatograms: (a) The TLC plate (P) is placed in a metal tray (D), insulated with expanded polystyrene (I), into which is poured liquid nitrogen. The flash-gun (G), 22 cm above the plate, had its plastic filter replaced by an infrared-absorbing filter (fl, and (optionally) a 313-nm interference filter (f2). (b) The programmer (w), controlled manually by switch 3 (L, live; N, neutral), is set up to close switch 5a, to fire the flash gun. The gun is swung out of the way immediately after the flash. After a 1-s delay, switch 5b is closed by the programmer for 14 s, operating the solenoid M-A and so opening the camera (C) shutter via the cable release (CR). The photographic film is exposed through the filter (f3). After the exposure, the opening of switches 2 and 3 enables the programmer to be re-adjusted prior to the next shot, without triggering unwanted flashes or photo-graphic exposures, respectively. (c) The camera may be replaced by a photomultiplier (PM) tube connected to a data-logger (DL) and an extrahigh-tension module (EHT). The flash can be manually activated by closing switch 2 with the PM lens (b) covered by a black shutter card (sh). The Metz Metablitz flash-gun may be replaced by a Bowens Mono-lite flash-gun mounted rigidly vertically above the tray (D). After the flash has been triggered by the programmer (w), the camera is swung through 90° into position above the dish and below the monoite flash-gun. (d) The flash-gun is replaced by the Oriel Photo-resist Illuminator (II); the tray (D) with the TLC place is moved on side (S) to a position under the camera during the 3-s delay. The activation time, delay time, and photographic exposure time are controlled by the programmer (w). (Reprinted with permission from A.G.W. Leigh and D.J. Leaback, *Anal. Chim. Acta* 1988, 212, 213..)

radiation for an interval of time such as 30 s. Immediately after irradiation was completed, the tray (D) was moved away manually on slide S to permit photographic exposure (Figure 5.5d). Phosphorescence lifetimes were also measured with the system. For these experiments, the camera/programmed shutter assembly

(Figure 5.5b) was replaced by a photomultiplier tube. The voltage output developed across a 1 megaohm load resistor was recorded on a data-logger (Figure 5.5c). For collecting routine quantitative data, they found that a simple (Metablitz) flash-gun proved to be more convenient and reproducible than a mercury discharge lamp. However, for the photography of the phosphorescent zones, the simple flash-gun did not produce a sufficiently uniform field of activating radiation. When they used a more powerful Monolite flash system, the field of illumination was much more uniform, and better results were obtained.

The instrumental systems, accessories, and sampling techniques discussed in this section illustrate the variety of approaches that have been developed for obtaining phosphorescence at low temperature from phosphors in solution and phosphors adsorbed on solid surfaces. Although there are disadvantages to measuring phosphorescence at low temperature, many times these disadvantages are outweighed by the sensitivity and spectral resolution that can be achieved. Usually, fluorescence can be measured at low temperature, and thus, both fluorescence and phosphorescence data can be obtained from a given sample. This is particularly important in the analysis of complex samples that contain several luminescent components.

5.3.   Applications

5.3.1. Solid Surfaces

Sawicki and Pfaff (24) developed methods for direct analysis of aromatic compounds on paper and thin-layer chromatograms by spectrophosphorimetry at liquid nitrogen temperature. They applied their methodology to air pollution analysis. The separated components were cut from filter paper, glass-fiber paper, or thin-layer chromatoplates. The solid matrix was put into a quartz sample tube, and the tube was positioned upside down in the cell compartment so the spot to be examined was in the source beam and perpendicular to it. They emphasized that drastic changes could occur in phosphorescence spectra for adsorbed compounds in a wet state compared to a dry state. Sawicki and Pfaff (24) readily characterized airborne particulate samples for phenanthrene and benzo[e]pyrene by obtaining phosphorescence spectra of the compounds adsorbed on glass-fiber paper.   In related work, Sawicki and Pfaff (25) introduced quenchophosphorimetry.   Components on a solid surface such as filter paper with and without a quencher were submerged in liquid nitrogen and then exposed to ultraviolet radiation.   When the exciting radiation was removed, the phosphorescence of the spots was noted quickly. A variety of compound types were investigated. The main advantage of this approach was the selectivity that could be obtained for certain compound types present in mixtures of compounds.   Several other researchers have used low-temperature phosphorescence for the characterization of compounds adsorbed on solid surfaces (26-30).

Leigh et al. (30) detected 4-methylumbelliferone and glycoside-I by fluorescence and phosphorescence on electrophero-grams, thin-layer chromatoplates, and paper chromatograms. They submerged the separated components, which were adsorbed on the solid matrix, in a thin layer of liquid nitrogen.   The components were excited with a Hanovia mercury strip lamp.   They photo-graphed the phosphorescence components after removing the mercury lamp. The limit of detection by eye for 4-methylumbelli-ferone was about the same with either the fluorescence or phospho-rescence procedures.   However, the limit of detection by eye for phosphorescence for glycoside-I was considerably lower with the phosphorescence approach compared to the fluorescence approach. For example, the limit of detection was 15 nmole by fluorescence, but 0.05 nmole by phosphorescence.

In later work, Leigh and Leaback (23) considered the low-temperature phosphorimetric determination of some 4-methyl-7-hydroxy coumarin (MU) derivatives on thin layer chromatoplates. The derivatives of MU were its 2-acetamido-2-deoxy-β-D-glucoside and the corresponding β-(1→4)-D-galacto-2- acetamido-2-deoxy-β-D-glucoside.   The apparatus they used was discussed in Section 5.2 and is illustrated in Figure 5.5. Earlier work by the authors showed that MU conjugates gave a purple fluorescence (23). However, on a solid surface at low temperature, a blue-green phosphorescence emission was observed, and phosphorescence life-times of about 1s were obtained. The low-temperature phosphores-cence approach permitted a sensitive and nondestructive way of easily visualizing the compounds on thin-layer chromatoplates. Reasonably good quantitative measurements of the MU conjugates were made, and the apparatus in Figure 5.5 permitted good quality photographs of the thin-layer chromatographs to be obtained.

As discussed in Section 5.2, Gifford et al. (21) constructed and evaluated a scanning thin-layer single-disc multi-slot phosphorimeter for the direct detection of phosphorescent components separated on thin-layer chromatograms. With an improved phosphorimeter system, Miller et al. (22) investigated solvent enhancement effects in low-temperature thin-layer phosphorimetry. They showed that the phosphorescence intensities of several adsorbed compounds were enhanced greatly by spraying the chromatography medium with a suitable solvent prior to examination of the chromatoplate. The enhancement of the phosphorescence was dependent on the stationary phase, the structure of the adsorbed material, and the solvent sprayed. Nanogram quantities of the separated solutes could be detected. Table 5.2 gives the detection limits of several compounds on thin-layer chromatoplates.

Using a cryogenic miniature refrigeration system (See Section 5.2), Lehotay et al. (20) reported analytical figures of merit for Rhodamine 6G, warfarin, and p-aminobenzoic acid adsorbed on filter paper. The phosphorescence of p-aminobenzoic acid adsorbed on filter paper could only be observed below 25°C. An important advantage of their system was the ability to study small volumes of sample. This would make the system attractive for clinical and biological applications.

The continuous cooling belt for low-temperature molecular

Table 5.2. Detection Limits of Compounds on TLC Plates (ng/spot)

Compound	TLC phosphorimetry 77 K	Visual detection limits, native phosphorescence 77 K[a]
Sulfanilamide	0.5	100
Sulfadiazine	2	1000
Sulfamethoxazole	3	1000
N-4-Acetylsulfanilamide	7	-
4-Aminobenzoic acid	0.6	100
Procaine hydrochloride	2	100

[a]de Silva, J.A.F.; Strojny, N. *Anal. Chem.* 1975, 47, 714. Reprinted with permission from J.N. Miller, D.L. Phillipps, D.T. Burns, and J.W. Bridges, *Anal. Chem.*, 50, 613. Copyright 1978 American Chemical Society.

luminescence spectrometry developed by Jones and Winefordner (18) (See Section 5.2) is essentially a low-temperature solid-surface luminescence technique. In their particular work, they measured either fluorescence or phosphorescence signals for a given compound. Limits of detection were reported for several polycyclic aromatic hydrocarbons, and the limits of detection ranged from 0.05 ng for 3,4-benzopyrene to 10 ng for phenanthrene. The linear dynamic range was 2 and 3 orders of magnitude in all cases. The relative standard deviation ranged from 3.6% to 10.7%. They demonstrated the applicability of the rotatable cooling device with a synthetic mixture of six polycyclic aromatic hydrocarbons. Table 5.3 gives the results they obtained (18). The % error in Table 5.3 was attributed to several factors: overlap of peaks, poor reproducibility of injections, prefilter and postfilter effects, and possible formation of excimers. For multicomponent analysis, such error is usually tolerable, especially when the analysis can be performed rapidly, which is the case for the continuous cooling belt system. In related work, Jones et al. (31) determined polycyclic aromatic hydrocarbons in cooked beef by low-temperature molecular fluorescence using the moving sample cooling belt. In principle, low-temperature phosphorescence signals could have been measured for the samples. However, phosphorescence data were not reported.

5.3.2. Shpol'skii Spectrometry

Shpol'skii spectrometry involves obtaining emission spectra of aromatic compounds dissolved in n-alkanes at low temperature (32,33). Very narrow band spectra are obtained that correspond to transitions involving vibrationally excited states of the electronic ground state. The reduction of spectral bandwidths is essentially matrix induced and frequently a xenon lamp is used to excite the luminescence. Shpol'skii spectrometry is normally associated with fluorescence spectrometry; however, several reports of phosphorescence spectra that show the Shpol'skii effect have been published (32,33).

Colmsjö et al. (34) investigated several sulfur heterocyclic polyaromatics and obtained well resolved fluorescence and phosphorescence Shpol'skii spectra. Figure 5.6 shows the phosphorescence spectrum of 7,8-epithiobenzo[ghi]perylene. The authors commented that the polycyclic aromatic hydrocarbon analog coronene also exhibits a strong quasi-linear phosphorescence.

Table 5.3. Quantitative Analysis of a Six-Component Mixture of PAHs in Hexane

PAH	Concn found, mg / L	Actual conc, mg / L	% Error
1,12-Benzoperylene	0.24 ± 0.14	0.30	20
1,2-Benzopyrene	0.40 ± 0.15	0.30	33
Coronene	0.42 ± 0.20	0.36	17
Perylene	0.41 ± 0.14	0.30	37
Phenanthrene	4.60 ± 2.82	3.30	39
Triphenylene	0.60 ± 0.45	0.70	14

Reprinted with permission from B. T. Jones and J. D. Winefordner, *Anal. Chem.*, <u>60</u>, 412. Copyright 1988 American Chemical Society.

Garrigues and Ewald (35) employed Shpol'skii spectrometry in n-alkanes frozen at 15 K for the identification of monomethylated isomers of pyrene, phenanthrene, and chrysene from crude oils. Best results were obtained for identification after fractionation of the crude oil by liquid chromatography. Figure 5.7 shows the emission spectra of monomethylphenanthrene in an equimolar

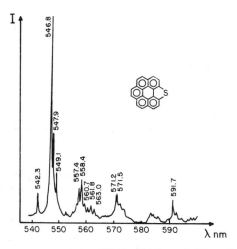

Figure 5.6. Phosphorescence spectrum of 7,8-epithiobenzo[ghi]perylene at 63 K (solvent, n-hexane). (Reprinted with permission from A.L. Colmsjo, Y.U. Zebuhn, C.E. Östoman, *Anal. Chem.* <u>54</u>, 1673. Copyright 1982 American Chemical Society.)

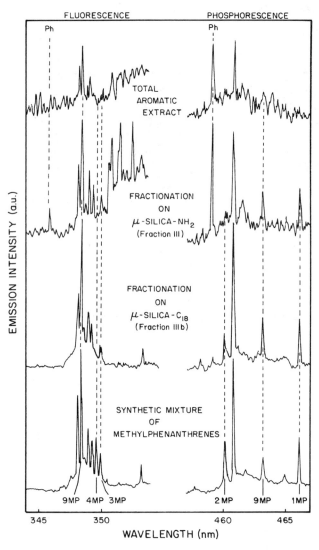

Figure 5.7. Emission spectra of monomethylphenanthrenes (MP) in an equimolar synthetic mixture (each MP at C = 2 x 10^{-7} M) and in the crude oil extracts, at different levels of fractionation, frozen in n-hexane at 15 K. Excitation at 298 nm for fluorescence spectra. Excitation at 297 nm for phosphorescence spectra. Peaks noted "Ph" indicate the emission of phenanthrene. The attributed peaks are only due to pure bands related respectively to identified compounds by reference with emission bands from synthesized molecules. (Reprinted with permission from P. Garrigues and M. Ewald, *Anal. Chem.* 55, 2155. Copyright 1983 American Chemical Society.)

synthetic mixture and in a crude oil extract.   By the use of appropriate excitation wavelengths, the authors were able to identify each methylphenanthrene in the synthetic mixture either from the fluorescence spectrum or from the phosphorescence spectrum. This clearly shows how both fluorescence and phosphorescence can be combined to characterize mixtures of compounds.   Figure 5.7 also illustrates how chromatography and luminescence spectrometry can be combined to identify components in complex mixtures.   Later, Garrigues et al. (36) reported high-resolution fluorescence and phosphorescence spectra of polycyclic aromatic hydrocarbons in Shpol'skii matrices.   They illustrated the analytical applicability of the technique for the identification of methylated polycyclic aromatic hydrocarbons in petroleum, sediment, and aerosol.

Elsaiid et al. (37) used laser excited Shpol'skii spectrometry to directly determine nitrogen-, oxygen- and sulfur-heterocyclic compounds in solvent refined coal, petroleum crude oil, and carbon black.   Both fluorescence and phosphorescence spectra were reported.   Characteristic quasi-linear Shpol'skii excitation and emission spectra were presented for the first time under site-selective excitation conditions.   In the site-selection technique, guest molecules occupy several different microenvironments or "sites" in a low-temperature matrix.   Thus, purely electronic energy levels of different molecules of the same solute are shifted to different extents.   Only those molecules with an energy difference between ground and excited state that exactly matches the photon energy can be excited.   Elsaiid et al. (37) demonstrated the potential for using the laser excited Shpol'skii technique in environmental and biological studies for the direct determination of nitrogen-, oxygen-, and sulfur-heterocyclic compounds and substitutional derivatives of the parent polycyclic aromatic compound.   The primary advantages of Shpol'skii spectra obtained by laser excitation are: (a) high selectivity for isomeric compounds; (b) direct analysis of mixtures with little sample pretreatment; (c) direct quantitation using deuterated analogs or standard additions; and (d) high sensitivity due to the combination of narrow excitation bands and the narrow-band, high spectral irradiance of the laser source (37,38).

The examples in this section show how fluorescence and phosphorescence Shpol'skii spectrometry complement one another. When both fluorescence and phosphorescence Shpol'skii spectra can be obtained from a sample, the amount of luminescence information is very substantial. This is a tremendous advantage for the identification and quantitation of luminescent components in mixtures.

5.3.3. Phosphorescence Line-Narrowing Spectrometry

Fluorescence line-narrowing or site-selection spectra have been reported for several compounds (33,39,40). The site-selection technique was briefly discussed in the previous section in reference to the work by Elsaiid et al. (37). With this approach, laser-excited narrow-line fluorescence spectra may be obtained from a low-temperature matrix from a well-defined distribution of excited molecules, denoted as an isochromat (33). Very few phosphorescence line-narrowing spectra have been reported. To obtain line-narrowed phosphorescence spectra, it is necessary to excite the T_1-S_0 transition with a monochromatic excitation source. The T_1-S_0 transition is a spin-forbidden transition, and thus, it is difficult to record phosphorescence line-narrowing spectra. Hofstraat et al. (33) have stated that because of the extremely low absorption coefficients of T_1-S_0 transitions, phosphorescence line-narrowing spectrometry probably has little analytical potential. However, additional work needs to be performed before detailed statements on the advantages and disadvantages of this technique can be given.

5.3.4. Low-Temperature Solution Phosphorescence

Some of the historical aspects of low-temperature solution phosphorimetry were given in Chapter 1. All of the earlier work in phosphorimetry was done with solutions at low temperature, and applications continue to appear today. Several reviews on low-temperature solution phosphorimetry have appeared (9,10,13,41-53). In this section, selected applications will be discussed to illustrate the variety of samples that can be analyzed by low-temperature solution phosphorimetry.

In 1957, Keirs et al. (54) reported several analytical uses of low-temperature solution phosphorimetry. They considered the potential for obtaining selectivity by spectral resolution of excitation and emission radiation and commented on phosphorescence time resolution. In 1958, Freed and Salmre (55) reported the construction of a phosphorimeter which was used for the determination of several drugs. Then, in 1962, Parker and Hatchard (56) modified a spectrofluorimeter for obtaining phosphorescence spectra. A year later Latz and Winefordner (57) described the construction of a phosphorimeter. Since that time, most of the work on low-temperature phosphorimetry has originated from Professor Winefordner's research group at the University of Florida.

Winefordner and Latz (57) applied phosphorimetry to the determination of aspirin in blood serum and plasma. The aspirin in the serum or plasma was extracted with chloroform, the chloroform was evaporated, and the residue was dissolved in EPA, which is a mixture of ethyl ether:isopentane:ethyl alcohol (5:5:2), and then, the appropriate phosphorescence data were obtained. None of the components normally present in serum or plasma resulted in serious interference.

McGlynn et al. (58) investigated the total luminescence of organic molecules of interest in petroleum samples. They obtained low-temperature fluorescence and phosphorescence data for naphthalene, phenanthrene, and 1,2,4,5-tetramethylbenzene. In addition, fluorescence data at room-temperature were obtained for the previous compounds. With the data obtained, they employed the fluorescence and phosphorescence for the analysis of mixtures of the three components. Their work serves to illustrate how fluorescence and phosphorescence complement one another.

Sheridan et al. (59) determined nanogram quantities of DNA by phosphorescence using Ag^+ as an external heavy atom. Boutilier et al. (60) used sodium iodide and silver nitrate as heavy atoms in the phosphorescence analysis of nucleosides naturally occurring in DNA with methanol:water (10:90) at 77 K. They investigated deoxyadenosine, deoxyguanosine, thymidine, and cytidine. The proper selection of pH, solvent matrix, and metal ion permitted the phosphorescence enhancement of one species while suppressing the phosphorescence of another species. Minimum detectable quantities ranged from 0.5 to 50 ng.

Aaron and Winefordner (61) reported phosphorescence data for purines in aqueous solution at 77 K. Purine derivatives are very important in nucleic acid chemistry. The authors used a rotating capillary tube as a sample cell and obtained the phosphorescence data from rigid aqueous solutions. Phosphorescence excitation and emission spectra, lifetimes, phosphorimetric analytical curves, and limits of detection were obtained. The phosphorescence spectra of the purines showed fine structure, which would be useful for identification purposes. In addition, the purines had high phosphorescence yields, and low detection limits were obtained. Absolute limiting detectable quantities of purine and the eight derivatives investigated were in the picogram range. Table 5.4 gives the

Table 5.4. Phosphorescence Analytical Characteristics of Substituted Purines[a]

Purine	Concentration range (M) of near linearity[b]	Slope of linear portion	Standard deviation[c]	Linear correlation coefficient	Limit of detection μg/mL[d]	Minimal detectable amount, ng[e]
Purine	2×10^3	0.88	0.04	0.999	0.01	0.2
6-Aminopurine	3×10^3	0.76	0.05	0.998	0.02	0.4
6-Methylpurine	3×10^3	0.96	0.02	1.000	0.01	0.2
6-Methylmercaptopurine	10^4	0.97	0.04	0.999	0.0006	0.01
6-Benzylaminopurine[f]	2×10^3	0.98	0.02	1.000	0.02	0.4
6-Chloropurine	10^4	0.91	0.07	0.999	0.001	0.03
6-Bromopurine	10^4	0.89	0.06	0.999	0.002	0.04
2-Amino-6-meyhyl-mercaptopurine[f]	10^5	0.94	0.03	1.000	0.0002	0.004
2,6-Diaminopurine	10^3	0.93	0.08	0.998	0.15	3.0

[a]In neutral methanol/water, 10/90 solution, v/v except otherwise noted.
[b]Near linearity means region over which slope of analytical curve is within 1% of the values designated in column 3.
[c]Standard deviation of phosphorescence signals taken on the linear portion of the analytical curve.
[d]Limit of detection is defined as the concentration giving a phosphorescence signal (located on the linear part of the analytical curve) two times greater than the background noise. The background signal was subtracted from the observed signal value or suppressed by means of the bias adjustment of the nanoammeter read-out.
[e]Absolute limiting quantity of compound detected by the method–calculated from the limit of detection with a volume of sample of 20 μL.
[f]Analytical curves obtained in H_2SO_4 0.1 N MeOH/H_2O 10/90 v/v solution.

Reprinted with permission from J.J. Aaron and J.D. Winefordner, *Anal. Chem.*, 44, 2127. Copyright 1972 American Chemical Society.

analytical data for purine and its substituted derivatives.

Phosphorescence versus pH titration curves for cytosine, cytidine, cytidine-5'-monophosphate, -diphosphate, and -triphosphate were obtained in methanol/water (10/90) and in a variety of aqueous sodium halide solutions frozen at 77 K (62). The shapes of the titration curves showed that molecular aggregates or "puddles" of cytosine and cytidine were present in relatively concentrated frozen solution (10^{-3} M). The aggregates were dissociated in dilute frozen solution ($\leq 10^{-4}$ M) or in ~1 M NaCl. No molecular aggregates were found for concentrated solutions of cytidine-5'-monophosphate, -diphosphate, and -triphosphate. Sodium bromide and sodium iodide concentrations of 0.1 M gave "reversed" sigmoidal phosphorescence titration curves for cytidine because of an anomalously large heavy-atom enhancement factor in acidic solution ranging between 30 and 50.

Rahn and Landry (63) investigated the fluorescence and phosphorescence (at 77 K) and the photochemistry (at 298 K) of several polynucleotides and DNA complexed with Ag^+. For all the samples investigated, they found that Ag^+ gave the heavy-atom effect which was indicated by the quenching of fluorescence, enhancement of phosphorescence, and the reduction of phosphorescence lifetime. Their results suggested a possible triplet precursor for thymine dimerization in Ag^+ complexes. Also, their studies revealed that the photochemistry of DNA was greatly altered when Ag^+ was bound to DNA. This suggested the possible application of Ag^+ binding in photochemical and photobiological studies.

The phosphorescence excitation and emission spectra and lifetimes of vitamins K_1, K_3, and K_5 were obtained at 77 K in n-hexane, methanol, ethanol, and mixtures of methanol-water (64). The limits of detection ranged from 0.07 to 1.5 µg/mL, but they were not influenced by the nature of the solvent. The authors considered the application of the phosphorimetric method for the determination of the K vitamins with previously published analytical methods. The limit of detection compared very favorably with the different analytical techniques for the determination of vitamins K_1 and K_3. For example, the limit of detection of vitamin K_1 was 200 µg/mL by gas chromatography and 1 µg/mL by phosphorimetry. Also, the phosphorimetric analytical curves were linear over about three orders of magnitude in concentration, whereas the linearity for the other methods was generally over a range of concentration of less than 2 orders of magnitude. Finally, the absolute minimal detectable amounts of vitamins K_3 and K_1 by phosphorimetry were 2 and 20 ng, respectively, which compared well with the respective values by chromatographic methods, namely, 4 and 500 ng.

Lukasiewicz et al. (15) studied the influence of methanol-water mixtures and of sodium chloride, sodium bromide, and sodium iodide aqueous solutions on the phosphorescence signals of several organic compounds at 77 K. The model compounds used were 3-indoleacetic acid, hippuric acid, and sulfacetamide. They found that only a few percent by weight of methanol in water, sodium iodide in a methanol-water mixture, and sodium chloride, sodium bromide, or sodium iodide in water yielded phosphorescence signals several orders of magnitude greater than the phosphorescence signal for the same sample in pure water. The salt solutions gave higher phosphorescence signals than methanolic solutions due to the heavy-atom effect. Iodide was the most effective heavy atom. They studied the systems with an open rotating quartz capillary cell and concluded that the optimum aqueous solvent for routine analytical phosphorimetric measurements was 5-30% aqueous solution of sodium iodide.

Mousa and Winefordner (65) employed phase-resolved phosphorimetry (See Chapter 3) and showed that phosphorescence emission and excitation spectra that severely overlapped could be phase resolved into the spectra of individual components. Fluorescence emission was also phase resolved from phosphorescence emission. The experiments were carried out with dilute solutions of pure compounds and synthetic binary mixtures. Examples of some of the compounds investigated are: benzophenone, 4-iodobiphenyl, and 4,4'dibromobiphenyl. The quantitation of the binary mixtures was found to be accurate and precise.

Mau and Puza (66) were interested in studying the triplet state of chlorophylls in relationship to the mechanism of photosynthesis. The triplet state of chlorophylls can be difficult to study because the phosphorescence emission is in the near infrared, and the phosphorescence efficiencies are low. They obtained the phosphorescence emission spectra of chlorophyll a, chlorophyll b, and their magnesium free analogs pheophytin a and pheophytin b in ether:isopentane:ethanol (5:5:2) at 77 K. Single emission bands were observed for the pure samples. For example, a typical emission band maxima was at 950 nm for chlorophyll a. The phosphorescence quantum yield for the compounds was 5×10^{-5}, and the phosphorescence lifetimes were in the range of 1.5-2.1 ms.

Aaron et al. (67) reported the room-temperature fluorescence and low-temperature phosphorescence analytical data for eight substituted quinolines in ethanol-water (10:90). The authors commented that the vibrational structure of the phosphorescence bands and phosphorescence lifetimes would be useful for the identification of the quinolines. The limits of detection for the quinolines were in the ppm or ppb range, depending on the specific quinoline derivative.

Acuna et al. (68) described the quantitative analysis of benzoic acid and o-, m-, and p-toluic acids by low-temperature phosphorimetry. These compounds have fluorescence quantum yields that were close to zero; thus, phosphorimetry was a useful approach for the analysis of these compounds. The four acids showed dimerization in nonpolar solvents. However, by using isopentane:ether (1:1), only monomeric species were present. In addition, this solvent system gave a transparent rigid glass at 77 K. The calibration curves were linear from 10^{-7} M to 10^{-3} M and the limit of detection was about 0.02 µg/mL.

The phosphorescence characteristics of fifty-two pesticides were obtained by Moye and Winefordner (69). Thirty-two of them gave sufficient phosphorescence so that excitation spectra, emission spectra, decay times, analytical curves, and limits of detection could be obtained. Moye and Winefordner (70) developed a method for the determination of urinary p-nitrophenol by thin-layer chromatography and phosphorimetry. p-Nitrophenol is a major metabolite of parathion. The time needed for the entire procedure was 40 min, and only 5 mL of urine was required for the analysis of samples containing 0.01 µg of p-nitrophenol. The average recovery of p-nitrophenol from urine samples was 88%, and the relative standard deviation of 2.5% was obtained. In later work, Aaron et al. (71) compared the low-temperature and solid-surface room-temperature phosphorescence of thirty-two pesticides. Low-temperature phosphorimetry was shown to be a sensitive technique, with limits of detection from 0.001 to 30 µg/mL. The solid-surface room-temperature phosphorescence approach was very simple and specific for some of the pesticides with absolute limits of detection between 10 and 50 ng. A detailed discussion of solid-surface techniques is given in Chapters 6-8.

Corfield et al. (72) assessed the use of low-temperature phosphorescence spectrometry for crude oil identification. They evaluated conventional phosphorimetry, synchronous excitation phosphorimetry, phosphorescence contour mapping, and phosphorescence lifetimes. Conventional phosphorescence spectra were not structured sufficiently to offer adequate discrimination among the crude oil samples investigated. Phosphorescence lifetime data were not very useful because of the complex nature of the observed decay rates. Synchronous excitation phosphorimetry has been discussed by Vo-Dinh and Gammage (73) and essentially involves scanning both the excitation and emission monochromators simultaneously with a fixed wavelength difference between the excitation and emission monochromators. With this technique, sharp phosphorescence emission bands are obtained which are useful for the characterization of mixtures of compounds. Corfield et al. (72) concluded that synchronous excitation phosphorimetry complemented synchronous excitation fluorometry for the identification of crude oil samples. The most successful technique they investigated was total-phosphorescence-contour spectra. These were two-dimensional representations of the three-dimensional dependence of phosphorescence intensity on excitation and emission wavelengths. The contour diagrams permitted differentiation among eight oil samples studied either through superposition or subtraction methods.

Wolfbeis et al. (74) obtained phosphorescence excitation and emission spectra and detection limits for twenty-two nitrated polynuclear aromatic hydrocarbons. Detection limits ranged from 0.3 ng/mL to 1 µg/mL. The spectral properties of 1-nitropyrene allowed its selective determination in the presence of other nitroaromatics. Scharf and Winefordner (75) investigated the effect of environment on the luminescence properties of p-aminoacetophenone in several matrices. Phosphorescence properties were reported for the compound adsorbed on filter paper at room temperature, in ethanol at 80 K, and in ether:pentane (4:1) at 80 K.

Khasawneh and Winefordner (76) described the room-temperature fluorescence and low-temperature phosphorescence characteristics of biphenyl and several polychlorinated biphenyls using hexane as a solvent. Analytical figures of merit and low-temperature phosphorescence lifetimes for biphenyl, eight congeners, and nine Aroclors were given (See Table 5.5). They pointed out that low amounts of each of the compounds could be determined either by room-temperature fluorescence or low-temperature phosphorescence. Limits of detection were reported to be between 0.08 ng/mL and 6.0 ng/mL using room-temperature fluorescence and 0.16 ng/mL and 7.5 ng/mL using low-temperature phosphorescence. They concluded that it was very difficult to determine one congener in the presence of others using either room-temperature fluorescence or low-temperature phosphores-

Table 5.5. Low-Temperature Phosphorescence Characteristics of Biphenyl and Some Polychlorinated Biphenyls

Compound	Limit of detection (ng / mL)	Lifetime[a] (s)	Relative standard deviation (%)	Blank relative standard deviation (%)
Biphenyl	0.80	2.5±0.21	2.5	9.0
4-Chlorobiphenyl	0.60	1.3±0.12	3.2	8.0
3,3'-Dichlorobiphenyl	0.65	1.45±0.12	9.5	3.5
3,4-Dichlorobiphenyl	0.30	1.65±0.18	4.0	3.5
4,4'-Dichlorobiphenyl	0.16	1.18±0.10	2.5	4.5
3,3',4,4'-Tetrachloro-biphenyl	0.75	1.10±0.13	1.0	9.5
3,3',5,5'-Tetrachloro-biphenyl	0.85	1.10±0.10	2.0	9.5
3,3',4,4',5-Penta-chlorobiphenyl	1.7	1.05±0.17	2.0	8.0
3,3',4,4'5,5'-Hexa-chlorobiphenyl	0.60	0.9±0.09	1.5	9.0
Aroclor 1016	0.50	0.80±0.07	3.0	4.0
Aroclor 1221	0.30	1.44±0.13	3.0	7.0
Aroclor 1232	0.30	1.24±0.13	2.5	8.0
Aroclor 1242	0.40	1.04±0.11	4.0	7.0
Aroclor 1248	1.6	1.0±0.10	2.0	9.0
Aroclor 1254	1.7	1.08±0.12	2.0	2.5
Aroclor 1260	2.0	1.15±0.11	2.5	10.0
Aroclor 1262	2.5	0.95±0.12	3.5	10.0
Aroclor 1268	7.5	0.82±0.095	4.0	6.5

[a]Lifetime is corrected for the instrument response.

Reprinted with permission from I. M. Khasawneh and J. D. Winefordner, *Microchem. J.* 1988, 37, 86.

cence. However, the data they obtained would be useful in conjunction with other techniques such as gas chromatography and liquid chromatography luminescence detection.

The luminescence characteristics of dibenzofuran and several polychlorinated dibenzofurans and dibenzo-p-dioxins were reported (77). These compounds are very stable and are the subject of concern by many environmentalists. The room-temperature fluorescence, low-temperature phosphorescence, and room-temperature phosphorescence properties of dibenzofuran and several polychlorinated dibenzofurans and dibenzo-p-dioxins were given by Khasawneh and Winefordner (77). Limits of detection as low as 0.02 ng/mL, 0.75 ng, and 0.60 ng/mL were listed for room-temperature fluorescence, room-temperature phosphorescence, and low-temperature phosphorescence, respectively. The luminescence data that they reported showed the suitability of the various luminescence approaches for the determination of these compounds.

The fluorescence and phosphorescence characteristics of fourteen vitamins were evaluated by Aaron and Winefordner (78). Analytically useful fluorescence signals were obtained for p-amino-benzoic acid, folic acid, calciferol, pyridoxine hydrochloride, riboflavine, α-tocopherol, and vitamin A. Useful phosphorescence signals were obtained for p-aminobenzoic acid, folic acid, niacin-amide, pyridoxine hydrochloride, and α-tocopherol. The other vitamins studied did not fluoresce or phosphorescence or could not be measured because of experimental problems. Limits of detection via phosphorescence were in the nanogram range. Their study showed the complementary nature of fluorescence and phosphorescence.

Miles and Schenk (79) examined the fluorescence and phosphorescence characteristics of seven phenylethylamines and ten barbiturates. Phenobarbital behaved very unusually because its phosphorescence emission maximum at 370 nm was at a lower wavelength than its fluorescence emission maximum at 415 nm in 0.1 N alcoholic sodium hydroxide.

Aaron et al. (80) carried out an analytical study of the phosphorescence of pyrimidine derivatives in frozen aqueous solution. Methanol:water (10:90) was used for seven pyrimidine derivatives. These compounds are important because of their presence in natural macromolecules and of their application in biochemistry and clinical chemistry. A basic medium (pH ~11) showed enhanced phosphorescence, and limits of detection between 10^{-5} to 10^{-8} M were obtained.

Morrison and O'Donnell (81) developed a method for the determination of diphenylhydantoin in plasma by phosphorescence spectrometry after permanganate oxidation. The oxidation product was extracted into methylcyclohexane, followed by phosphores-cence measurement at 446 nm. A detection limit of 0.05 µg/mL was obtained, with an error of less than 10%. Sample sizes considerably less than 1 mL were employed, and the presence of phenobarbital or other anticonvulsants did not interfere with the determination.

Bridges et al. (82) reported the luminescence properties of several sulfonamide drugs. They emphasized the fluorescence characteristics of these drugs, but phosphorescence data were reported for sixteen sulfonamides. They commented that when fluorescence yields were low for the sulfonamides, phosphorescence could be used for an analysis. The ultraviolet absorption spectra, fluorescence and phosphorescence excitation and emission spectra of seven thiobarbiturates were examined by King and Gifford (83). Their work showed primarily the complementary nature of fluorescence and phosphorescence. Gifford et al. (84) considered the luminescence characteristics of several classes of drugs affecting the central nervous system. The excitation, fluorescence and phosphorescence properties of twenty-nine compounds of psychopharmacological interest were studied at 77 K. They concluded that phosphorimetric analysis could often be used to obtain information concerning the mechanism of psychotropic drug interactions not obtainable because of insensitive analytical methods. Twenty anti-inflammatory and antipyretic drugs were examined for their fluorescence and phosphorescence characteristics. Room-temperature fluorescence, low-temperature fluorescence, and low-temperature phosphorescence excitation and emission wavelengths were reported. In addition, the phosphorescence lifetimes were reported for several compounds. Their results showed that of the twenty compounds studied, all but six could be determined at trace levels. Also, fluorescence and phosphorescence complemented one another. For example, phenazine gave a very weak fluorescence at 370 nm, but a strong phosphorescence with two maxima at 405 nm and 420 nm.

O'Donnell and Winefordner (52) reviewed the potential of phosphorescence spectrometry in clinical chemistry with emphasis on low-temperature phosphorescence techniques. They commented that low-temperature would be very useful for those molecular species difficult or impossible to measure by conventional methods. Also, they provided nine tables of phosphorescence data for a variety of compounds and reviewed the literature through about 1974.

de Silva et al. (85) carried out an extensive study of luminescence properties of pharmaceuticals of the tetrahydrocarbazole, carbazole, and 1,4-benzodiazepine classes. Preliminary examination of the luminescence characteristics of the model compounds was made on thin-layer chromatoplates at room temperature and at liquid nitrogen temperature. Low-temperature fluorescence and/or phosphorescence data were obtained for fifty-seven compounds in different solvents. Nanogram detection limits were obtained in several cases. They modified a commercial spectrofluorometer to accommodate a phosphoroscope, which permitted the use of a sample tube spinning apparatus (11), and open-ended capillary cells (14), which extended the use of the system to cracked glasses, snowed matrices, and aqueous solutions.

Sternberg et al. (86) were interested in the solvent effects on the phosphorescence and fluorescence of cocaine and methyl benzoate. Cocaine contains the benzoate chromophore. They found that there was virtually no solvent effect on the fluorescence of the two compounds. However, the phosphorescence spectrum of methyl benzoate exhibited several long wavelength bands in acetonitrile that were absent in nonpolar solvents. The phosphorescence emission spectrum of cocaine differed from methyl benzoate both in acetonitrile and dimethyl sulfoxide. In all solvents, the phosphorescence emission of cocaine was less resolved than that of methyl benzoate. The main conclusion from their work was that there was a clear difference in the triplet states of cocaine and methyl benzoate.

de Lima and Nicola (87) reported the room-temperature and low-temperature phosphorescence properties for five 1,8-naphthyridine derivatives. Some of the 1,8-naphthyridine derivatives are highly effective against gram negative pathogens. For both the room-temperature and low-temperature results, dilute alkaline solutions were used. Calibration curves, limits of detection, phosphorescence lifetimes, and the effect of various parameters such as the NaOH concentration, effect of irradiation, and effect of temperature in the sample compartment were reported. Detection limits in the nanogram and microgram ranges were obtained by both phosphorescence techniques.

Baeyens et al. (88) reported the low-temperature phosphorescence determination of mebendazole and flubendazole in anthelmintic preparations. Both tablets and suspensions were used in the development of the analytical methods. The average percent recovery and percent relative standard deviation for mebendazole in pharmaceutical preparations were 98.99% and 2.76%, respectively. For flubendazole, the average percent recovery was 98.08% and the percent relative standard deviation was 2.14%. In related work, Baeyens et al. (89) reported the luminescence properties of benzoylbenzimidazoles and related benzimidazoles.

Khasawneh et al. (90) investigated the low-temperature phosphorescence properties of sixty-six different pharmaceutical drugs. Analytical figures of merit were presented for thirty compounds using a mixture of ethanol/water as a solvent. The limits of detection were in the range of 2 to 200 ppb. Table 5.6 gives a partial listing of the analytical data for fifteen of the

Table 5.6. Analytical Figures of Merit of Low-Temperature (77 K) Phosphorimetry for Some Pharmaceutical Compounds

Compound	Limit of detection[a] (LOD) (ng / mL)	Upper concentration of linear dynamic range[b] (μg / mL)	Blank relative standard deviation[c] (%RSD)
Anafranil[a]	100	10	29.4
Antrenyl[a]	170	90	12.6
Apresoline hydro- chloride[a]	160	10	17.0
Brethine[a]	240	7	6.0
Butazolidin[a]	440	9	18.8
Chloropheniramine maleate[b]	200	3	4.5
Coramine[a]	3,500	870	11.0
DBI	250	95	6.1
Dexchloropheniramine maleate[b]	40	4	36.6
Diazoxide[b]	10	2	4.3
Esidrix	2	1	4.5
Estradiol[b]	200	4	5.9
Ethyl estradiol[b]	200	4	5.9
Fluphenazine[b]	60	2	13.6
Griseofulvin[b]	20	1	14.0

[a]Limit of detection is the concentration that gives a signal three times the standard deviation of 16 blanks.

[b]Linear dynamic range extends from the limit of detection to the concentration where the slope has decreased 5% (the slope of log intensity vs log concentration).

[c]Blank relative standard deviation is based on 16 blank measurements.

Reprinted with permission from I. Khasawneh, J. Kerkhoff, D. Siegel, A. Jurgensen E. Inman, and J.D. Winefordner, *Microchem. J.* 1985, 31, 281.

pharmaceutical compounds. In other work, Khasawneh et al. (91) obtained the phosphorescence spectral characteristics, lifetimes and limits of detection of thirty pharmaceutical compounds. Low- and room-temperature data were obtained using ethanol:water (80:20) glass and filter paper, respectively.

Warner and co-workers (92-94) have discussed multidimensional phosphorimetry. They used the rapid scanning capability of a video fluorometer. The video fluorometer was originally developed at the University of Washington (95). The video fluorometer used by Warner and co-workers has been described in the literature (96), and modifications to the system to obtain low-temperature phosphorescence data have been discussed (93). The system is capable of rapidly obtaining a phosphorescence emission-excitation matrix (PEEM). The PEEM is essentially a mapping of phosphorescence intensity as a function of multiple excitation and emission wavelengths simultaneously. A block diagram of the video phosphorimeter is shown in Figure 5.8. They demonstrated the applicability of multidimensional phosphorimetry by acquiring and analyzing three

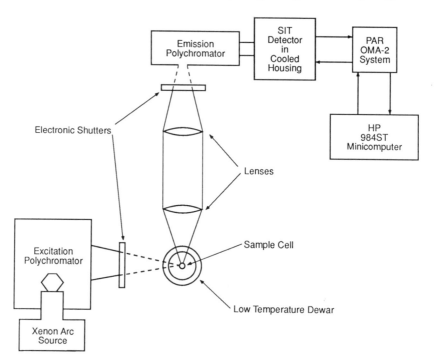

Figure 5.8. Block diagram of video phosphorimeter. (Printed with permission from C.N. Ho and I.M. Warner, *Anal. Chem.*, 54, 2386. Copyright 1982 American Chemical Society.)

sets of synthetic data. Three polycyclic aromatic hydrocarbons were employed, namely, coronene (10^{-8} M), phenanthrene (10^{-7} M), and triphenylene (10^{-8} M). From these three compounds, three samples, consisting of a single component system, a binary mixture, and a ternary mixture were analyzed. In addition, a burned oil residue sample was investigated. Their research demonstrated the general usefulness of multidimensional phosphorimetry for multicomponent analyses. Also, they showed how a complex sample can often be reduced to a single component using time resolution.

In principle, derivative phosphorescence, synchronous phosphorescence, and constant energy synchronous phosphorescence spectrometry could be used to characterize or quantitate phosphorescent components at low temperature. However, these approaches have not been used at low temperature to any great extent. Luminescence derivative spectrometry involves taking the derivatives of either the excitation or emission spectra, which can result in a significant improvement in resolution of the luminescence bands (97). In synchronous luminescence spectrometry, both the excitation and emission monochromators are scanned at the same time at a fixed wavelength difference ($\Delta\lambda$). Vo-Dinh (98) has considered the details of synchronous luminescence spectrometry and discussed the conditions for choosing $\Delta\lambda$. The main advantages of the synchronous luminescence technique are its simplicity and enhanced selectivity achieved for relatively complex mixtures. With synchronous luminescence spectrometry, a single luminescent band is frequently obtained for a pure component. Vo-Dinh (98) has discussed selectivity in synchronous phosphorimetry with emphasis on room-temperature phosphorescence. The concepts in Vo-Dinh's work should be directly applicable to low-temperature phosphorimetry. For constant-energy synchronous luminescence spectroscopy, a constant energy difference is maintained between the excitation and emission wavelengths as each monochromator is scanned through the spectral region of interest (99,100). Files et al. (101) considered gasoline and crude oil finger-printing using low-temperature and room-temperature constant energy synchronous luminescence spectrometry. Gasoline engine exhaust has also been analyzed using the constant energy approach at low temperature (102).

If two or more compounds possess similar absorption spectra and phosphorescence spectra, it may be possible to determine the concentrations of the species in a mixture by using time-resolved phosphorimetry, if the compounds have different phosphorescence lifetimes (103). With this technique, the phosphorescence is measured at short time intervals after excitation is terminated. By the correct choice of the time delay, namely, the time between the end of excitation and the start of measurement, it is possible to resolve the phosphorescence signals in a mixture. Barnes and Winefordner have considered the details of the optimization of time-resolved phosphorimetry (104).

Time-resolved phosphorimetry was first demonstrated as an approach for chemical analysis by Keirs et al. (54). They resolved a mixture of acetophenone and benzophenone at concentrations in the range of 10^{-3} and 10^{-6} M. Harbaugh et al. (105) used pulsed-source phosphorimetry for the measurement of phosphorescence lifetimes of several structurally- and spectrally-similar organic compounds. They emphasized that with a heavy atom present, the phosphorescence lifetime was shortened for all the compounds investigated.

McDuffie and Neely (106) employed time-resolved phosphorimetry for the determination of griseofulvin in the presence of dechlorogriseofulvin. The phosphorescence lifetimes of griseofulvin and dechlorogriseofulvin were 0.11 s and 1.16 s, respectively. Figure 5.9 gives the log decay curve for a mixture of griseofulvin and dechlorogriseofulvin. The linear concentration range for griseofulvin was between 5×10^{-4} to 1×10^{-6} M.

Harbaugh et al. (107) used pulsed-source time-resolved phosphorimetry for the qualitative and quantitative analysis of mixtures of drugs. For the compounds they investigated, no physical separation was needed, and temporal resolution was sufficient to resolve mixtures of the phosphors. Generally, because phosphorescence lifetimes of organic compounds can vary significantly with structure and with the environment, temporal resolution can be much more selective than spectral resolution. In their work, they investigated morphine, ethylmorphine, codeine, quinine, procaine, phenobarbital, amobarbital, cocaine, phetamine, and methamphetamine. They also emphasized that three important considerations in time-resolved phosphorimetry are phosphorescence spectra, phosphorescence lifetime, and phosphorescence intensity. For the best results, the phosphorescence signals of the compounds should be about the same because species with weak signals would be lost in the noise of the observed signal. They presented quantitative results for binary and ternary mixtures of drugs. For example, for a mixture of codeine and morphine that contained 0.023 mg/mL and 0.11 mg/mL, respectively, they found

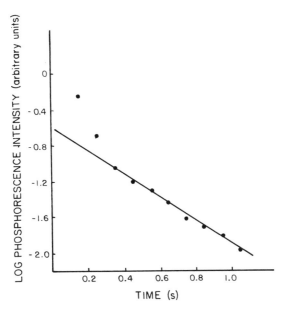

Figure 5.9. Logarithmic decay curve for mixture of griseofulvin (0.25×10^{-6} M, T=0.11 s) and dechlorogriseofulvin (0.75×10^{-6} M, T = 1.16 s). (Printed with permission from J.R. McDuffie and W.C. Neely, *Anal. Biochem.* 1973, $\underline{54}$, 507.)

0.022 mg/mL and 0.12 mg/mL, with a relative error of -4.3% and +9.0%, respectively.

Boutilier and Winefordner (108,109) used a nitrogen laser or a Chromatix flashlamp pumped dye laser to excite organic compounds at 77 K using time-resolved measurements. The instrumental system they used was discussed in Chapter 3 (Figure 3.6). They investigated heavy atoms such as I^-, Tl^+, and Ag^+ with regard to their effect on both phosphorescence lifetimes and detection limits. Some of the compounds they studied were anturane, benzophenone, butazolidine, carbazole, cocaine, ethylmorphine, morphine, phenanthrine, phenobarbital, phenylcyclidine, procaine, and quinine. In general, the compounds investigated were found to have both long (seconds) and short (milliseconds) phosphorescence lifetimes. The short lifetime component appeared primarily in the presence of a heavy atom. The solvents used were ethanol:water (10:90), and the detection limits with the N_2 laser (337 nm) were frequently superior to those with the doubled flashlamp pumped dye laser (270 nm). Table 5.7 compares the limits of detection for four compounds using two different lasers.

Constant-energy synchronous scanning was combined with

Table 5.7. Comparison of Limits of Detection Using Avco Nitrogen Laser and Chromatix CMX-4 Laser

Compound	Delay time ms	Gate time ms	Solvent[a]	Limits of detection,[b] ng / mL	
				Avco N_2[c]	CMX-4[d]
Benzophenone	0.2	2.0	E/W	0.92	0.22
440 nm[f]	0.2	2.0	KI	2.4	0.41
	4.0	2.0	$AgNO_3$	0.62	0.10
Quinine,	10.0	20	E/W	1.9	13
515 nm	0.2	2.0	KI	0.40	9.2
	4.0	2.0	$AgNO_3$	0.21	2.7
Phenanthrene,	10.0	20	E/W	5.5	8.6
500 nm	10.0	20	KI	4.3	14
	0.2	2.0	$AgNO_3$	1.9	0.26
Carbazole,	9.0	10	E/W	0.55	e
440 nm	9.0	10	KI	0.34	e
	0.2	1.0	$AgNO_3$	0.038	e

[a]E /W is 10 / 90 v/v ethanol/water, KI is 0.75 M KI, $AgNO_3$ is 0.1 M $AgNO_3$.
[b]Based on signal-to-noise of 3, observed time constant of 3.3 s.
[c]Avco nitrogen laser, 337.1 nm, 15 Hz, except carbazole, 20 Hz.
[d]Chromatix CMS-4 laser, 270 nm, 15 Hz.
[e]Only Avco laser used to determine carbazole detection limits.
[f]Emission wavelengths measured.
Reprinted with permission for G.D. Boutilier and J.D. Winefordner, *Anal. Chem.*, 51, 1384. Copyright 1979 American Chemical Society.

time resolution to increase the spectral selectivity in low-temperature phosphorimetry (110). A personal computer-controlled flashlamp phosphorimeter consisted of a conventional spectrophosphorimeter with computer control of the scanning excitation and emission monochromators. Source pulse, repetition rate, delay time between termination of the source pulse and the gate opening, and detector gate width were under a variety of forms of electronic control.   Laserna et al. (110) commented that time-resolved constant-energy synchronous phosphorimetry is a more complicated technique than time-resolved phosphorimetry and constant-energy synchronous phosphorimetry. Thus, the previous two techniques should be attempted prior to using time-resolved constant-energy synchronous phosphorimetry. Their results showed that the simplification capability of spectral and temporal parameters can be

successfully employed to extract intensity information of single
analytes and increase discrimination power against undesirable
components in mixtures. In their work, they employed pharmaceuticals as model compounds that showed severe spectral overlap
within a given spectral region. Figure 5.10 shows the results
obtained for diphenylpyraline and sulfanilamide, two relatively

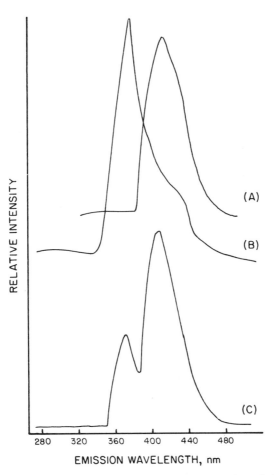

Figure 5.10. TRCESP spectra obtained at a constant energy value of $\Delta\bar{V} = 10{,}000\ \mathrm{cm}^{-1}$,
delay time of $t_d = 1$ ms, and gate time of $t_g = 6$ ms for (A) a pure solution of sulfanilamide
(5 ppm) with a gain of 10^6, (B) a pure solution of diphenylpyraline (50 ppm) with a gain of
10^7, and (C) a mixture of the two--sulfanilamide (1 ppm) and diphenylpyraline (50 ppm).
(Printed with permission from J.J. Laserna, M.A. Mignardi, R. von Wandruszka, and J.D.
Winefordner, *Appl. Spectrosc.* 1988, 42, 1112.)

long-lived phosphors with phosphorescence lifetimes of 4.0 s and 1.3 s, respectively. These two phosphors were resolved by mainly spectral means. In another case, they investigated a synthetic mixture of nupercaine and amobarbital, and both spectral and temporal discrimination were successful.

References

1. Tsai, S.C.; Robinson, G.W. *J. Chem. Phys.* 1968, 49, 3184.
2. Parker, C.A.; Joyce, T.A. *Trans. Faraday Soc.* 1969, 65, 2823.
3. Clark, W.D.; Litt, A.D.; Steel, C. *Chem. Commun.* 1969, 1087.
4. Saltiel, J.; Curtis, H.C.; Metts, L.; Miley, J.W.; Winterle, J.; Wrighton, M. *J. Am. Chem. Soc.* 1970, 92, 410.
5. Vander Donckt, E.; Matagne, M.; Sapir, M. *Chem. Phys. Lett.* 1973, 20, 81.
6. Turro, N.J.; Liu, K.C.; Chow, M.F.; Lee, P. *Photochem. Photobiol.* 1978, 27, 523.
7. Donkerbroek, J.J.; Elzas, J.J.; Gooijer, C.; Frei, R.W.; Velthorst, N.H. *Talanta* 1981, 28, 717.
8. Turro, N.J. *Modern Molecular Photochemistry*; Benjamin/ Cummings: Menlo Park, CA, 1978; pp 129-130.
9. Ward, J.L.; Walden, G.L.; Winefordner, J.D. *Talanta* 1981, 28, 201.
10. Winefordner, J.D.; St. John, P.A.; McCarthy, W.J. In *Fluorescence Assay in Biology and Medicine*; Udenfriend, S., Ed.; Academic Press: New York, 1969; Vol. II, pp 86-87.
11. Hollifield, H.C.; Winefordner, J.D. *Anal. Chem.* 1968, 40, 1759.
12. Zweidinger, R.; Winefordner, J.D. *Anal. Chem.* 1970, 42, 639.
13. Aaron, J.J.; Winefordner, J.D. *Talanta* 1975, 22, 707.
14. Lukasiewicz, R.J.; Rozynes, P.A.; Sanders, L.B.; Winefordner, J.D. *Anal. Chem.* 1972, 44, 237.
15. Lukasiewicz, R.J.; Mousa, J.J.; Winefordner, J.D. *Anal. Chem.* 1972, 44, 963.
16. Ward, J.L.; Bateh, R.P.; Winefordner, J.D. *Appl. Spectrosc.* 1980, 34, 15.
17. Ward, J.L.; Walden, G.L.; Bateh, R.P.; Winefordner, J.D. *Appl. Spectrosc.* 1980, 34, 348.

18. Jones, B.T.; Winefordner, J.D. *Anal. Chem.* 1988, <u>60</u>, 412.
19. Scharf, G.; Smith, B.W.; Winefordner, J.D. *Anal. Chem.* 1985, <u>57</u>, 1230.
20. Lehotay, S.J.; Jones, B.T.; Mignardi, M.A.; Files, L.A.; Winefordner, J.D. *Microchem. J.* 1987, <u>36</u>, 235.
21. Gifford, L.A.; Miller, J.N.; Burns, D.T.; Bridges, J.W. *J. Chromatogr.* 1975, <u>103</u>, 15.
22. Miller, J.N.; Phillipps, D.L.; Burns, D.T.; Bridges, J.W. *Anal. Chem.* 1978, <u>50</u>, 613.
23. Leigh, A.G.W.; Leaback, D.J. *Anal. Chim. Acta* 1988, <u>212</u>, 213.
24. Sawicki, E.; Pfaff, J.D. *Anal. Chim. Acta* 1965, <u>32</u>, 521.
25. Sawicki, E.; Pfaff, J.D. *Mikrochim. Acta* 1966, 322.
26. Szent-Gyorgyi, A. *Science* 1957, <u>126</u>, 751.
27. Randerath, K. *Anal. Biochem.* 1967, <u>21</u>, 480.
28. Mayer, R.T.; Holman, G.M.; Bridges, A.C. *J. Chromatogr.* 1974, <u>90</u>, 390.
29. Isenberg, I.; Smerdon, M.J.; Cardenas, J.; Miller, J.; Schaup, H.W.; Bruce, J. *Anal. Biochem.* 1975, <u>69</u>, 531.
30. Leigh, A.G.W.; Creme, S.; Leaback, D.H. *J. Chromatogr.* 1979, <u>178</u>, 592.
31. Jones, B.T.; Glick, M.R.; Mignardi, M.A.; Winefordner, J.D. *Appl. Spectrosc.* 1988, <u>42</u>, 850.
32. de Lima, C.G. *Crit. Rev. Anal. Chem.* 1986, <u>16</u>, 177.
33. Hofstraat, J.W.; Gooijer, C.; Velthorst, N.H. In *Molecular Luminescence Spectroscopy: Methods and Applications - Part II*; Schulman, S.G., Ed.; Wiley: New York, 1988; Chapter 4.
34. Colmsjö, A.L.; Zebuhn, Y.U.; Östman, C.E. *Anal. Chem.* 1982, <u>54</u>, 1673.
35. Garrigues, P.; Ewald, M. *Anal. Chem.* 1983, <u>55</u>, 2155.
36. Garrigues, P.; de Sury, R.; Bellocq, J.; Ewald, M. *Analysis* 1985, <u>13</u>, 81.
37. Elsaiid, A.E.; Walker, R.; Weeks, S.; D'Silva, A.P.; Fassel, V.A. *Appl. Spectrosc.* 1988, <u>42</u>, 731.
38. D'Silva, A.P.; Fassel, V.A. *Anal. Chem.* 1984, <u>56</u>, 985A.
39. Brown, J.C.; Edelson, M.C.; Small, G.J. *Anal. Chem.* 1978, <u>50</u>, 1394.
40. Brown, J.C.; Ducanson, J.A.; Small, G.J. *Anal. Chem.* 1980, <u>52</u>, 1711.
41. *Fluorescence and Phosphorescence Analysis*; Hercules, D.M.. Ed.; Wiley: New York, 1966.

42. Konev, S.V. *Flourescence and Phosphorescence of Proteins and Nucleic Acids*; Plenum Press: New York, 1967.

43. Zander, M. *Phosphorimetry*; Academic Press: New York, 1968.

44. Parker, C.A. *Photoluminescence of Solutions*; Elsevier: New York, 1968.

45. Becker, R.S. *Theory and Interpretation of Fluorescence and Phosphorescence*; Wiley: New York, 1969.

46. Winefordner, J.D.; Schulman, S.G.; O'Haver, T.C. *Luminescence Spectrometry in Analytical Chemistry*; Wiley: New York, 1972.

47. Schulman, S.G. *Fluorescence and Phosphorescence Spectroscopy: Physicochemical Principles and Practice*; Pergamon Press: New York, 1977.

48. Reference 10, Chapter 2.

49. Winefordner, J.D.; McCarthy, W.J.; St. John, P.A. In *Methods of Biochemical Analysis*; Glick, D., Ed.; Wiley: New York, 1967; Vol. XV, pp 369-483.

50. Baeyens, W.R.G. In *Molecular Luminescence Spectroscopy: Methods and Applications - Part I*; Schulman, S.G., Ed.; Wiley: New York, 1985; Chapter 2.

51. Lott, P.F.; Hurtubise, R.J. *J. Chem. Educ.* 1974, 51, A315, A357.

52. O'Donnell, C.M.; Winefordner, J.C. *Clin. Chem.* 1975, 21, 285.

53. Hurtubise, R.J. *Anal. Chem.* 1983, 55, 669A.

54. Keirs, R.J.; Britt, R.D.; Wentworth, W.E. *Anal. Chem.* 1957, 29, 202.

55. Freed, S.; Salmre, W. *Science* 1958, 128, 1341.

56. Parker, C.A.; Hatchard, C.G. *Analyst* 1962, 87, 664.

57. Winefordner, J.D.; Latz, H.W. *Anal. Chem.* 1963, 35, 1517.

58. McGlynn, S.P.; Neely, B.T.; Neely, C. *Anal. Chim. Acta* 1963, 28, 472.

59. Sheridan, R.E.; O'Donnell, C.M.; Pautler, E.L. *Anal. Biochem.* 1973, 53, 657.

60. Boutilier, G.D.; O'Donnell, C.M.; Rahn, R.O. *Anal. Chem.* 1974, 46, 1508.

61. Aaron, J.J.; Winefordner, J.D. *Anal. Chem.* 1972, 44, 2127.

62. Aaron, J.J.; Spann, W.J.; Winefordner, J.D. *Talanta* 1973, 20, 855.

63. Rahn, R.O.; Landry, L.C. *Photochem. Photobiol.* 1973, 18, 29.

64. Aaron, J.J.; Winefordner, J.D. *Anal. Chem.* 1972, 44, 2122.
65. Mousa, J.J.; Winefordner, J.D. *Anal. Chem.* 1974, 46, 1195.
66. Mau, A.W.H.; Puza, M. *Photochem. Photobiol.* 1977, 25, 601.
67. Aaron, J.J.; Ward, J.L.; Winefordner, J.D. *Analysis* 1982, 10, 98.
68. Acuna, A.U.; Ceballos, A.; Molera, M.J. *Anales de Quimica* 1976, 72, 410.
69. Moye, H.A.; Winefordner, J.D. *J. Agr. Food Chem.* 1965, 13, 516.
70. Moye, H.A.; Winefordner, J.D. *J. Agr. Food Chem.* 1965, 13, 533.
71. Aaron, J.J.; Kaleel, E.M.; Winefordner, J.D. *J. Agr. Food Chem.* 1979, 27, 1233.
72. Corfield, M.M.; Hawkins, H.L.; John, P.; Soutar, I. *Analyst* 1981, 106, 188.
73. Vo-Dinh, T.; Gammage, R.B. *Anal. Chem.* 1978, 50, 2054.
74. Wolfbeis, O.S.; Posch, W.; Gubitz, G.; Tritthart, P. *Anal. Chim. Acta* 1983, 147, 405.
75. Scharf, G.; Winefordner, J.D. *Spectrochim. Acta* 1985, 41A, 899.
76. Khasawneh, I.M.; Winefordner, J.D. *Microchem. J.* 1988, 37, 86.
77. Khasawneh, I.M.; Winefordner, J.D. *Talanta* 1988, 35, 267.
78. Aaron, J.J.; Winefordner, J.D. *Talanta* 1972, 19, 21.
79. Miles, C.I.; Schenk, G.H. *Anal. Chem.* 1973, 45, 130.
80. Aaron, J.J.; Fisher, R.; Winefordner, J.D. *Talanta* 1974, 21, 1129.
81. Morrison, L.D.; O'Donnell, C.M. *Anal. Chem.* 1974, 46, 1119.
82. Bridges, J.W.; Gifford, L.A.; Hayes, W.P.; Miller, J.N.; Burns, D.T. *Anal. Chem.* 1974, 46, 1010.
83. King. L.A.; Gifford, L.A. *Anal. Chem.* 1975, 47, 17.
84. Gifford, L.A.; Miller, J.N; Bridges, J.W.; Burns, D.T. *Talanta* 1977, 24, 273.
85. de Silva, J.A.F.; Strojny, N.; Stika, K. *Anal. Chem.* 1976, 48, 144.
86. Sternberg, V.I.; Singh, S.P.; Narain, N.K. *Spectrosc. Lett.* 1977, 8, 639.
87. de Lima, C.G.; de M. Nicola, E.M. *Anal. Chem.* 1978, 50, 1658.
88. Baeyens, W.R.G.; Fattah, F.A.; De Moerloose, P. *Anal. Lett.*

1985, <u>18</u>, 2105.

89. Baeyens, W.R.G.; Fattah, F.A.; De Moerloose, P. *Anal. Lett.* 1985, <u>18</u>, 2143.

90. Khasawneh, I.; Kerkhoff, J.; Siegel, D.; Jurgensen, A.; Inman, E.; Winefordner, J.D. *Microchem. J.* 1985, <u>31</u>, 281.

91. Khasawneh, I.M.; Alvarez-Coque, M.C.G.; Ramos, G.R.; Winefordner, J.D. *J. Pharm. Biomed. Anal.* 1989, <u>7</u>(11), 29.

92. Ho, C.N.; Warner, I.M. *Anal. Chem.* 1982, <u>54</u>, 2486.

93. Ho, C.N.; Warner, I.M. *Trends Anal. Chem.* 1982, <u>1</u>, 159.

94. Warner, I.M.; Patonay, G.; Thomas, M.P. *Anal. Chem.* 1985, <u>57</u>, 463A.

95. Johnson, D.W.; Callis, J.B.; Christian, G.D. *Anal. Chem.* 1977, <u>49</u>, 747A.

96. Warner, I.M.; Fogarty, M.P.; Shelly, D.C. *Anal. Chim. Acta* 1979, <u>109</u>, 361.

97. O'Haver, T.C. *Anal. Chem.* 1979, <u>51</u>, 91A.

98. Vo-Dinh, T. *Anal. Chem.* 1978, <u>50</u>, 396.

99. Inman, E.L.; Winefordner, J.D. *Anal. Chem.* 1982, <u>54</u>, 2018.

100. Inman, E.L.; Files, L.A.; Winefordner, J.D. *Anal. Chem.* 1986, <u>58</u>, 2156.

101. Files, L.A.; Moore, M.; Kerkhoff, M.J.; Winefordner, J.D. *Microchem. J.* 1987, <u>35</u>, 305.

102. Files, L.A.; Jones, B.T.; Hanamura, S.; Winefordner, J.D. *Anal. Chem.* 1986, <u>58</u>, 1440.

103. Winefordner, J.D. *Acc. Chem. Res.* 1969, <u>2</u>, 361.

104. Barnes, C.G.; Winefordner, J.D. *Appl. Spectrosc.* 1984, <u>38</u>, 214.

105. Harbaugh, K.F.; O'Donnell, C.M.; Winefordner, J.D. *Anal. Chem.* 1973, <u>45</u>, 381.

106. McDuffie, J.R.; Neely, W.C. *Anal. Biochem.* 1973, <u>54</u>, 507.

107. Harbaugh, K.F.; O'Donnell, C.M.; Winefordner, J.D. *Anal. Chem.* 1974, <u>46</u>, 1206.

108. Boutilier, G.D.; Winefordner, J.D. *Anal. Chem.* 1979, <u>51</u>, 1384.

109. Boutilier, G.D.; Winefordner, J.D. *Anal. Chem.* 1979, <u>51</u>, 1391.

110. Laserna, J.J.; Mignardi, M.A.; von Wandruszka, R.; Winefordner, J.D. *Appl. Spectrosc.* 1988, <u>42</u>, 1112.

CHAPTER 6

SOLID-SURFACE ROOM-TEMPERATURE PHOSPHORESCENCE

6.1. Introduction

Analytical measurements of the fluorescence and phosphorescence of compounds adsorbed on solid materials constitute the general area of solid-surface luminescence analysis. A variety of solid materials has been used in solid-surface luminescence such as filter paper, silica gel, sodium acetate, polymers, and cyclodextrins. Several of the arguments presented for sensitivity and selectivity in solution luminescence analysis can also be applied to solid-surface luminescence analysis. One major difference between solid-surface luminescence and solution luminescence is that in solid-surface luminescence, the luminescent molecules are adsorbed on small particles or fibers, whereas in solution luminescence, the molecules are dissolved in a solvent. The source radiation impinging on the solid matrix and the luminescence from the adsorbed molecules will be scattered. The scattered source radiation and scattered luminescent radiation is reflected from the surface of the solid material and can be transmitted through the solid material, if the experimental conditions allow for the transmission of the radiation. Wendlandt and Hecht (1) discussed the difference between specular reflection and diffuse reflection, and these phenomena are important in solid-surface luminescence analysis. Specular reflection or mirror reflection is defined by Fresnel equations and occurs from smooth surfaces. Diffuse reflection of exciting radiation results by penetration of the incident radiation into the interior of the solid substrate, and multiple scattering occurs at the boundaries of individual particles or fibers of the solid matrix. Ideal diffuse reflection takes place when the angular distribution of the reflected radiation is independent of the angle of incidence of source radiation (2). Several theories of diffuse reflection have been presented with varying degress of success. Körtum (2) has emphasized that specular reflection and diffuse reflection are two important limiting cases, and all possible variations are found in practice between these two extremes. Usually, with solid-surface luminescence analysis, diffuse luminescence is measured.

Commercial and laboratory-constructed instruments are used to measure solid-surface luminescence. Commercial instruments for solid-surface fluorescence measurements became available about 1968. Laboratory-constructed instruments and modifications to commercial instruments have been described in the literature for measuring both fluorescence and phosphorescence from compounds adsorbed on solid surfaces. Hundreds of applications have been presented in areas such as environmental research, forensic science, pesticide analysis, food analysis, pharmaceutical analysis, biochemistry, medicine, and clinical chemistry. Most of the applications have been with solid-surface fluorescence. However, numerous applications in which room-temperature phosphorescence is employed for an analysis are now appearing in the literature (3-5).

Roth (6) first suggested the use of room-temperature phosphorescence (RTP) from his detection of RTP from several organic compounds adsorbed on cellulose. Lloyd and Miller (7) have commented on the first observations of RTP. Schulman and Walling (8,9) independently measured RTP from several compounds. Paynter et al. (10) analytically developed the phenomena of RTP observed by Walling and Schulman, with filter paper as a solid surface.

In this chapter, the practical aspects, types of phosphorescence data, and instrumentation for solid surface RTP will be considered. Hurtubise (3,4) has discussed luminescence from solid surfaces in some detail, and Vo-Dinh (5) has given an excellent treatment on room-temperature phosphorimetry from solid surfaces. In Chapter 7 some of the physicochemical interactions responsible for solid-surface phosphorescence will be discussed, and in Chapter 8 several applications of solid-surface phosphorescence will be covered.

6.2. Practical Considerations

In this section, the various solid substrates used in RTP

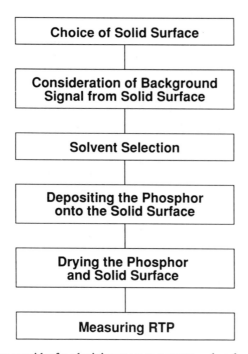

Figure 6.1. Steps to consider for obtaining room-temperature phosphorescence from solid-surfaces.

work, background signals from the solid surfaces, solvent selection, application of the sample to the surface, drying conditions, and other experimental aspects will be considered. Hurtubise (3) has reviewed several procedural aspects in solid-surface luminescence analysis, and Vo-Dinh (5) has discussed, in detail, the practical aspects of RTP. Figure 6.1 gives a general outline of the steps involved prior to the measurement of RTP.

6.2.1. Available Solid Substrates

Only certain solid materials are useful for obtaining RTP from adsorbed organic compounds. Filter paper is the most widely used solid surface for inducing RTP. However, a variety of other solid substrates are available for RTP, which permits the analysis of numerous organic compounds (Table 6.1). Vo-Dinh (5) has discussed, in detail, the characteristics of several of the solid substrates available for RTP.

No detailed guidelines have been developed in selecting a

solid-surface for RTP, primarily because the physicochemical interactions involved with RTP are not fully understood. As mentioned earlier, filter paper is the most widely employed solid matrix for RTP. Normally one is concerned with obtaining a strong RTP signal from an adsorbed compound. Filter paper is very versatile in this respect, although frequently it is necessary to optimize the experimental conditions to obtain strong RTP signals (5). Recently, α-cyclodextrin/NaCl and β-cyclodextrin/NaCl mixtures have been employed in RTP work (25-27). The cyclodextrin-salt mixtures yielded both room-temperature fluorescence and RTP from numerous organic compounds under essentially the same experimental conditions for all the phosphors. For example, a heavy atom was not needed to obtain relatively strong RTP signals (25,26). Also, the luminescence spectra from phosphors adsorbed on cyclodextrin-salt mixtures are usually better defined than the luminescence spectra from phosphors adsorbed on filter paper. One disadvantage of cyclodextrin-salt mixtures is the sample preparation time. For example, generally, it is necessary to add the powdered mixture to a test tube, then add a solution of the phosphor to the test tube, evaporate the solvent and dry the solid mixture, and finally, add the solid mixture to a sample holder. With filter paper, a solution of the phosphor is added to the filter paper, and then, the filter paper and sample are dried. However, the disadvantage of sample preparation time for cyclodextrin-salt mixtures could be outweighed by better defined luminescence spectra and the same experimental conditions for obtaining RTP signals for a variety of compounds. In the remaining part of this section, a general discussion will be given for several of the materials listed in Table 6.1. Vo-Dinh (5) has given an extensive treatment for most of the materials listed in Table 6.1, and the interested reader can consult Reference 5 for more details.

    Vo-Dinh et al. (28) compared over twenty different kinds of commercial filter paper using pyrene as a standard compound. Their criteria for selection of a solid surface was the relative RTP signal of the sample compared to the RTP background of the filter paper. Schleicher and Schuell filter papers yielded the highest signal-to-background ratio; however, Whatman filter papers gave very close to the same signal-to-background ratio. Bateh and Winefordner (29) conducted an extensive study of cotton-linter pulps, wool pulps and several filter papers as substrates for RTP. They concluded that the difference in performance between the poorest paper and the best paper was substantially less than an order of magnitude.

Table 6.1. Several Solid Surfaces for Inducing RTP

Solid Surface	References
Several Brands of filter paper	5,10-12
Polyacrylic acid-treated filter paper	13
Ion-exchange filter paper	14,15
Sodium acetate	
Powder	5,16
Pellets	5,17
Impregnated paper	5,17,18
Silica gel chromatoplates with a salt of polyacrylic acid as a binder	5,19
Polyacrylic acid-sodium chloride or sodium bromide mixtures	5,20-23
Chalk, H_3BO_3 / T-7 clay / NaOH, $CaHPO_4$ / T-7 clay / cornstarch / NaOH	5,24
α-Cyclodextrin / salt mixtures and β-cyclodextrin / salt mixtures	25-27

Ford and Hurtubise (19,30) investigated several brands of silica gel chromatoplates and column chromatographic silica gel as substrates for obtaining RTP from adsorbed compounds. The chromatoplates that contained a polyacrylate binder yielded the strongest RTP signals from the phosphors investigated. The polymer itself was essential for obtaining strong RTP signals (19,30). Comparison of the RTP of eight different nitrogen heterocycles on the silica gel chromatoplates and filter paper showed that the phosphors adsorbed on filter paper yielded somewhat stronger RTP signals than on silica gel chromatoplates (19). For example, under a nitrogen atmosphere the RTP signal of benzo(f)quinoline was 1.6 times greater on filter paper, and the RTP of phenanthridine was 1.1 times greater on filter paper. The reproducibility of the RTP signals and the RTP linear dynamic ranges for the chromatoplates were comparable to similar analytical data obtained on filter paper. However, not as many compounds yielded RTP signals on the silica gel chromatoplates relative to filter paper. The use of chromatoplates does permit the measurement of RTP in conjunction with thin-layer chromatography (31). For example, Ford and Hurtubise (31) separated the three isomers of phthalic acid using a silica gel chromatoplate with a polyacrylate binder. After development of the chromatoplate, it was dried and then viewed under shortwave UV radiation. Three RTP signals were observed that corresponded to the phthalic acid isomers. Interestingly, there has been very little use of RTP with thin-layer chromatography. This is most likely the result of the special conditions that are needed to obtain RTP signals with silica gel. However, the use of silica gel chromatoplates with a polyacrylate binder to separate components and characterization of the separated components by RTP should find use in the future.

Various powders have also been used to obtain RTP signals from adsorbed compounds. One such mixture, α-cyclodextrin/ NaCl, was discussed earlier in this section. Another material that has found use in RTP work is sodium acetate (16-18). Von Wandruszka and Hurtubise (16,32) showed that a variety of phosphors gave RTP signals with sodium acetate. Sodium acetate has essentially no phosphorescence background signal, and is available in very pure form. Fewer compounds yield RTP on sodium acetate than on filter paper. However, this aspect would give sodium acetate an added degree of selectivity for certain phosphors compared to filter paper. In addition, the RTP of p-aminobenzoic acid adsorbed on sodium acetate showed a relative insensitivity to ambient moisture, and thus, may prove useful in situations where moisture has to be strictly controlled to obtain a RTP signal from a phosphor (16).

Polymer-salt mixtures were investigated as substrates for RTP (20). Polyacrylic acid-salt mixtures were found to be effective in obtaining RTP signals from adsorbed phosphors. The RTP linear dynamic range and limits of detection for nine model compounds were compared for filter paper and 1% polyacrylic acid-NaBr. On the average, the linear dynamics ranges were greater, and the limits of detection lower, for filter paper compared to the polymer-salt mixture. The RTP relative average deviation for 4-phenylphenol adsorbed on 1% polyacrylic acid-NaBr was 4.2%, whereas the relative average deviation for the same compound on filter paper was 5.5%. Also, no RTP signal was obtained for chrysin adsorbed on filter paper; however, the 1% polyacrylic acid-NaBr mixture yielded an analytically useful RTP signal from chrysin. Senthilnathan et al. (13) spotted polyacrylic acid solutions of 4-phenylphenol, p-aminobenzoic acid, 1,2-benzocarbazole, and 5,6-benzoquinoline on filter paper, and the RTP results obtained were compared with similar samples spotted on filter paper without polyacrylic acid. Improvement in sensitivity ranged from 26 times for 5,6-benzoquinoline to 1.1 for 1,2-benzocarbazole, and limits of detection improved from 100 times for 5,6-benzoquinoline to 1.1 times for p-aminobenzoic acid with polyacrylic acid adsorbed on filter paper compared to filter paper without polyacrylic acid. The relative standard deviations for the samples with polyacrylic acid added were also improved.

Citta and Hurtubise (33) examined several model aromatic carbonyl compounds on various surfaces for RTP and room-temperature fluorescence. The results showed that it was important to investigate several surfaces to obtain optimal luminescence conditions for this class of compounds. In addition, both RTP and room-temperature fluorescence were shown to be analytically useful. Table 6.2 presents analytical data for RTP from four aromatic carbonyl compounds. It can be seen that 2-acetonaphthone only gave useful RTP signals on filter paper treated with polyacrylic acid. The best limits of detection were obtained with 0.5% polyacrylic acid-NaCl, and the lowest percent relative standard deviation was obtained with silica gel chromatoplates. The results in Table 6.2 indicate that no one surface gave the optimum RTP analytical data for the compounds studied.

RTP and room-temperature fluorescence analytical figures of merit were obtained for 5,6-benzo(f)quinoline, p-aminobenzoic acid, 4-phenylphenol, and phenanthrene on four different surfaces (34). The four solid surfaces investigated were silica gel with a polyacrylate binder, filter paper, 1% polyacrylic acid-NaBr, and 80% α-cyclodextrin-NaCl. Generally, filter paper and 80% α-cyclodextrin-NaCl gave better analytical figures of merit. The use of 80% α-cyclodextrin-NaCl involved more sample preparation time; however, it gave the lowest average limits of detection for RTP. Another advantage of 80% α-cyclodextrin-NaCl compared to filter paper is that the luminescence emission bands are frequently better defined than the emission bands of compounds adsorbed on filter paper. In comparing all the analytical data for the four model compounds, filter paper gave the best overall results (34). Richmond and Hurtubise (27) showed that 1% α-cyclodextrin gave essentially the same luminescence results as 80% α-cyclodextrin-NaCl. The 1% α-cyclodextrin-NaCl mixture was easier to handle and was also more cost effective.

Table 6.2. Analytical Data for RTP from Model Compounds Adsorbed on Several Surfaces

Compound and solution	Silica gel chromatoplate			Filter paper			Filter paper with polyacrylic acid			0.5% Polyacrylic acid		
	Linear range,ng	LOD[a]	RSD,%	Linear range,ng	LOD[a]	RSD,%	Linear range,ng	LOD[a]	RSD,%	linear range,ng	LOD[a]	RSD,%
2-Acetonaphthone (Ethanol)	—[c]	—[c]	—[c]	—[c]	—[c]	—[c]	0.8-75	0.8	4.6[d]	—[c]	—[c]	—[c]
5,12-Naphthacenequinone[b] (1,2-Dichloroethane-ethanol)	0.4-60	0.4	1.1[f]	1.7-55	1.7	4.1[f]	0.5-55	0.5	2.5[f]	0.2-45	0.2	3.3
α-Naphthoflavone (0.1 M HCl-ethanol)	0.8-75	0.8	3.2[d]	1.7-50	1.7	3.5[d]	0.9-100	0.9	4.2[d]	0.5-50	0.5	5.4[d]
β-Naphthoflavone (0.1 M HBr-ethanol)	0.4-50	0.4	1.4[d]	0.4-50	0.4	4.7[d]	2.1-100	2.1	3.5[d]	0.2-75	0.2	5.7[d]

[a] LOD = limit of detection (ng); amount of sample needed to give a S/N of 3.
[b] Phosphoroscope was not used.
[c] No RTP observed.
[d] Relative standard deviation (RSD) calculated from 6 spots of 100 ng each.
[e] No data reported due to nonreproducible results.
[f] Relative standard deviation (RSD) calculated from 9 spots of 100 ng each.
Reprinted with permission from L.A. Citta and R.J. Hurtubise, Microchem. J. 1986, 34, 56.

6.2.2. Background Signals from Solid Substrates

Almost all of the surfaces used to obtain RTP from adsorbed compounds give phosphorescence background signals. However, sodium acetate has little, if any, phosphorescence background.

Ward et al. (35) discussed various attempts to reduce the phosphorescence background of filter paper. They investigated Schleicher and Schuell 604 and 903 filter papers. They soaked and eluted the filter paper samples with solvents to reduce the background signals. The solvents investigated were acetone, benzene, 0.1 M HCl, 0.1 M H_2SO_4, chloroform, ethanol, hexane, 0.1 M HNO_3, methanol, methylene chloride, and 1 M sodium hydroxide. They also heated filter paper samples at 120°C and 250°C in an attempt to reduce background signals. With the various approaches they used, no significant reduction in phosphorescence background was noted.

Bateh and Winefordner (29) evaluated a variety of chemical treatments to reduce phosphorescence background in filter paper. The various treatments were designed to deal with trace metals and hemicelluloses and/or lignin in cellulose pulp. Diethylenetriamine-pentaacetic acid was employed to remove trace amounts of transition metals that might contribute to the phosphorescence background. They found that the background was only slightly reduced, but the phosphor signal improved considerably. Other treatments were evaluated to test for the presence of extractable hemicelluloses and/or lignin in cellulose pulp. None of the treatments confirmed the presence of such materials. The authors indicated that hemicellulose and/or lignin may be responsible for the background signal.

McAleese and Dunlap (36) investigated the possibility of reducing the phosphorescence background in filter paper by illumination of the filter paper with light from a 150-W xenon lamp. As Figure 6.2 shows, the background emission at 495 nm decreased 87% during the first 3 hours of illumination with 285-nm light from the 150-W xenon lamp. In other experiments, several filter paper samples were subjected to illumination by white light from the xenon lamp. After 24 hours of exposure, the samples of filter paper were dried in a glovebag for 2 hours. Compared to control samples, the background intensities were diminished by an average of 10.3-fold. The authors commented that by using efficient illumination of the sample, completely drying the sample, and reducing the background emission from filter paper, it would be

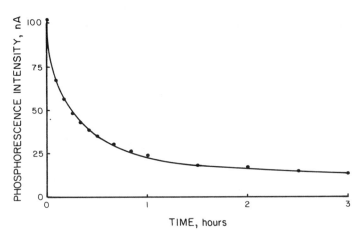

Figure 6.2. Phosphorescence intensity at 495 nm plotted as a function of time for paper exposed to 285 nm light. (Reprinted with permission from D.L. McAleese and R.B. Dunlap, *Anal. Chem.*, <u>56</u>, 600. Copyright 1984 American Chemical Society.)

possible to detect 0.5 pg of p-aminobenzoic acid adsorbed on filter paper.

    Su et al. (37) discussed a computer approach for background correction for RTP measurements. An Apple computer was interfaced to an Aminco-Bowman spectrophotofluorometer with a phosphorescence attachment. The RTP spectra were obtained from the computer-interfaced spectrometer, retrieved and then corrected against the background emission. The computer approach did not improve sensitivity; however, the spectral quality and the precision for quantitative analysis were improved (5).

6.2.3. Solvent Selection

    The solvent employed to deposit the phosphor on the solid substrate can be very important in obtaining a strong RTP signal. Some aspects to consider are the solubility of the phosphor and the possibility of any chemical interactions of the solvent with the solid matrix. A general discussion of the different solvent systems that have been used in RTP work has been given (5). Acidic, alkaline, and neutral solutions have been used in a number of applications. One finds that ethanol-water solvents have been extensively employed in RTP work. The trial-and-error approach is used in optimizing the solvent or solvent system that will yield strong RTP signals. However, the relatively large body of literature available on RTP applications can be consulted for general guidelines on choosing a solvent for a specific compound or a compound with closely related properties. Below, some examples are given that illustrate how the solvent can influence the final RTP signal.

Sodium acetate has been used to induce RTP from several compounds. For one such compound, p-aminobenzoic acid, it was shown that the anion of p-aminobenzoic acid was adsorbed on sodium acetate with ethanol as a solvent (38). In the initial solution chemistry, some dissolved sodium acetate reacted with the acid to form the anion of p-aminobenzoic acid, and thus, the anion was adsorbed on the surface and yielded an RTP signal. However, if an acetone or ether solution of p-aminobenzoic acid was adsorbed onto sodium acetate, no RTP was observed from the p-aminobenzoic acid (16). It was determined that sodium acetate was insoluble in acetone and ether; thus, the partial neutralization of p-aminobenzoic acid could not occur prior to adsorption with these solvents (38). The results with acetone and ether also indicated that p-amino-benzoic acid adsorbed on sodium acetate did not give RTP. In other work, it was shown that a 0.1 M HBr-methanol solution of benzo(f)-quinoline adsorbed onto 0.5% polyacrylic acid-sodium chloride yielded an RTP signal 3.5 times greater than a comparable sample adsorbed from a 0.1 M HBr ethanol solution (21). De Lima and de M. Nicola (39) carried out a study of the effect of NaOH concentration on the RTP of a variety of compounds adsorbed on filter paper. For each phosphor, there existed a certain optimum range of NaOH concentrations that gave the strongest RTP signal.

6.2.4. Adsorption of Phosphor onto the Solid Surface

For flat surfaces such as filter paper and chromatoplates, sample volumes in the range of 1 to 10 µL are delivered to the surface with a microsyringe. The size of the initial spot should be as small and uniform as possible. Also, when a single volume of solution can be delivered to the surface, then normally better reproducibility is obtained compared to the application of multiple volumes to the surface. If a syringe is used, one source of error during the application of the sample is "creep back" on the tip of the syringe (40). Part of the drop can curl back around the tip of the syringe and remain after the sample is deposited on the surface. This source of error can be minimized by using a very fine tip or by coating the external portion of the stem of the syringe with silicone (3). However, these special precautions are not needed in most applications and only would be necessary for very high precision.

The technique used to adsorb various samples on powders for RTP studies is briefly described as follows. A small volume of organic solvent is introduced into a test tube from a micropipet, and then 1-10 μL of standard or sample solution is added from a micropipet. After this, a constant amount of powder is added to the test tube, and the contents are mixed. Then, the test tube is placed in an oven at 80-100°C until the organic solvent is evaporated. The dry solid can be transferred quantitatively to a small mortar and pestle, which is used to gently break up conglomerate particles. Alternatively, the conglomerate particles can be broken up in the test tube with a small spatula. The powder is then transferred to a special sample holder.

Spotting a sample on a flat surface with a microsyringe or a micropipet is more rapid than handling powders. However, in RTP work, a powdered substrate may offer greater selectivity or have a lower luminescence background than flat surface materials. Vo-Dinh (5) has given an extensive comparison of the approaches that have been used to apply the phosphor to the solid substrate.

Ford and Hurtubise (30) spotted several samples of benzo(f)-quinoline and 4-azafluorene onto a silica gel chromatoplate and obtained RTP signal reproducibilities of 1.7% and 3.0%, respectively. They also investigated plate-to-plate variation of the RTP signal. Three aluminum-backed silica gel chromatoplates were each spotted with five 100 ng spots of benzo(f)quinoline, and after drying the chromatoplates, the RTP signals were obtained. The average relative RTP intensities obtained for the three different chromatoplates were 100, 88.4, and 92.9. Because of plate-to-plate variation in the RTP signals, any quantitative method which employs RTP with chromatoplates should include RTP measurements of the appropriate standards along with the unknowns on the same chromatoplate.

Bateh and Winefordner (29) compared various lots of diethylenetriaminepentaacetic (DTPA) acid treated Schleicher and Schuell filter papers. The filter papers were treated with DTPA because the DTPA enhanced the RTP signals of adsorbed p-amino-benzoic acid. Table 6.3 compares the results of relative signals and relative standard deviations for several different lots of treated filter paper. The percent relative standard deviation for the blanks ranged from 3.5% to 4.7%, and for sample signals from p-aminobenzoic

Table 6.3. Results of "Lot"-Analysis of DTPA-Treatment S & S 903[a]

	Blank		PABA		
	Mean relative signal	RSD,%	Mean relative signal	RSD,%	S_A / S_B
W94 (1980)	6.3	3.7	360	3.2	57
W94 (1981)	5.8	4.4	405	2.2	70
W93	5.4	4.3	423	3.5	78
W92	5.2	3.4	408	2.3	78
W12	6.3	3.6	399	2.7	63
W02	5.3	4.7	407	3.3	77
W01	5.2	3.5	400	2.4	77

[a]Mean relative signals and relative standard deviations calculated from 16 measurements of blank and PABA. S_A (mean relative signal of PABA); S_B (mean relative signal of blank). Reprinted with permission from R.P. Bateh and J.D. Winefordner, *Talanta* 1982, 29, 713.

acid, the percent relative standard deviation was from 2.2% to 3.5%. As indicated in Table 6.3, the ratios of analyte to blank signals are relatively high, which shows that strong RTP signals can be obtained from the adsorbed phosphor. In general, the problem of lot-to-lot variation for filter paper is normally not noticed because filter paper is inexpensive, and it takes a large number of RTP measurements to consume one lot (5).

By contrasting the percent relative standard deviation results for the four surfaces in Table 6.2, one can obtain an idea of how the different surfaces compare as far as sample application and overall sample preparation. Of the three flat surfaces, the silica gel chromatoplates gave the lowest relative standard deviation. For the 0.5% polyacrylic acid-NaCl (powder sample), the overall percent relative standard deviation was somewhat higher than for the flat surfaces.

6.2.5. Drying Conditions

After the phosphor is adsorbed on the solid surface, it is necessary to dry the adsorbed sample prior to the RTP measurement step. Moisture is one of the main factors that can diminish an RTP signal. The sample can be dried by blowing hot air onto the sample with an air blower, placing the sample inside a desiccator, heating the sample inside an oven, or placing the sample under an infrared heating lamp.   In general, a hot air blower is not recommended because of the possibility of contaminating the sample with trace impurities in the forced air flow (5).

With an oven, the sample is normally dried for 15-30 min at 100°C. However, other temperatures have been employed. In some cases, it is necessary to allow the sample to dry in ambient air for about 5 min after spotting the sample.   For example, Ford and Hurtubise (30) indicated that it was necessary to dry adsorbed nitrogen heterocycles under ambient conditions before placing the samples in an oven. It was found that some of the samples would discolor if placed in an oven immediately after spotting.

The use of a desiccator requires about 1-5 hr to dry a sample. However, this approach is mild and is important for samples that undergo thermal decomposition.   McAleese et al. (41) used a vacuum desiccator when studying the effects of moisture and oxygen quenching on RTP. They routinely dried the samples for 1 hr. They also found that by treating filter paper with sodium citrate, the RTP signals of several phosphors were not affected by humidity when exposed to a relative humidity of 60% or less.

Drying the phosphor with an infrared heating lamp has been found to be very practical, and only about 3-5 min is needed to dry filter paper samples.   This method of drying samples is rapid, convenient, and usually is nondestructive (5).

After the sample has been dried by one of the methods discussed above, the sample is ready for RTP measurement. Normally, the sample is kept dry by passing a flow of air or other gas such as nitrogen or argon into the sample compartment. However, in routine work, during the measurement step, a drying gas is frequently not employed because the signal enhancement is less than an order of magnitude (5). However, one has to decide on a case-by-case basis whether a drying gas is needed.

6.2.6. Precision, Accuracy, and Limits of Detection

Various aspects of precision, accuracy, and limits of detection were considered in some of the previous sections of this chapter. However, it is important to summarize some of the more important aspects of these parameters.   The relative standard deviation of RTP measurements for pure compounds ranges from 1.0 to 10%. For complex samples, the relative standard deviation is from 10 to 30%. The previous relative standard deviation range is considered very good when one considers the complexity of some of the samples investigated (5). The accuracy of the RTP approach has been shown to be very good to excellent. In a round-robin analysis of a coal-liquid sample for polycyclic aromatic hydrocarbons, the RTP approach compared well with the data from several other laboratories in which other analytical techniques were used to obtain compositional data (5,42). As another example, p-aminobenzoic acid was determined by RTP in multicomponent vitamin tablets without separating the p-aminobenzoic acid (32). The amount of p-aminobenzoic acid determined was identical to that reported by the manufacturer.

The limits of detection that have been reported for RTP are approximately in the range of 1 pg to 200 ng. Low nanogram limits of detection are easily obtained. If the background signals from the solid surfaces can be reduced, then the limits of detection at the sub-picogram level, and possibly lower, could be routinely achieved. Also, assuming the background signal from the solid surface were minimized, then by the use of laser excitation and a heavy atom to enhance intersystem crossing, significant improvements in the limits of detection should result.

6.3. Types of Phosphorescence Data

The luminescence data obtained for compounds adsorbed on solid surfaces is similar to that acquired for compounds in the liquid phase. Fluorescence and phosphorescence excitation and emission spectra, fluorescence and phosphorescence lifetimes, fluorescence and phosphorescence quantum yields, and fluorescence and phosphorescence polarization data are obtained. However, very little data has been reported on phosphorescence polarization data for compounds adsorbed on solid surfaces. In this section, a discussion will be given of various types of phosphorescence data that can be obtained from compounds adsorbed on solid surfaces.

6.3.1. Excitation and Emission Spectra

One can readily obtain fluorescence and phosphorescence excitation and emission spectra from organic compounds adsorbed on solid surfaces. Vo-Dinh (5) has devoted an entire chapter to RTP spectra of organic compounds. In addition, he has provided a table listing excitation and emission wavelengths and limits of detection for about 200 organic compounds adsorbed on various surfaces.

In general, it has been found that RTP excitation and emission spectra are somewhat more diffuse than the low-temperature solution phosphorescence excitation and emission spectra. Also, the room-temperature spectra frequently show a red shift compared to the solution low-temperature phosphorescence spectra. Recently, in this laboratory, we have compared the RTP solid-surface emission spectra, low-temperature solid-surface phosphorescence spectra, and low-temperature solution phosphorescence spectra of various phosphors. The phosphors were adsorbed on 30% β-cyclodextrin-NaCl for the solid-surface spectra and dissolved in ethanol for the low-temperature solution spectra. Figure 6.3 compares the phosphorescence emission spectra under room-temperature conditions and low-temperature conditions. The best defined spectrum was obtained at low temperature in solution (Figure 6.3A). However, the low-temperature solid-surface phosphorescence spectrum (Figure 6.3B) compared favorably with the low-temperature solution spectrum in that several bands were well defined. The room-temperature solid-surface phosphorescence spectrum (Figure 6.3C) was somewhat broadened and not as well defined compared to the two low-temperature spectra.

6.3.2. Phosphorescence Polarization

Very little work has been done in measuring phosphorescence polarization from compounds adsorbed on solid surfaces because one deals with scatter of the source and phosphorescence radiation. Dalterio and Hurtubise (22) attempted to measure the extent of polarization of phosphors adsorbed on polyacrylic acid-salt mixtures and filter paper. However, within experimental error, a value of zero polarization was measured. They indicated that scattering of exciting light and emitted light caused depolarization. Recent work in our laboratory has indicated that strongly emitting phosphors adsorbed on sodium acetate, filter paper, and 30% β-cyclodextrin-NaCl do yield phosphorescence polarization. More experimental data is needed to fully develop the applicability of phosphorescence polarization from scattering surfaces. The approach should prove useful in assessing the orientation of the phosphor adsorbed on the surface and how rigidly the phosphor is held to the surface.

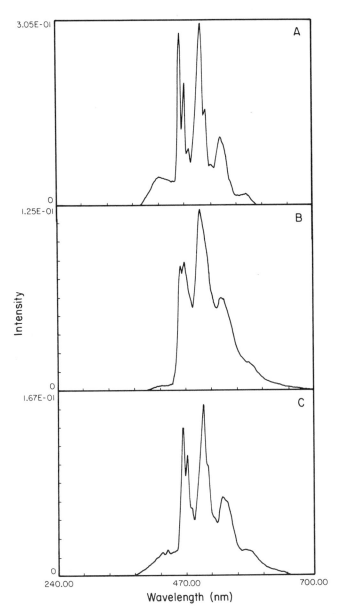

Figure 6.3. Low-temperature solution phosphorescence spectrum of phenanthrene (A); low-temperature phosphorescence spectrum of phenanthrene adsorbed on 30% β-cyclodextrin-NaCl (B); room-temperature phosphorescence spectrum of phenanthrene absorbed on 30% β-cyclodextrin-NaCl (C).

6.3.3. Phosphorescence Lifetimes

Phosphorescence lifetimes have been measured for several phosphors adsorbed on solid matrices at room temperature. Generally the RTP lifetimes are shorter than the corresponding phosphorescence lifetime at low temperature. In addition, in some cases, two decaying components have been observed for a pure compound adsorbed on a solid surface. The references cited can be consulted for solid-surface phosphorescence lifetime data (38,43-58). Table 6.4 lists selected values of phosphorescence lifetimes from the literature.

Aaron et al. (15) reported the RTP lifetimes of indole and several of its derivatives adsorbed on filter paper in the presence of 1 M iodide. The phosphorescence decay of the substituted indoles included both short and long decaying components. The short-component values ranged from 0.8 to 1.5 ms, and the long-component values ranged from 2.6 to 6.2 ms. They indicated that the RTP of the long-lived component could be attributed to the decay of the excited triplet state. For the short-lived component, they suggested that it could be due to a close lying triplet state, or to a matrix effect.

Senthilnathan and Hurtubise (38) described two decaying components for the anion of p-aminobenzoic acid adsorbed on sodium acetate, and the anion adsorbed on sodium acetate/NaCl mixtures. For the anion of p-aminobenzoic acid adsorbed on pure sodium acetate, the lifetime of the short component was 1.1 s, whereas the lifetime of the long component was 1.5 s. They suggested that inhomogeneous matrix effects were responsible for the short and long decaying components. Additional work is needed to explain why, in some cases, short and long decaying components are observed for phosphors adsorbed on solid surfaces.

Phosphorescence lifetimes are useful in the characterization of mixtures of phosphors and in the calculation of fundamental rate constants related to triplet state molecules. The rate constants are, in turn, important in mechanistic studies of the interactions of phosphors with solid surfaces. (See Chapter 7.)

6.3.4. Luminescence Quantum Yield

There have been several reports on methods for the determination of fluorescence and phosphorescence quantum yields in

Table 6.4. Phosphorescence Lifetimes of Phosphors Adsorbed on Solid Surfaces

Phosphor	Surface	Lifetime, s		Reference
		RT[a]	-180°C	
p-Aminobenzoic acid + I⁻	Anion exchange	0.019	--	43
p-Aminobenzophenone	Filter paper	0.002[b]	--	44
	Filter paper	0.019[c]		
Pyrene	Filter paper	0.018	--	46
4,4′-Dichlorobiphenyl	Filter paper	0.006	--	47
1,2,3,4-Tetrachlorodibenzofuran	Filter paper	0.001	--	49
p-Aminobenzoic acid	Sodium acetate	1.1[b]	--	38
		1.5[c]		
p-Aminobenzoic acid	50% Sodium Acetate / NaCl	1.2	2.4	55
4-Phenylphenol	80% α-Cyclodextrin / NaCl	1.3	2.5	52
Benzo(f)quinoline	80% α-Cyclodextrin / NaCl	0.87	2.5	53
Pronated form of benzo(f)quinoline	Silica gel chromatoplate with a polyacrylate binder	1.4	--	56

[a] Room temperature
[b] Short decaying component
[c] Long decaying component

solution (59-61). Less work has been reported for determining the luminescence quantum yield for solid materials. Obtaining quantum yields for solid materials is more difficult than similar measurements in solution. Usually factors such as experimental geometry, sample reflectivity, and sample preparation are very exacting in obtaining accurate values of luminescence quantum yield from solid materials.

The quantum yield of sodium salicylate was measured by Kristianpoller (62) with a second phosphor which had a constant quantum yield for both the exciting radiation and emitted radiation. This permitted highly efficient collection of the fluorescence emission of sodium salicylate. Wrighton et al. (63) developed a technique for the determination of the absolute quantum yields of powdered samples employing a conventional scanning emission spectrofluorometer. Their approach was applied to the determination of luminescence yields of National Bureau of Standards phosphors, sodium salicylate, and several metal complexes. Photoacoustic spectroscopy was used by Adams et al. (64,65) for the determination of the absolute quantum efficiencies of tetraphenylbutadiene, yellow liumogen, and sodium salicylate. Kirkbright et al. (66) determined the fluorescence quantum yield of several 2-substituted benzthiazole solids by both optical measurements and photoacoustic measurements.

Only recently has there been a rather detailed study on the determination of room-temperature solid-surface fluorescence and phosphorescence quantum yields of organic compounds adsorbed on surfaces (67). However, Wrighton et al. (63) reported the quantum yield of sodium salicylate mixed with inorganic salts. It is necessary to have methods for the determination of quantum yields for compounds adsorbed on surfaces for both theoretical and practical luminescence work. For a fundamental understanding of the interactions in solid-surface luminescence, it is important to measure quantum yield values under several experimental conditions. For analytical development work, it is necessary to have some fundamental measure for deciding which surfaces and conditions give the largest quantum yields. The remaining part of this section will consider a method for the determination of the fluorescence and phosphorescence quantum yields for compounds adsorbed on solid surfaces. Also, a general comparison of the magnitude of the quantum yield values on different surfaces will be given.

Ramasamy et al. (67) have described, in detail, a method for the determination of the room-temperature fluorescence (RTF) and RTP quantum yields for organic compounds adsorbed on different surfaces. Their approach involved several modifications to the method originally developed by Wrighton et al. (63). For accurate quantum yield determinations, careful consideration had to be given to the sample holder, sample preparation, positioning of the sample, and various instrumental aspects. The main advantage of the approach developed by Ramasamy et al. (67) is that a conventional scanning fluorescence spectrophotometer could be used. Critical to the reproducibility of the measurements were primarily sample alignment and the amount of source radiation impinging on the sample. Sample alignment was standardized by first setting the excitation monochromator at 600 nm and then measuring the scattered source radiation from the standard, sample, or blank by scanning the emission monochromator ±10 nm on either side of 600 nm. The scattered relative intensity was set to the same level for the standard, sample, and blank. It was necessary to suspend a lead weight from the bottom of the cell compartment to minimize movement of the sample due to vibrations caused by the rotating chopper during solid-surface RTP measurements. It was also found that old source lamps with weak output resulted in low quantum yield values. Therefore, it was important that the lamp be checked routinely. In addition, it was necessary for the sample holder to be frequently painted black to obtain reproducible results. Below is a general discussion of the quantum yield method developed by Ramasamy et al. (67).

Sodium salicylate was chosen as a solid reference standard because it is easy to handle, it is commercially available in pure form, and its fluorescence quantum yield has been well character-ized (62-65). In the work by Ramasamy et al. (67), the absolute quantum yield of sodium salicylate was determined. This was accomplished by obtaining the corrected fluorescence emission spectrum for sodium salicylate and the relative amount of radiation absorbed by sodium salicylate. The relative amount of radiation absorbed by sodium salicylate was obtained as follows. The blank solid surface was scanned from 10 nm below to 10 nm above the excitation wavelength employed, and the area of the recorded reflectance band was obtained. Next, the sample adsorbed on the solid surface was scanned in the same fashion, and the area of the resulting reflectance band was recorded. The difference then between the area of the blank reflectance band and the sample reflectance band was proportional to the amount of radiation

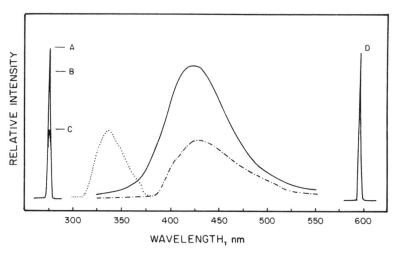

Figure 6.4. Reflectance bands and RTF and RTP spectra from PABA adsorbed on sodium acetate for calculating quantum yields: (A) reflectance band for NaOAc; (B) reflectance band for PABA adsorbed on NaOAc; (C) reflectance band for sodium salicylate adsorbed on NaOAc; (D) reference reflectance band with NaOAc, RTF spectrum (•••) of PABA adsorbed on NaOAc, RTP spectrum (-•-) of PABA adsorbed on NaOAc, and RTF spectrum (—) of sodium sallicylate adsorbed on NaOAc. (Reprinted with permission from S. M Ramasamy, V.P. Senthilnathan, and R.J. Hurtubise, *Anal. Chem.*, <u>58</u>, 612. Copyright 1986 American Chemical Society.)

absorbed. Figure 6.4 shows the reflectance bands (A and B) for sodium acetate and p-aminobenzoic acid adsorbed on sodium acetate. Once the corrected fluorescence emission spectra and the reflectance bands were obtained, it was possible to calculate the absolute fluorescence quantum yield for sodium salicylate using Equation (6.1).

$$\phi_F = \frac{\text{(fluorescence area)}}{\text{(blank reflectance area)} - \text{(standard reflectance area)}} \qquad (6.1)$$

The fluorescence yield of sodium salicylate was used to check the instrument and procedure for the determination of quantum yields. In addition, it was employed for the calculation of the relative quantum yields of unknown samples. This aspect is discussed more fully below (70).

Figure 6.4 shows the corrected RTF and RTP spectra for p-aminobenzoic acid adsorbed on sodium acetate, and the reflectance band obtained for adsorbed p-aminobenzoic acid. The relative fluorescence and phosphorescence quantum yields were calculated using Equation (6.2).

$$\phi_{\text{RTF or RTP}} = \frac{(\phi_{\text{std}})(P_{\text{std}})(\text{analyte RTF and RTP area})}{(P_{\text{analyte}})(\text{RTF standard area})} \quad (6.2)$$

In Equation 6.2, "P" refers to the amount of radiation absorbed by the adsorbed sample. The amount of radiation absorbed by standard or analyte was calculated by the following three methods: (a) the difference in areas of diffuse reflectance bands of the appropriate blank and sample; (b) the difference in the relative diffuse reflectance values, R_{∞} (R_{∞}, reflectance of the sample/reflectance of $BaSO_4$), of the blank and the sample; and (c) the difference in the Kubelka-Munk function values $F(R_{\infty}) = (1 - R_{\infty})^2/2R_{\infty}$ of the sample and blank. The different methods were used to calculate the amount of radiation absorbed by the samples because some of the solid surfaces absorbed the exciting radiation. Sodium acetate essentially absorbed no radiation in the range used to excite the samples; thus, the absolute quantum yield of sodium salicylate was determined while adsorbed on sodium acetate, and the fluorescence quantum yield of sodium salicylate thus obtained was used in calculating the quantum yields of the samples adsorbed on other solid matrices.

Table 6.5 gives the RTF and RTP quantum yield values of compounds adsorbed on solid supports with and without nitrogen passing over the surface. Higher quantum yields were obtained with nitrogen present for both RTF and RTP except for the RTF of 5,6-benzoquinoline adsorbed on filter paper. Generally, higher quantum yields are expected with nitrogen present because of the absence of oxygen. Table 6.5 shows that p-aminobenzoic acid gave the largest quantum yield values of the three samples investigated.

It was necessary to correct for the absorption due to the solid support when determining the quantum yields for samples adsorbed on polyacrylic acid-NaBr mixtures and filter paper (67). As discussed earlier in this section, the correction was made by one of three means: (a) difference in the area of reflectance bands; (b) difference in relative reflectance band with $BaSO_4$ as a reference; and (c) difference in the Kubelka-Munk function values of the sample and the blank with $BaSO_4$ as a reference. The details

Table 6.5 RTF and RTP Quantum Yields of Compounds Adsorbed on Solid Supports With and Without Nitrogen[a,b]

Compound	λ_{Ex}, nm	ϕ_{RTF}		ϕ_{RTP}	
		With N_2	Without N_2	With N_2	Without N_2
p-Aminobenzoic acid on sodium acetate	277	0.22 ± 0.05	0.15 ± 0.02	0.51 ± 0.09	0.29 ± 0.10
4-Phenylphenol on PAA-NaBr	270	0.10 ± 0.01	0.087 ± 0.004	0.11 ± 0.01	0.050 ± 0.01
5,6-BQ (0.1 M HBr) on filter paper	290	0.076 ± 0.004	0.084 ± 0.002	0.015 ± 0.002	0.0024 ± 0.001

a Average of three determinations at 95% confidence limit.
b For the powdered samples, each sample was packed in the depression twice for each run, and thus, two values were obtained for each run.
Reprinted with permission from S.M. Ramasamy, V.P. Senthilnathan, and R.J. Hurtubise, *Anal. Chem.*, 58, 612. Copyright 1986 American Chemical Society.

of the use of $BaSO_4$ as a reflectance standard and the use of the Kubelka-Munk function in correcting for absorption by substrates have been discussed by Körtum (2). As Table 6.6 shows, essentially identical quantum yield values were obtained by using correction methods (a) and (b) with or without nitrogen flowing over the surface. By use of the Kubelka-Munk function for correction due to substrate absorption, similar quantum yield values were obtained for all three correction methods for nitrogen gas flowing over the surface as indicated in Table 6.6. However, without nitrogen gas, somewhat different results were obtained using method (c) vs. methods (a) and (b). The results indicated that nitrogen gas should be flowing over the surface for good correlation between methods (a), (b), and (c). Ramasamy et al. (67) commented that additional work would be needed to explain the differences in Table 6.6 by method (c) without nitrogen compared to methods (a) and (b). However, the method developed by them is applicable to compounds on a variety of surfaces with quite different characteristics. The approximate minimum quantum yield value that could be determined for a sample would be 0.001 (67).

Table 6.7 lists several additional quantum yield values for the fluorescence and phosphorescence of compounds adsorbed on different surfaces at 23°C and -180°C. In general, the ϕ_f values show no change or a smaller change from 23°C to -180°C relative to the corresponding ϕ_p values. Most of the ϕ_p values show a rather large change from 23°C to -180°C. The data in Table 6.7 are important practically in that the quantum yield values give a relative measure of the sensitivity of an analytical method at room temperature as well as at low temperature. For example, the ϕ_f of p-aminobenzoic acid adsorbed on sodium acetate at room temperature is 0.19, whereas at -180°C the ϕ_f is 0.22. Thus, little would be gained by measuring the fluorescence at low temperature. For 4-phenylphenol adsorbed on filter paper, the ϕ_p at room temperature is 0.02; however, at low temperature the ϕ_p is 0.53. Thus, a significant enhancement would be achieved in sensitivity for 4-phenylphenol at -180°C (Table 6.7). The previous discussion is somewhat oversimplified because sensitivity is also affected by other factors such as background signals and the efficiency of absorption of exciting radiation.

It is of interest to compare solution phase quantum yield values with solid-surface quantum yield values. For example, the solution room-temperature fluorescence, low-temperature fluorescence, and low-temperature phosphorescence of p-aminobenzoic

Table 6.6 Comparison of Quantum Yield Values Corrected for Substrates That Absorb Exciting Radiation[a]

	ϕ_{RTF}		ϕ_{RTF}	
	Method a or b	Method c	Method a or b	Method c
4-Phenylphenol Adsorbed on 1% PAA-NaBr				
With N	0.10 ± 0.00	0.10 ± 0.03	0.11 ± 0.01	0.11 ± 0.03
Without N$_2$	0.087 ± 0.009	0.022 ± 0.025	0.050 ± 0.025	0.012 ± 0.008
5,6-Benzoquinoline (0.1 M HBr) Adsorbed on Filter Paper				
With N	0.076 ± 0.009	0.091 ± 0.016	0.014 ± 0.009	0.018 ± 0.009
Without N$_2$	0.084 ± 0.015	0.14 ± 0.05	0.0024 ± 0.005	0.0040 ± 0.004

[a]Average of triplicate determinations at 95% confidence level. Reprinted with permission from S. M. Ramasamy, V. P. Senthilnathan, R. J. Hurtubise, *Anal. Chem.*, 58, 612. Copyright 1986 American Chemical Society.

Table 6.7 Fluorescence and Phosphorescence Quantum Yields of Compounds Adsorbed on Solid Surfaces

Compound	Solid Surface	ϕ_f		ϕ_p		Reference
		23°C	-180°C	23°C	-180°C	
Anion of p-aminobenzoic acid	Sodium acetate	0.19	0.22	0.40	0.61	54
Anion of p-aminobenzoic acid	50% Sodium acetate / NaCl	0.20	---	0.14	---	55
Phenanthrene	80% α-Cyclodextrin / NaCl	0.29	0.29	0.13	0.26	53
Benzo(f)quinoline	80% α-Cyclodextrin / NaCl	0.41	0.40	0.09	0.22	53
Benzo(f)quinoline	0.05% α-Cyclodextrin / NaCl	0.22	0.36	0.05	0.12	53
p-Aminobenzoic acid	80% α-Cyclodextrin / NaCl	0.47	0.45	0.22	0.60	52
p-Aminobenzoic acid	0.05% α-Cyclodextrin / NaCl	0.22	0.46	0.07	0.32	52
4-Phenylphenol	80% α-Cyclodextrin / NaCl	0.33	0.34	0.08	0.29	52
4-Phenylphenol	0.05 α-Cyclodextrin / NaCl	0.27	0.36	0.03	0.21	52
4-Phenylphenol	Filter paper	0.33	0.51	0.02	0.53	57
Benzo(f)quinoline	Filter paper	0.50	0.65	0.012	0.11	58

acid have been reported as 0.15, 0.28, and 0.39, respectively (67). The corresponding data from Table 6.7 for p-aminobenzoic acid adsorbed on 80% α-cyclodextrin/NaCl gives 0.47, 0.45, and 0.60, respectively. It is seen that the ϕ_f and ϕ_p values are greater on the solid matrix. For 4-phenylphenol, the solution room-temperature fluorescence, low-temperature fluorescence, and low-temperature phosphorescence were 0.25 ± 0.03, 0.59 ± 0.03, and 0.43 ± 0.04, respectively (67). The equivalent solid-surface quantum yield values taken from Table 6.7 for 4-phenylphenol adsorbed on filter paper are 0.33, 0.51, and 0.53, respectively. The pooled standard deviations for these fluorescence and phosphorescence quantum yields were 0.04 and 0.01, respectively (57). In this case, the solid-surface room-temperature ϕ_f and the solid-surface low-temperature phosphorescence ϕ_p values are higher than the corresponding solution quantum yield values. However, the low-temperature solution quantum yield value (0.59) is somewhat higher than the low-temperature solid-surface fluorescence quantum yield value (0.51). Other quantum yield data are needed to make meaningful correlations between solid-surface and solution quantum yield values. In Chapter 7, a discussion is given on the use of quantum yield values in solid-surface luminescence to calculate rate constants related to the triplet state and the construction of energy level diagrams from quantum yield values.

6.4.   Sample Holders

Sample holders are important devices in RTP work and are simple to construct (5). They are designed to hold powders or flat surfaces such as filter paper. Paynter et al. (10) described a "finger-type" sample holder which was designed to fit into the phosphorescence accessory of an Aminco-Bowman spectrofluorometer. A filter paper disc with the phosphor adsorbed on it was mounted on the flat section of a rod and held in place by a flat brass plate with a circular window. The height of the sample inside the cell compartment could be adjusted by threaded brass bushings that held the steel rod to the spectrometer cell cover. With this type of sample holder, it is necessary that every sample be aligned prior to each RTP measurement. A modified version of the "finger-type" sample holder was developed to reduce the handling time (68). Individual holder tips were designed to be rapidly interchangeable. Each holder tip was constructed from segments of an aluminum rod with half of the low part of the cylinder removed so that it was semicircular. A slit plate with a hole in the center served to hold the paper disc in place.

Vo-Dinh and Martinez (42) used a similar sample holder to the one described in the preceeding paragraph. The primary advantage of their sample holder was that it could be inserted directly in the top of a phosphoroscope assembly without any mounting and positioning procedure. Figure 6.5 shows the sample holder as it would function during the measurement step.

A bar-type sample holder was designed by Ward et al. (69) for multiple sampling (Figure 6.6). The bar-type holder could be inserted into the sample compartment of a Aminco-Bowman spectrophosphorimeter and held four 0.6-cm-diameter filter paper discs. The main advantage of the multiple-sample bar holder was the decrease in sample handling time.

The sample holders described above have been designed primarily to hold samples adsorbed on filter paper discs. Powder samples could be employed with these sample holders; however, the circular depression in the sample holder would probably have to have a greater depth for the holder to accommodate the sample more effectively. For the measurement of RTP from p-aminobenzoic acid adsorbed on sodium acetate, the samples were placed in a

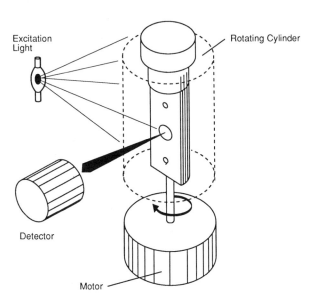

Figure 6.5. Sample holder as used in the spectroscopic measurement step. (Reprinted with permission from T. Vo-Dinh and P.R. Martinez, *Anal. Chim. Acta* 1981, 125, 13.)

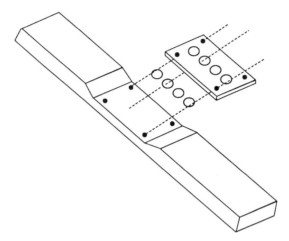

Figure 6.6. Multiple sample bar: 11 in x 7/8 in x 1/4 in aluminum. (Reprinted with permission from J.L. Ward, R.P. Bateh, and J.D. Winefordner, *Analyst* 1982, <u>107</u>, 335.)

series of 12 circular depressions that were milled into a brass plate along a line parallel to the edge (32). The plate was blackened, which gave it a dull, nonreflecting finish. The sample holder with samples was positioned on the moving stage of a densitometer, and the RTP intensity of twelve samples could be measured within three minutes. In later work the brass plate was modified to accommodate 22 powder or filter paper circle samples (20). A similar sample holder, which was designed to be used with a densitometer and α-cyclodextrin/NaCl mixtures, was described by Bello and Hurtubise (26). This sample holder had a degassing assembly for the samples, which was detachable and composed of two brass blocks with four gas channels carved in each block. The gas channels were equally spaced so that each channel was aligned with one of the circular depressions containing the samples. The width of the gas channel covered entirely the width of the depression so that a stream of gas could be passed onto the entire surface of the powder sample. The sample holder shown in Figure 6.7 fit into the cell compartment of a spectrofluorimeter and was used to obtain luminescence data from phosphors adsorbed onto 80% cyclodextrin/NaCl. For both of the sample holders described by Bello and Hurtubise (26), it was not necessary to place a quartz plate over the samples to prevent the samples from coming out of the depressions. The gentle flow of nitrogen over the surface did not disrupt the sample in the circular depression.

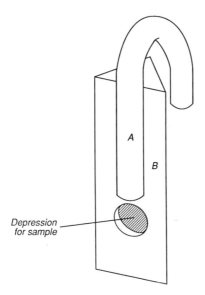

Figure 6.7. Sample holder used with cyclodextrin-salt mixtures: (A) Fixed degassing assembly made from copper tubing; (B) Triangular brass sample holder. (Reprinted with permission from J.M. Bello and R.J. Hurtubise, *Anal. Lett.* 1986, <u>19</u>, 775.)

For quantum yield measurements, a triangular brass sample holder, which was painted black, fit into the cell holder of a spectrofluorimeter (67). The triangular sample holder had a 6.5 mm diameter circular depression, and powder samples and circular pieces of filter paper were held in place by the sample holder. In temperature variation experiments, a semicircular, black Delrin plastic piece with a 0.5 in diameter depression was used as a sample holder. It was attached to one end of a fiber glass rod. For powder samples, it was necessary to cover the sample with a small quartz plate because the sample and holder were placed in a Dewar into which cold gaseous nitrogen was passed. The quartz plate prevented the sample from coming out of the sample holder as the cold nitrogen gas flowed into the Dewar.

Burrell and Hurtubise (70) designed a novel sample holder for solid and liquid luminescent samples that was used with a Perkin-Elmer LS-5 spectrofluorimeter. Figure 6.8A and 6.8B show the general design and sample positioning capabilities of the assembly. As indicated in Figure 6.8A, the accessory could accommodate a 1-cm-square quartz cell containing a solution or a

View from LS-5: **A**

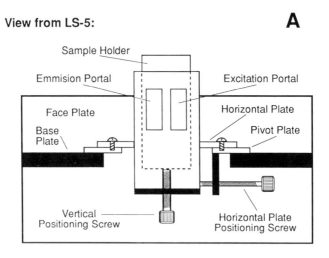

Top View: **B**

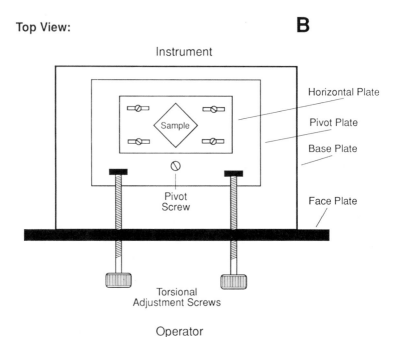

Figure 6.8. View of the sample accessory from the instrument (A). Top view of sample accessory (B). (Reprinted with permission from G. J. Burrell and R. J. Hurtubise, *Appl. Spectrosc.* 1988, 42, 173.

section of a chromatoplate, a triangular brass block with a powdered sample in a circular depression in the brass block, or a suitably supported filter paper strip spotted with adsorbate. The room-temperature fluorescence and RTP signals from compounds adsorbed on chromatoplate sections were very dependent on position. However, positioning of the sample was facilitated by the external control features of the sample accessory (Figure 6.8). The horizontal and vertical positioning of the samples were accomplished with the adjustment screws in Figure 6.8A. In Figure 6.8B, the torsional sample positioning screws are shown. The external controls permitted rapid and reliable sample positioning and optimization of room-temperature fluorescence and RTP signals. With the accessory in Figure 6.8, several samples were analyzed quickly with very good precision.

6.5.   Instrumentation

The major instrumentation for phosphorimetry was discussed in Chapter 3. In this Section, instrumentation developed for solid-surface phosphorimetry will be considered. A variety of instrumental aspects for solid-surface phosphorescence measurements have been reported (3-5).

Commercial instruments can be readily used for RTP measurements; however, special sample holders have to be used. Many types of sample holders were discussed in the preceding section. Modifications to commercial instruments can be relatively minor or sophisticated depending on one's needs. Ford and Hurtubise (30) designed a phosphoroscope and reflection mode assembly for use with a Schoeffel SD 3000 spectrodensitometer to obtain RTP signals from compounds adsorbed on silica gel and filter paper. The same instrument could be used to measure RTP from powdered samples. The reflection mode assembly permitted the distance from the source exit to the photomultiplier tube to be varied by means of an adjustable slide. In addition, the angle of the photomultiplier tube housing could be adjusted to maximize the reflected RTP striking the lens of the photodetector system. The phosphoroscope assembly was constructed with a variable speed dc motor (0-12 V) and a rotating disk phosphoroscope, and an assembly was designed for use with the modified reflection mode unit of the spectrodensitometer. The phosphoroscope was constructed of thin sheet aluminum that was painted flat black to minimize scattered radiation. When the compound on the solid surface was excited by the source radiation, the photomultiplier tube was in a position so the phosphoroscope blocked the emitted radiation, and no luminescence was detected by the photomultiplier tube. As the phosphoroscope rotated, it reached a position so that the adsorbed compound was no longer excited, and the detector system detected any delayed luminescence.

Vo-Dinh et al. (28) described an automatic phosphorimetric instrument for RTP measurements with a continuous filter paper device. Figure 6.9 shows a block diagram of the instrumental system. The detection unit of the system was an Aminco-Bowman spectrophotofluorometer. The sample compartment was modified and equipped with a laboratory-constructed rotating mirror assembly for detection of phosphorescence. The rotating mirror and general operating aspects are shown in Figure 6.10. A diagonally

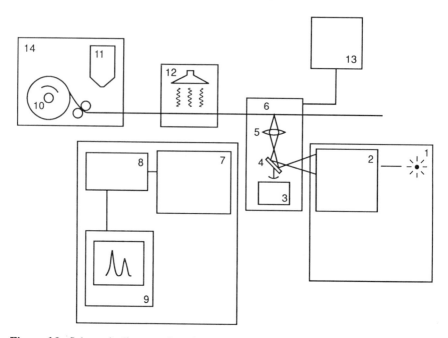

Figure 6.9. Schematic diagram of an AutoAnalyzer continuous filter with the room temperature phosphorescence detection system: (1) light source; (2) excitation monochromator; (3) rotation motor-phosphoroscope; (4) reflecting surface; (5) optics; (6) filter paper; (7) emission monochromator; (8) detection unit; (9) recorder; (10) filter paper roll; (11) spotting syringe; (12) drying IR lamp; (13) dry air supply; (14) AutoAnalyzer continuous filter (Reprinted with permission from T. Vo-Dinh, G. L. Walden, and J. D. Winefordner, *Anal Chem.*, 49, 1126. Copyright 1977 American Chemical Society.)

A

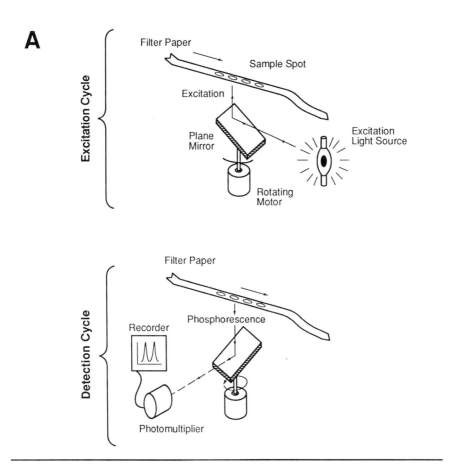

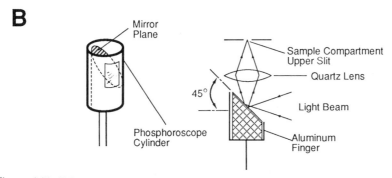

Figure 6.10. Principle of phosphorimetric excitation and detection with the rotating mirror phosphorimeter (A); Design of the rotating mirror assembly (B). (Reprinted with permission from T. Vo-Dinh, G.L. Walden, and J.D. Winefordner, *Anal. Chem.*, 49, 1126. Copyright 1977 American Chemical Society.)

cut section of an aluminum cylindrical rod was used for the mirror and reflecting surface. The well polished surface permitted good reflection of ultraviolet radiation. Excitation radiation from the excitation monochromator was reflected onto the surface of the filter paper by the reflecting surface (Figure 6.10A). The filter paper moved horizontally across a slit located at the top of the sample compartment. The reflection plane moved into the emission path as the cylindrical reflecting surface rotated, and the phosphorescence emitted by the sample was reflected back into the detection system. The excitation radiation was not observed by the detector during the excitation period, and scattered radiation was decreased substantially by inserting the reflecting cylindrical surface into an Aminco-Keirs phosphoroscope attachment (Figure 6.10B). At the upper part of the sample compartment, an aperture allowed a continuous warm-air flow over the filter paper. A Technicon Continuous Filter paper roll was used as a solid surface and was drawn into the drying chamber and the cell compartment (Figure 6.9). The samples were manually spotted drop by drop with a hypodermic syringe onto the moving filter paper. After spotting, the filter paper was fed into the drying chamber where the sample remained for about 2 min. After the drying step, the paper was passed over the sample compartment of the spectrophotofluorometer (Figure 6.9). Two important experimental variables had an influence on the RTP measured. These were drying the samples before measurement and the continuous flushing of dry gas through the sample compartment during measurements. Optimal predrying time was between 5 and 10 min for the compounds they investigated. Also, several series of 10 to 15 identical samples of different materials were measured. The relative standard deviation for most series of measurements was less than 5%. The limit of detection was found to be in the nanogram and subnanogram ranges. Later, Lue-Yen Bower and Winefordner (71) designed an improved version of the experimental system just discussed with a new filter paper guide that permitted continuous sampling of organic phosphors adsorbed on filter paper. Generally the analytical data showed that the new system would be very useful in areas where several samples were handled routinely.

Walden and Winefordner (72) compared ellipsoidal and parabolic mirror systems in fluorometry and room-temperature phosphorimetry. With the mirrors, the collection efficiency of sample luminescence for small volume samples improved. The authors commented that one problem with most commercial luminescence instruments is that only a small fraction of the total 4π Sr of the emitted luminescence is collected and measured. An instrument with f/4 collection optics collects approximately 0.015π Sr of the total emitted luminescence. Therefore, collection of a larger fraction of the total luminescence would permit either measurement of a smaller concentration of sample or the measurement of a smaller sample volume. Walden and Winefordner indicated that with RTP on filter paper, emission from the front surface was considerably greater than emission levels from the back surface. In addition, they showed that high collection efficiency mirrors can be very useful for the analysis of samples of small area. Table 6.8 gives a comparison of the limits of detection by RTP from an Aminco spectrofluorimeter, a parabolic mirror-monochromator system, and an ellipsoidal mirror filter system.

Goeringer and Pardue (73) reported the development of a silicon intensified target vidicon camera system for phosphorescence studies and presented applications of RTP to organic salts deposited on filter paper. The instrument allowed time-resolved spectra to be recorded with a minimum scan time of 8 ms/scan. Spectral decay data were processed by different regression methods to obtain rate constants, lifetimes, and initial intensity. The methods permitted the handling of multiple data rates, multiwavelength data for each component, and for an internal standard procedure that reduced the effects of experimental variables. The

Table 6.8 Room-Temperature Phosphorimetry Absolute Limits of Detection

Compound	Amico SPF	Parabolic mirror system monochromator	Ellipsoidal mirror system filter
Penanthrene	0.9	2	0.02
Carbazole	0.7	0.2	1
Tryptophan	0.2	1	8
1,2,5,6-Dibenzanthracene	8	0.6	0.09
8-Amino-1-naphthol-3,6-disulfonic acid	0.5	0.2	0.09

Reprinted with permission from G.L. Walden and J.D. Winefordner, *Appl. Spectrosc.* 1979 <u>33</u>, 166.

internal standard method reduced imprecision by factors of 2 to 5 depending upon the component of interest. Figure 3.7 in Chapter 3 gives a block diagram of their time-resolved phosphorimeter.

Warren et al. (74) discussed quantitative RTP with internal standard and standard addition techniques. Their experimental measurement system consisted of a 200-w Xe-Hg source, excitation and emission monochromators, a reference beam splitter and photomultiplier tube, a sample module, and an emission photomultiplier tube. They employed a rotating sample holder which functioned as both a sample holder and phosphoroscope. The system was controlled by a master microprocessor, which ran programs written in BASIC and assembly language. Separate microprocessors, which communicated over serial lines with the master microprocessor, controlled the excitation and emission monochromators. Input and output devices included a teletypewriter for communication between the operator and the master microprocessor, a thermal plotter/printer, and a video monitor for display of experimental data and calculated results.

Su et al. (75) constructed a continuous sampling system that was suitable for RTP, low-temperature fluorescence, low-temperature phosphorescence, and room-temperature fluorescence. Samples of powders, paper discs, and liquids could be handled with the system. The sampling system essentially consisted of a circular chopper disc with 20 equally spaced concentric circular depressions. Twenty copper rings were inserted into the depressions to hold paper discs firmly in the correct position for RTP observation. To the bottom of the copper disc, a smaller diameter concentric hollow copper cylinder was welded to facilitate the conduction of a cold stream from liquid nitrogen to the copper disc and to the sample for low-temperature studies. A laboratory-built chopper was used to block fluorescence signals from phosphorescence signals.

McAleese and Dunlap (76) investigated the problem of sample sizes larger than the excitation beam in conventional instruments. A loss in phosphorescence sensitivity can occur for samples outside the illumination area. In their work, they modified a spectrophotofluorometer by shifting the collimating mirror closest to the light source toward the excitation grating. They also reduced the size of the filter paper so that complete front surface illumination was accomplished. In general, improved sensitivity and precision were observed for the RTP technique.

Jones et al. (48) constructed a Becquerel-disc phosphoroscope for the measurement of lifetimes in RTP. Lifetimes determined with the Becquerel-disc phosphoroscope were compared with those obtained by using a pulsed-source system. There was good agreement between the RTP lifetimes of the individual samples using the two methods for the determination of RTP lifetimes. However, the precision was better by a factor of about 2 for the Becquerel-disc phosphoroscope. In addition, the Becquerel-disc system was capable of measuring longer lifetimes than the flashlamp system.

Scharf et al. (43) designed a sample compartment for use in RTP and for samples at -65°C adsorbed to solid surfaces. The sample compartment assembly was designed for the Perkin-Elmer LS-5 fluorometer and contained four positions for samples and blanks, which permitted convenient subtraction of the blank from the sample spectra. In addition, the system was designed so that the angle between the exciting beam and the solid-surface sample could be easily adjusted. By using an immersion heater for heating and a dry ice-acetone mixture for cooling, they were able to work in the temperature range of -65°C to 70°C.

Gifford et al. (77) constructed and evaluated a thin-layer, single-disk, multi-slot phosphorimeter for the direct measurement of phosphorescence from separated components on thin-layer chromatograms near liquid nitrogen temperature. The phosphorimeter could also be used to measure RTP of samples adsorbed on solid surfaces. Later, Miller et al. (78) described the construction and evaluation of an improved thin-layer phosphorimeter. In a typical experiment, an aluminum backed thin-layer chromatoplate or filter paper was affixed with elastic bands to the outside of a hollow copper sample drum which could be filled with liquid nitrogen. The rate of rotation of the turntable was controlled by a variable-output transformer, and a scanning rate of 3-40 cm min^{-1} was obtainable. Gifford et al. (77) discussed the ability of their phosphoroscope to resolve short- and long-lived phosphorescence. Also, they compared disks of different slot dimensions and showed under what design conditions the phosphorescence intensity became constant.

References

1. Wendlandt, W.W.; Hecht, H.G. *Reflectance Spectroscopy*; Wiley: New York, 1966.
2. Körtum, G. *Reflectance Spectroscopy*; Springer-Verlag: New York, 1969.

3.   Hurtubise, R.J. *Solid-Surface Luminescence Analysis*; Marcel Dekker: New York, 1981.
4.   Hurtubise, R.J. In *Molecular Luminescence Spectroscopy: Methods and Applications - Part II*; Schulman, S.J., Ed.; Wiley: New York, 1988; Chapter. 1.
5.   Vo-Dinh, T. *Room-Temperature Phosphorimetry for Chemical Analysis*; Wiley: New York, 1984.
6.   Roth, M. *J. Chromatogr.* 1967, 30, 276.
7.   Lloyd, J.B.F.; Miller, J.N. *Talanta* 1979, 26, 180.
8.   Schulman, E.M.; Walling, C. *Science* 1972, 178, 53.
9.   Schulman, E.M.; Walling, C. *J. Phys. Chem.* 1973, 77, 902.
10.   Paynter, R.A.; Wellons, S.L.; Winefordner, J.D. *Anal. Chem.* 1974, 46, 736.
11.   Lue-Yen Bower, E.; Ward, J.L.; Walden, G.; Winefordner, J.D. *Talanta* 1980, 27, 380.
12.   Schulman, E.M.; Parker, R.T. *J. Phys. Chem.* 1977, 81, 1932.
13.   Senthilnathan, V.P.; Hurtubise, R.J. *Anal. Chim. Acta* 1984, 157, 203.
14.   Su, S.Y.; Winefordner, J.D. *Can. J. Spectrosc.* 1983, 28, 21.
15.   Aaron, J.J.; Andino, M.; Winefordner, J.D. *Anal. Chim. Acta* 1984, 160, 171.
16.   von Wandruszka, R.M.A.; Hurtubise, R.J. *Anal. Chem.* 1977, 49, 2164.
17.   Lue-Yen Bower, E.; Winefordner, J.D. *Anal. Chim. Acta* 1978, 102, 1.
18.   Parker, R.T.; Freedlander, R.S.; Schulman, E.M.; Dunlap, R.B. *Anal. Chem.* 1979, 51, 1921.
19.   Ford, C.D.; Hurtubise, R.J. *Anal. Chem.* 1980, 52, 656.
20.   Dalterio, R.A.; Hurtubise, R.J. *Anal. Chem.* 1982, 54, 224.
21.   Ramasamy, S.M.; Hurtubise, R.J. *Anal. Chem.* 1982, 54, 2477.
22.   Dalterio, R.A.; Hurtubise, R.J. *Anal. Chem.* 1983, 55, 1084.
23.   Dalterio, R.A.; Hurtubise, R.J. *Anal. Chem.* 1984, 56, 336.
24.   Su, S.Y.; Winefordner, J.D. *Microchem. J.* 1982, 27, 151.
25.   Bello, J.M.; Hurtubise, R.J. *Appl. Spectrosc.* 1986, 40, 790.
26.   Bello, J.M.; Hurtubise, R.J. *Anal. Lett.* 1986, 19, 775.
27.   Richmond, M.D.; Hurtubise, R.J. *Appl. Spectrosc.* 1989, 43, 810.
28.   Vo-Dinh, T.; Walden, G.L.; Winefordner, J.D. *Anal. Chem.* 1977, 49, 1126.
29.   Bateh, R.P.; Winefordner, J.D. *Talanta* 1982, 29, 713.
30.   Ford, C.D.; Hurtubise, R.J. *Anal. Chem.* 1979, 51, 659.

31. Ford, C.D.; Hurtubise, R.J. *Anal. Chem.* 1978, 50, 610.
32. von Wandruszka, R.M.A.; Hurtubise, R.J. *Anal. Chem.* 1976, 48, 1784.
33. Citta, L.A.; Hurtubise, R.J. *Microchem. J.* 1986, 34, 56.
34. Purdy, B.B.; Hurtubise, R.J. *Microchem. J.* 1989, 39, 330.
35. Ward, J.L.; Lue Yen-Bower, E.; Winefordner, J.D. *Talanta* 1981, 28, 119.
36. McAleese, D.L.; Dunlap, R.B. *Anal. Chem.* 1984, 56, 600.
37. Su, S.Y.; Bolton, D.L.; Winefordner, J.D. *Chem. Biomed. Environ. Instr.* 1982, 12, 55.
38. Senthilnathan, V.P.; Hurtubise, R.J. *Anal. Chem.* 1985, 57, 1227.
39. de Lima, C.G.; de M. Nicola, E.M. *Anal. Chem.* 1978, 50, 1658.
40. Kirchner, J.G. *J. Chromatogr.* 1973, 82, 101.
41. McAleese, D.L.; Freedlander, R.S.; Dunlap, R.B. *Anal. Chem.* 1980, 52, 2443.
42. Vo-Dinh, T.; Martinez, P.R. *Anal. Chim. Acta* 1981, 125, 13.
43. Scharf, G.; Smith, B.W.; Winefordner, J.D. *Anal. Chem.* 1985, 57, 1230.
44. Scharf, G.; Winefordner, J.D. *Talanta* 1986, 33, 17.
45. Khasawneh, I.M.; Alvarez-Coque, M.C.G.; Ramos, G.R.; Winefordner, J.D. *J. Pharm. Biomed. Anal.* 1989, 7, 29.
46. Ramos, G.R.; Alvarez-Coque, M.C.G.; O'Reilly, A.M.; Khasawneh, I.M.; Winefordner, J.D. *Anal. Chem.* 1988, 60, 416.
47. Khasawneh, I.M.; Winefordner, J.D. *Microchem. J.* 1988, 37, 77.
48. Jones, B.T.; Smith, B.W.; Berthod, A.; Winefordner, J.D. *Talanta* 1988, 35, 647.
49. Khasawneh, I.M.; Winefordner, J.D. *Talanta* 1988, 35, 267.
50. Ramos, G.R.; Khasawneh, I.M.; Alvarez-Coque, M.C.G.; Winefordner, J.D. *Talanta* 1988, 35, 41.
51. Khasawneh, I.M.; Chamsaz, M.; Winefordner, J.D. *Anal. Lett.* 1988, 21, 125.
52. Bello, J.M.; Hurtubise, R.J. *Appl. Spectrosc.* 1988, 42, 619.
53. Bello, J.M.; Hurtubise, R.J. *Anal. Chem.* 1988, 60, 1291.
54. Ramasamy, S.M.; Hurtubise, R.J. *Anal. Chem.* 1987, 59, 432.
55. Ramasamy, S.M.; Hurtubise, R.J. *Anal. Chem.* 1987, 59, 2144.
56. Burrell, G.J.; Hurtubise, R.J. *Anal. Chem.* 1987, 59, 965.
57. Ramasamy, S.M.; Hurtubise, R.J. *Talanta* 1989, 36, 315.

58. Ramasamy, S.M.; Hurtubise, R.J. *Appl. Spectrosc.* 1989, 43, 616.
59. Melhuish, W.H. *J. Opt. Soc. Am.* 1964, 54, 183.
60. Demas, J.N.; Crosby, G.A. *J. Phys. Chem.* 1971, 75, 991.
61. Rhys Williams, A.T.; Winfield, S.A.; Miller, J.N. *Analyst* (London) 1983, 108, 1471.
62. Kristianpoller, N. *J. Opt. Soc. Am.* 1964, 54, 1285.
63. Wrighton, M.S.; Ginley, D.S.; Morse, D.L. *J. Phys. Chem.* 1974, 78, 2229.
64. Adams, M.J.; Highfield, J.G.; Kirkbright, G.F. *Anal. Chem.* 1980, 52, 1260.
65. Adams, M.J.; Highfield, J.G.; Kirkbright, G.F. *Analyst* (London) 1981, 106, 850.
66. Kirkbright, G.F.; Spillane, D.E.M.; Anthony, K.; Brown, R.G.; Hepworth, J.D.; Hodgson, K.W.; West, M.A. *Anal. Chem.* 1984, 56, 1644.
67. Ramasamy, S.M.; Senthilnathan, V.P.; Hurtubise, R.J. *Anal. Chem.* 1986, 58, 612.
68. Lue-Yen Bower, E.; Winefordner, J.D. *Anal. Chim. Acta* 1978, 101, 319.
69. Ward, J.L.; Bateh, R.P.; Winefordner, J.D. *Analyst* 1982, 107, 335.
70. Burrell, G.J.; Hurtubise, R.J. *Appl. Spectrosc.* 1988, 42, 173.
71. Lue-Yen Bower, E.; Winefordner, J.D. *Appl. Spectrosc.* 1979, 33, 9.
72. Walden, G.L.; Winefordner, J.D. *Appl. Spectrosc.* 1979, 33, 166.
73. Goeringer, D.E.; Pardue, H.L. *Anal. Chem.* 1979, 51, 1054.
74. Warren, M.W.; Avery, J.P.; Malmstadt, H.V. *Anal. Chem.* 1982, 54, 1853.
75. Su, S.Y.; Asafu-Adjaye, E.; Ocak, S. *Analyst* 1984, 109, 1019.
76. McAleese, D.L.; Dunlap, R.B. *Anal. Chem.* 1984, 56, 836.
77. Gifford, L.A.; Miller, J.N.; Burns, D.T.; Bridges, J.W. *J. Chromatogr.* 1975, 103, 15.
78. Miller, J.N.; Phillips, D.L.; Burns, D.T.; Bridges, J.W. *Anal. Chem.* 1978, 50, 613.

CHAPTER 7

PHYSICOCHEMICAL INTERACTIONS IN SOLID-SURFACE PHOSPHORESCENCE

There are several solid materials that have been used to obtain room-temperature phosphorescence (RTP) from adsorbed compounds. However, a detailed understanding of the specific physical and chemical interactions required for RTP from adsorbed compounds is not fully developed. This chapter is organized according to several of the solid surfaces that have been used in obtaining RTP from adsorbed phosphors. A discussion is given about the important interactions that have been discovered for each surface. A general summary is presented for the interactions and models that have been developed for solid-surface RTP. Finally, in the last section, a comparison is made of the general properties, interactions, and experimental conditions for solid surfaces that are important in obtaining RTP from adsorbed compounds.

7.1. Sodium Acetate

Von Wandruszka and Hurtubise (1,2) first reported the use of sodium acetate as a material that would stimulate RTP from certain organic compounds. In their investigations of the interactions of compounds adsorbed on sodium acetate, comparison of molecular structures and considerations of reflectance, fluorescence, and infrared spectra were used in addition to surface-area data, solvent interactions, and various molecular criteria (1). Ethanol solutions of the compounds were used to deposit the compounds on the sodium acetate. Their data implied that the differences in RTP intensities for a given compound were not due mainly to inherent molecular effects, but rather to differences in rigidity of the absorbed compound. p-Aminobenzoic acid was adsorbed on sodium acetate the most efficiently because it gave the largest room-temperature to low-temperature phosphorescence ratio compared to the other compounds investigated. It was found that certain molecular requirements were needed to obtain strong RTP signals for the compounds they investigated. For example, 3-methyl-4-aminobenzoic acid gave a 11.5-fold reduction in RTP compared to p-aminobenzoic acid. It was concluded that the presence of a carboxyl group bonded to the 1-position was one requirement for compounds with the benzene nucleus. Also, attached to the 4-position on the ring, an electron-donating, hydrogen-bonding substituent appeared to be necessary. They also found that the organic solvent used to adsorb the compounds on sodium acetate was important in obtaining RTP, but water gave a three-fold reduction in RTP for p-aminobenzoic acid. Aprotic solvents such as ether, acetone, dimethylformamide, and cyclohexane gave no RTP from p-aminobenzoic acid on sodium acetate. Von Wandruszka and Hurtubise (1) showed that the adsorption of p-aminobenzoic acid on sodium acetate was preceded by partial neutralization with dissolved sodium acetate in alcoholic solutions. In later RTP work, it was reported that sodium acetate was insoluble in acetone and ether (3). Thus, the partial neutralization of p-aminobenzoic acid could not occur prior to adsorption with these solvents. The results with acetone and ether indicated that the acid form of p-aminobenzoic acid, which was adsorbed on sodium acetate, did not give RTP. The p-aminobenzoic acid anion formed in ethanolic solution had a strong tendency to adsorb on the surface of sodium acetate, forming the sodium salt. This conclusion was supported by the strong RTP of the sodium salt of p-aminobenzoic acid when it adsorbed on suspended sodium acetate in an ethanolic solution of the sodium salt of p-aminobenzoic acid (2). Two compounds with the indole nucleus showed RTP on sodium acetate, 5-hydroxyindoleacetic acid and 5-hydroxytryptophan. Analytically useful RTP signals were obtained only when alkaline ethanolic solutions of the compounds were evaporated on sodium acetate. For the ethanol solutions of 5-hydroxyindoleacetic acid and 5-hydroxytryptophan, the dissolved sodium acetate did not give the required neutralization of the solutes because the indole compounds are very weak acids. Thus, it was necessary to add sodium hydroxide to ethanol solutions of these compounds.

Von Wandruszka and Hurtubise (1) investigated the mode of adsorption of p-aminobenzoic acid and other compounds on sodium acetate, talc, and starch, using reflectance spectroscopy. The compounds studied did not give RTP on talc and starch. There was a blue shift of about 35 nm in the reflectance maximum for p-aminobenzoic acid adsorbed on sodium acetate compared to the compound adsorbed on talc or starch. The reflectance spectral results indicated strong interactions between sodium acetate and adsorbed p-aminobenzoic acid, showing the formation of the sodium salt of p-aminobenzoic acid upon adsorption. Other supporting reflectance and fluorescence data were also reported (1).

Infrared spectroscopy was used to study the adsorption of p-aminobenzoic acid on sodium acetate (1). Infrared spectra of p-aminobenzoic acid, of sodium acetate, and of p-aminobenzoic acid adsorbed on sodium acetate were obtained. Strong sodium acetate infrared bands obscured much of the adsorbate spectra, but several observations and conclusions were reported. The N-H stretching vibrations of p-aminobenzoic acid at 3350-3450 cm^{-1} actually disappeared for p-aminobenzoic acid adsorbed on sodium acetate. This result showed that the bands were shifted to longer wavelengths and broadened due to hydrogen bonding between the amino group of p-aminobenzoic acid and the carboxyl group of the sodium acetate surface. o-Aminobenzoic acid did not give RTP when adsorbed on sodium acetate. Similar infrared experiments were carried out with o-aminobenzoic acid adsorbed on sodium acetate, and the o-aminobenzoic acid retained its N-H infrared bands. Because o-aminobenzoic acid undergoes intramolecular hydrogen bonding between the amino group and the carboxyl group in o-aminobenzoic acid, this compound was not held to the surface by strong intermolecular hydrogen bonding.

Von Wandruszka and Hurtubise (1) assumed that chemisorbed p-aminobenzoic acid molecules on the sodium acetate were distinguished from physically adsorbed molecules by their RTP characteristics. Only those molecules that strongly and directly interacted with the sodium acetate were held rigidly enough to yield RTP. They postulated that p-aminobenzoic acid molecules in the second and subsequent adsorbed layers would not give RTP, but would decrease the signal by absorbing exciting radiation and possibly absorbing emitted phosphorescence from the chemisorbed molecules. It was shown that the maximum RTP signal was obtained at 6100 ng p-aminobenzoic acid on 10 mg sodium acetate. It was assumed that the maximum RTP signal corresponded to complete monolayer coverage of the sodium acetate. Von Wandruszka and Hurtubise (1) used a method developed by Snyder

(4) to calculate the surface area occupied by a flatly adsorbed p-aminobenzoic acid molecule. They assumed that p-aminobenzoic acid was adsorbed flatly, based on spectral and other data that they obtained. Then they calculated the surface area of sodium acetate as 1.8 m^2/g. The previous surface area value was identical to the surface area of sodium acetate obtained from a commercial source. The results they obtained showed that p-aminobenzoic acid was adsorbed flatly on sodium acetate. Additional calculations showed that two sodium acetate molecules were necessary to hold one p-aminobenzoic acid molecule to the surface. A similar investigation was carried out with 5-hydroxyindoleacetic acid, and it was found that three sodium acetate molecules were needed to hold one 5-hydroxyindoleacetic acid molecule.

The RTP lifetimes and relative RTP intensities of the anion of p-aminobenzoic acid adsorbed on sodium acetate-sodium chloride mixtures ranging from 0.1% sodium acetate to 100% sodium acetate were investigated (5). The luminescence intensity and phosphorescence lifetime values varied over a wide range, and short and long decaying phosphorescent components were detected from the anion of p-aminobenzoic acid. It was speculated that inhomogeneous matrix effects were responsible for the slow and fast phosphorescence decay. The lifetime of both decaying components showed the same general pattern as the sodium acetate content in the mixtures increased. A plot of the log(RTP) versus log(mole fraction of sodium acetate) gave an approximate S-shaped curve with RTP intensity increasing as the mole fraction of sodium acetate increased. The maximum RTP signal was obtained with pure sodium acetate. In their work, it was shown that it is important to consider the initial wet chemistry and the dried solid matrix when dealing with solid-surface RTP. The solubility of sodium acetate from the sodium acetate-sodium chloride mixtures in the ethanol solutions used to adsorb the anion of p-aminobenzoic acid onto the solid matrix was shown to be an important factor in obtaining a high RTP signal. A simple calculation showed that a mixture with about 1.2% sodium acetate would give a saturated solution of sodium acetate. Based on solubility data, RTP intensity data, and phosphorescence lifetime data, the following explanation was presented for some of the factors that yielded RTP for the anion of p-aminobenzoic acid adsorbed on sodium acetate. The RTP lifetime of the anion of p-aminobenzoic acid became approximately constant at about the same point at which the ethanol solution became saturated with sodium acetate. Before adsorption then, p-aminobenzoic acid reacted in a saturated solution with dissolved sodium acetate to form the p-aminobenzoic acid anion. When the solvent was evaporated, the p-aminobenzoic acid anions had a greater probability of interacting with the closest sodium acetate species, namely, the dissolved sodium acetate. The abrupt increase in the phosphorescence lifetime near 1.4% sodium acetate was an indication of the fact that no more sodium acetate could dissolve in the adsorbing solvent. As the undissolved sodium acetate increased in the mixtures, the molecules of sodium acetate would pack the dry matrix more efficiently and either protect the p-aminobenzoic acid anion from collisions with oxygen molecules or hold the anion in a more rigid state, as indicated by the increase in RTP beyond about 1.4% sodium acetate. Maximum RTP was achieved with pure sodium acetate partly because the pure sodium acetate could pack more effectively than sodium acetate-sodium chloride mixtures. From the work of Senthilnathan and Hurtubise (6), two major aspects for the RTP of the p-aminobenzoic acid anion on sodium acetate-sodium chloride mixtures and pure sodium acetate were indicated. The p-aminobenzoic acid anions achieved a certain rigidity at approximately 1.4% sodium acetate by strong interaction with sodium acetate molecules. However, protective matrix effect and/or increased rigidity of the matrix caused by increasing sodium acetate content in the solid matrix allowed for enhanced RTP intensity, with maximum RTP obtained with pure sodium acetate.

Solid-surface fluorescence (ϕ_f) and phosphorescence (ϕ_p) quantum yield values were obtained from 23° to -180°C for the anion of p-aminobenzoic acid adsorbed on sodium acetate (7). Phosphorescence lifetime values were also obtained for the adsorbed anion from 23° to -196°C. The fluorescence quantum yield values remained essentially constant as a function of temperature. However, the phosphorescence quantum yield values changed considerably with temperature. Table 7.1 gives the values obtained. The phosphorescence lifetime experiments showed two decaying components, and each component showed a gradual increase in phosphorescence lifetime and then appeared to level off at lower temperatures. Ramasamy and Hurtubise (7) calculated various fundamental solid-surface luminescence parameters for the anion of p-aminobenzoic acid adsorbed on sodium acetate for the temperature range of 23° to -180°C. The parameters calculated were the rate constants for phosphorescence (k_p), for radiationless transition from the triplet state (k_m), for bimolecular quenching (k_q), and triplet formation efficiency (ϕ_t).

Table 7.1. Fluorescence and Phosphorescence Quantum Yield Values for the Anion of p-Aminobenzoic Acid Adsorbed on Sodium Acetate Over a Wide Temperature Range[a]

Temperature, °C	Fluorescence quantum yield, ϕ_f	Phosphorescence quantum yield, ϕ_p
23	0.19 ± 0.06[b]	0.40 ± 0.07
0	0.18 ± 0.07	0.37 ± 0.05
-40	0.20 ± 0.10	0.43 ± 0.18
-80	0.21 ± 0.05	0.55 ± 0.16
-120	0.20 ± 0.05	0.63[c,d] (0.05)[e]
-140	0.24 ± 0.04	0.59[c,d] (0.02)[e]
-160	0.22 ± 0.04	0.48[d] ± 0.16[b]
-180	0.22 ± 0.06	0.61[c,d] (0.05)[e]

[a] The results are the average of at least four determinations, except for three of the samples. The ϕ_p value at -160°C is the average of three determinations.
[b] 95% confidence limits.
[c] Average of duplicate determinations.
[d] Temperature held for two hours prior to quantum yield measurements.
[e] Range.

Reprinted with permission from S.M. Ramasamy and R. J. Hurtubise, *Anal. Chem.*, 59, 432. Copyright 1987 American Chemical Society.

The triplet formation efficiency was calculated with Equation (7.1), where ϕ_f is fluorescence quantum yield.

$$\phi_f + \phi_t = 1.0 \qquad (7.1)$$

Equation (7.1) has been discussed by Turro (8) and Parker (9). Parker has stated that if the ratio of phosphorescence quantum yield to phosphorescence lifetime (ϕ_p/τ_p) is approximately constant with temperature, then Equation (7.1) is valid and can be used to calculate ϕ_t. Ramasamy and Hurtubise (7) showed that ϕ_p/τ_p was constant with temperature and thus used Equation (7.1) to calculate ϕ_t. Because the values of ϕ_p and τ_p were obtained experimentally and ϕ_t was calculated with Equation (7.1), it was then possible to calculate k_p from Equation (7.2).

$$\phi_p = k_p \phi_p \tau_p \qquad (7.2)$$

Turro (8) has shown that under conditions in which the rate constant for internal conversion from the singlet state (k_i) is zero, the quantum yield of nonradiative triplet to ground-state transitions (ϕ_{ts}) is given by the difference ($\phi_t - \phi_p$). In turn, ϕ_{ts} is given by Equation (7.3).

$$\phi_{ts} = \phi_t k_m \tau_p \qquad (7.3)$$

Ramasamy and Hurtubise (7) assumed that k_i was zero and also the rate constant for bimolecular quenching (k_q) was zero. Their temperature variation experiments showed that k_i was zero or very small relative to the other rate constants. Also, because their samples were dried and all measurements were done under a nitrogen atmosphere, they assumed that oxygen and water had a small, if any, contribution to quenching of phosphorescence. In addition, the sodium acetate was essentially free of luminescent impurities, and thus, the quenching from impurities in sodium acetate was minimal. Therefore, with the assumptions above, Equation (7.3) could be used to calculate k_m values for the anion of p-aminobenzoic acid. Because two phosphorescent decaying components were obtained, Ramasamy and Hurtubise (7) calculated k_p and k_m for both components. Table 7.2 gives ϕ_t values and average k_p and k_m values at the various temperatures for each of the decaying components. Table 7.2 shows that ϕ_t values and k_p values are essentially constant with temperature. In general, ϕ_t is not always constant with temperature; however, k_p is considered a fundamental constant and in most cases is independent of temperature. It is clear from Table 7.2 that k_m undergoes large changes from 23° to -180°C. Ramasamy and Hurtubise (7) interpreted the data in Table 7.2 as follows. Equation (7.2) was rewritten to give Equation (7.4).

$$\phi_p = (\phi_t) \frac{k_p}{k_p + k_m} \qquad (7.4)$$

Because ϕ_t and k_p are essentially constant with temperature, the change in k_m will have the greatest effect on ϕ_p. Obviously, k_m is

Table 7.2. Luminescence Parameters for the p-Aminobenzoic
Acid Anion Adsorbed on Sodium Acetate

Temperature	ϕ_t	k_p	k_m
23	0.81	0.38	0.39
0	0.82	0.29	0.35
-40	0.80	0.31	0.26
-80	0.79	0.36	0.16
-120	0.80	0.35	0.096
-140	0.76	0.33	0.095
-160	0.78	0.30	0.10
-180	0.78	0.29	0.080

Reprinted with permission from S.M. Ramasamy and R.J. Hurtubise, *Anal. Chem.*, 59, 432.
Copyright 1987 American Chemical Society.

a function of temperature, and the more rigid the matrix at low
temperature, the smaller the value of k_m. This indicated that the
solid-surface phosphorescence for the anion of p-aminobenzoic acid
adsorbed on sodium acetate was a function of how rigidly the anion
was held by the solid matrix. Oxygen and moisture would not be
major contributing factors to the decrease of ϕ_p because of the
experimental conditions used in the work. From a fundamental
analytical viewpoint, the data for k_m in Table 7.2 shows that it
would be necessary to minimize k_m at room temperature to
maximize ϕ_p.

In related work by Ramasamy and Hurtubise (7), the curves
for phosphorescence lifetime as a function of temperature for the
anion of p-aminobenzoic acid adsorbed on sodium acetate yielded
additional information about RTP. Oelkrug and co-workers (10-12)
reported similar solid-surface phosphorescence curves for polycyclic aromatic hydrocarbons adsorbed on γ-alumina. Their solid-
surface phosphorescence lifetime curves followed Equation (7.5).

$$\frac{1}{\tau_p} - \frac{1}{\tau_p^o} = k_1 \exp(-E_a/T) \qquad (7.5)$$

The meaning of the preexponential factor (k_1) and activation energy
term (E_a) is not fully developed (13). In Equation (7.5), the term τ_p
is the phosphorescence lifetime at a given temperature and τ_p^o is the

phosphorescence lifetime when the exponential term is negligible. In the work by Ramasamy and Hurtubise (7), graphs of $\ln(1/\tau_p - 1/\tau_p^\circ)$ vs. $1/T$ were prepared from which the preexponential and activation energy terms were obtained for the anion of p-aminobenzoic acid adsorbed on sodium acetate. The average preexponential term for the short and long decaying components was $2.1 \, s^{-1}$, and the average activation energy term for the short and long decaying components was $392 \, cm^{-1}$. Oelkrug et al. (12) compared the k_1 and E_a values for naphthalene in durene, plastics, and adsorbed on γ-alumina. They concluded that thermal quenching was considerably smaller when adsorbed on alumina than in environments such as durene and plastics. They reported a k_1 value of $15 \, s^{-1}$ and an E_a value of $800 \, cm^{-1}$ for naphthalene adsorbed on highly activated alumina. These values are roughly comparable to the corresponding values obtained for the anion of p-aminobenzoic acid adsorbed on sodium acetate. In general, the more strongly the compound interacts with the surface, the smaller are the k_1 and E_a values (11,12).

Room-temperature fluorescence and phosphorescence quantum yield, triplet formation efficiency, and phosphorescence lifetime values were obtained for the anion of p-aminobenzoic acid adsorbed on sodium acetate and several sodium acetate-sodium chloride mixtures (13). From these data, rate constants for phosphorescence and for radiationless transition from the triplet state were obtained. Table 7.3 gives quantum yield data for the anion of p-aminobenzoic acid adsorbed on sodium acetate-sodium chloride mixtures at room temperature and at -180°C. As shown in Table 7.3, for the room-temperature samples, the maximum room-temperature fluorescence and RTP quantum yields were obtained with 100% sodium acetate. However, as indicated in Table 7.3, the low-temperature phosphorescence quantum yield values for the 1.4%, 5.0%, 50% and 100% sodium acetate samples are considerably greater than those for the corresponding sodium acetate samples at room temperature. The RTP lifetime values for the p-aminobenzoic acid anion ranged from 0.006 s for the 0.1% sodium acetate mixture to 1.36 s for 100% sodium acetate. At -180°C, the anion of p-aminobenzoic acid adsorbed on 100% sodium acetate gave a phosphorescence lifetime of 2.66 s (13). Ramasamy and Hurtubise (13) calculated several fundamental luminescence parameters for the anion of p-aminobenzoic acid adsorbed on the sodium acetate-sodium chloride mixtures, using approaches that were similar for the anion of p-aminobenzoic acid

Table 7.3. RTF and RTP Quantum Yields for the Anion of PABA Adsorbed on Sodium Acetate-Sodium Chloride Mixtures[a]

Sodium Acetate, %	ϕ_{RTF}[a,b]	ϕ_{RTP}[a,b]	ϕ_f[b,c]	ϕ_p[b,c]
0.1	0.04	0.02		
0.4	0.07	0.04		
0.5	0.07	0.05		
1.0	0.07	0.06		
1.4	0.10	0.05	0.21	0.80
5.0	0.10	0.09	0.20	0.81
50	0.20	0.14	0.17	0.77
100	0.24	0.36	0.19[d]	0.61[d]

[a]Results are the average of at least two determinations.
[b]Overall average reproducibility at the 95% confidence level $\pm$ 0.02.
[c]Values obtained at -180°C. Duplicate runs.
[d]Values taken from Reference 7.

Reprinted with permission from S.M. Ramasamy and R.J. Hurtubise, *Anal. Chem.*, 59, 2144. Copyright 1987 American Chemical Society.

adsorbed on sodium acetate (7). As discussed earlier in this section, if the ratio ϕ_p/τ_p is constant with temperature, then Equation (7.1) can be used to calculate ϕ_t. However, with the sodium acetate-sodium chloride mixtures, it was shown that the ratio of ϕ_p/τ_p was not constant with temperature; thus Equation (7.6) was used to calculate ϕ_t (9).

$$\phi_t = \frac{\phi_p \tau'_p}{\phi'_p \tau_p} \qquad \phi'_t \sim \frac{\phi_p \tau'_p}{\phi'_p \tau_p} (1 - \phi'_f) \qquad (7.6)$$

The primed terms in Equation (7.6) refer to low temperature (-180°C). The substitution of $1-\phi'_f$ for ϕ'_t in Equation (7.6) was reliable because at low temperature ϕ'_t is approximately equal to $1-\phi'_f$ (13). The expression to the right of the approximation sign in Equation (7.6) was used to calculate ϕ_t at room temperature for the anion of p-aminobenzoic acid adsorbed on sodium acetate-sodium chloride mixtures. Two decaying components were also observed for the anion of p-aminobenzoic acid adsorbed on the sodium acetate-sodium chloride mixtures; however, average phosphorescence lifetimes were employed to calculate k_p and k_m. Table 7.4 gives the calculated luminescence parameters for the p-amino-benzoic acid anion adsorbed on sodium acetate-sodium chloride mixtures. Table 7.4 shows that at room temperature the ϕ_t values increase with sodium acetate content, whereas at -180°C the ϕ_t values are essentially constant. All the k_p values in Table 7.4 are about the same for the different sodium acetate mixtures whether at room temperature or at -180°C. The somewhat small values for k_m for the 50% and 100% sodium acetate samples and the approximately zero values for k_m for the 1.4% and 5.0% mixtures at low temperature show that nonradiative transitions from the triplet state are not important at low temperature. At room temperature, the k_m values for the four mixtures are considerably higher with the 1.4% mixture giving the highest k_m value. This shows that the nonradiative transition from the triplet state at room temperature for the mixtures makes an important contribution to the loss of phosphorescence emission.

The rate constants related to the singlet state were not obtained for the anion of p-aminobenzoic acid adsorbed on sodium acetate-sodium chloride mixtures. However, the percentages of absorbed photons were calculated that were involved in radiative and nonradiative transitions (8,13). By using ϕ_t, ϕ_f, and ϕ_p values, the diagrams in Figure 7.1 were constructed. It is shown in Figure 7.1 that the major loss of energy for the 1.4, 5.0, and 50% mixtures is from internal conversion from the singlet excited state to the singlet ground state. For the 100% sodium acetate sample, there was no loss of energy by internal conversion from the excited singlet state. In Figure 7.1, it is seen that the ϕ_f and ϕ_p values for 1.4% and 5.0% sodium acetate are relatively low, which suggests that the anion of p-aminobenzoic acid is not held rigidly. This indicated that the anion would be accessible to vibrational modes at a lower percentage of sodium acetate that would not be available at a higher percentage of sodium acetate. The previous statement is supported by the relatively constant ϕ_f and ϕ_p values at low temperature for the four sodium acetate mixtures (Table 7.3). At low temperature, the p-aminobenzoic acid anion would not be held rigidly, and vibrational modes that cause nonradiative transitions would not be permitted. Figure 7.1 shows the relative complexity of the luminescence processes and how the various radiative and nonradiative processes are interrelated. For example, Figure 7.1 shows that the 100% sodium acetate gave the highest RTP yield. However, it also gave the greatest percentage nonradiative loss of

Table 7.4. Luminescence Parameters for the PABA Anion Adsorbed on Sodium Acetate-Sodium Chloride Mixtures[a]

Sodium Acetate, %	ϕ_t^b	k_p^b, s^{-1}	k_m^b, s^{-1}	ϕ_t^c	k_p^c, s^{-1}	k_m^c, s^{-1}
1.4	0.17 ± 0.07	0.41 ± 0.24	0.98 ± 0.25	0.79 ± 0.03	0.40 ± 0.02	-0.01 ± 0.05
5.4	0.18 ± 0.04	0.43 ± 0.14	0.43 ± 0.15	0.80 ± 0.03	0.44 ± 0.03	-0.01 ± 0.06
50	0.30 ± 0.05	0.39 ± 0.09	0.45 ± 0.10	0.83 ± 0.03	0.39 ± 0.02	0.03 ± 0.05
100	0.76 ± 0.03	0.35 ± 0.03	0.39 ± 0.05	0.78 ± 0.03	0.29 ± 0.02	0.09 ± 0.05

[a] Error based on pooled standard deviation and propagation of error equations.
[b] Room temperature.
[c] -180°C.

Reprinted with permission from S.M. Ramasamy and R.J. Hurtubise, *Anal. Chem.*, 59, 2144. Copyright 1987 American Chemical Society.

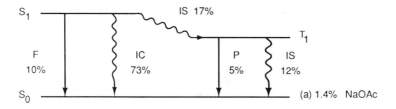

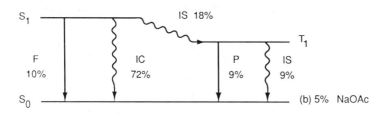

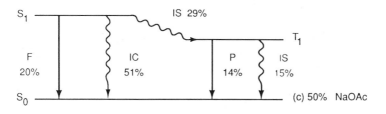

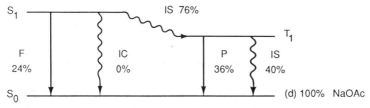

Figure 7.1. Energy diagrams for the anion of PABA adsorbed on sodium acetate and sodium acetate-sodium chloride mixtures: (a) 1.4% sodium acetate-sodium chloride; (b) 5.0% sodium acetate-sodium chloride; (c) 50% sodium acetate-sodium chloride; (d) sodium acetate. (S_0) ground state; (S_1) singlet state; (T_1) triplet state; (F) fluorescence; (IC) internal conversion; (IS) intersystem crossing; (P) phosphorescence. (Reprinted with permission from S. M. Ramasamy and R. J. Hurtubise, *Anal. Chem.*, <u>59</u>, 2144. Copyright 1987 American Chemical Society.)

energy from the excited triplet state. Therefore, if experimental conditions could be adjusted to minimize the nonradiative transition process, then the RTP signal could be enhanced by about a factor of two.

As discussed earlier in this section, the adsorption of the p-aminobenzoic acid anion involved the formation of the sodium salt of p-aminobenzoic acid on the sodium acetate surface as well as hydrogen bonding (1). Also, it was discussed that sodium acetate molecules that were undissolved in ethanol can pack the matrix efficiently in the dry state and either protect the p-aminobenzoic acid anion from collisions with oxygen molecules or hold the p-aminobenzoic acid anion in a rigid state (6). Also, temperature variation experiments supported a mechanism for RTP in which the p-aminobenzoic acid anion was held rigidly on sodium acetate (7). In earlier work, von Wandruszka and Hurtubise (2) showed that the p-aminobenzoic acid anion adsorbed from an ethanol solution onto a suspension of sodium acetate in the ethanol solvent. Thus, a portion of the p-aminobenzoic acid anions adsorb onto the surface of the undissolved sodium acetate prior to evaporation of the ethanol solvent. This indicated that as the ethanol evaporated, more and more p-aminobenzoic acid anions would be deposited on the surface of undissolved sodium acetate. This condition would favor a rather homogeneous distribution of p-aminobenzoic acid anions on the dry sodium acetate surface. Also, prior to a saturated solution of sodium acetate, the solid-surface RTP was less than the RTP of samples prepared with solutions that were saturated with sodium acetate. Table 7.3 shows the increase in both room-temperature fluorescence and RTP quantum yields of the p-aminobenzoic acid anion as the amount of sodium acetate increases. By plotting ϕ_f and ϕ_p versus the log of the ratio of millimoles of dissolved sodium acetate to the millimoles of dissolved p-aminobenzoic acid anion, some interesting graphs are obtained in Figure 7.2. As indicated in this Figure, there are two major parts to both graphs, namely, the region before saturation of the ethanol with sodium acetate and the region beyond the saturated ethanol solution. On the abscissa in Figure 7.2, the ethanol solution is saturated with sodium acetate at a log value of 3.36. For the region prior to saturation of the solution used for sample preparation, there is an approximate linear increase in the ϕ_f and ϕ_p values as the log [mmol of dissolved sodium acetate/mmol of p-aminobenzoic acid anion] increases. Even though there is a gradual linear increase in ϕ_f and ϕ_p before saturation, a more dramatic increase in ϕ_f and ϕ_p occurs beyond a

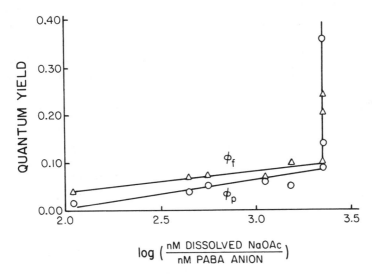

Figure 7.2. Graphs of fluorescence quantum yield (ϕ_f) and phosphorescence quantum yield (ϕ_p) vs. log of the ratio of millimoles of dissolved sodium acetate to millimoles of PABA anion. (Reprinted with permission from S.M. Ramasamy and R.J. Hurtubise, *Anal. Chem.*, 59, 2144, Copyright 1987 American Chemical Society.)

saturated solution of sodium acetate (Figure 7.2). As discussed earlier, the undissolved sodium acetate can pack the solid matrix more effectively than NaCl, and thus, RTP should increase for the p-aminobenzoic acid anion. Ramasamy and Hurtubise (13) made the conclusion that for the anion of p-aminobenzoic acid adsorbed on sodium acetate, ϕ_p is partially a function of the effective packing of the matrix with sodium acetate beyond the saturated solution of sodium acetate used in sample preparation. For example, Table 7.3 shows that ϕ_p increases 4-fold from 5.0% sodium acetate to pure sodium acetate. In addition, the matrix packing effect also contributes to the high room-temperature fluorescence quantum yield as indicated in Figure 7.2.

A summary of the conditions and interactions for the RTP of the anion of p-aminobenzoic acid is as follows (1,2,6,7,13).

1. In the sample preparation step, the p-aminobenzoic acid anion is formed by the reaction of p-aminobenzoic acid with sodium acetate.

2. If the solution is saturated with sodium acetate, the p-aminobenzoic acid anion can adsorb directly on the surface of the undissolved sodium acetate. This allows a rather homogeneous distribution of the p-aminobenzoic acid anion on the sodium acetate.

3. In the dried matrix, other interactions occur such as hydrogen bonding of the amino functionality with sodium acetate; however, how effectively the matrix is packed is a major factor in producing a high RTP quantum yield.

All of the aspects considered in items 1, 2, and 3 above contribute to or provide the appropriate conditions so the phosphor can be held rigidly in the solid matrix. In addition, the matrix would protect the phosphor from interacting with atmospheric constituents. However, for the experiments discussed in this section, oxygen and moisture were not a problem because the experiments were performed under dry nitrogen conditions.

7.2. Silica Gel

Silica gel has been shown to be useful for obtaining RTP from organic compounds under certain conditions (14-20). Ford and Hurtubise (14) tested several brands of silica gel for their ability to yield RTP from benzo(f)quinoline. Table 7.5 lists the brands of silica gel investigated. As shown in Table 7.5, acidic ethanol solutions of benzo(f)quinoline adsorbed on some brands of silica gel gave strong RTP signals from benzo(f)quinoline. However, benzo-(f)quinoline exhibited moderate RTP to no RTP on other brands and types of silica gel when ethanol or acidic ethanol solutions were used. Various experiments were carried out to determine to what extent the RTP enhancement could be attributed to adsorbing benzo(f)quinoline on EM silica gel chromatoplates in the protonated form (14). For example, the hydrochloride of the compound was prepared by passing HCl gas through an ether solution of benzo(f)-quinoline. The product of the reaction was shown to be the hydrochloride by solution fluorescence spectroscopy. A comparison was made of the RTP relative intensity values for equimolar amounts of benzo(f)quinoline and the hydrochloride spotted from ethanol, and of benzo(f)quinoline spotted from 0.1 M HCl in ethanol onto an EM chromatoplate. The relative RTP signals were in the order 1.0, 1.03, and 11.6, respectively. These data showed

Table 7.5. Silica Gel Brands Tested as RTP Supports for B(f)Q

Brand	Description	RTP (neutral)[b]	RTP (acid)[c]
EM[a]	Al backed TLC chromatoplate	moderate	strong
EM	Glass backed	moderate	strong
EM	Plastic backed TLC chromatoplate	moderate	strong
EM	Glass backed (HPTLC) chromatoplate	moderate	strong
Brinkmann	Plastic backed N-HR (TLC) chromatoplate	moderate	strong
Brinkmann	Plastic backed Sil-G (TLC) chromatoplate	none	none
S & S[d]	Glass backed (TLC) chromatoplate	none	moderate
Applied Science Labs	Glass backed Permakotes I (TLC) chromatoplate	none	weak
EM	Silica Gel 40, cloumn chromatography	none	none
EM	Silica Gel 60, column chromatography	none	none
EM	Silica Gel 100, column chromatography	none	none
MN[e]	Silica Gel 60, column chromatography	none	weak
MN	Kieselgel 60 for TLC chromatography	none	weak
Mallinckrodt	SilicAR TLC-7G chromatography	none	none

[a]EM Laboratories.
[b]B(f)Q spotted from ethanol.
[c]B(f)Q spotted from 0.1 M HCl ethanol.
[d]Schleicher & Schuell.
[e]Macherey, Nagel & Co.
Reprinted with permission from C.D. Ford and R. J. Hurtubise, *Anal. Chem.*, 52, 656. Copyright 1980 American Chemical Society.

that the RTP enhancement was more than the result of adsorbing the protonated benzo(f)quinoline on the chromatoplate. They assumed that the HCl interacted with the chromatoplate in some manner which allowed strong adsorbate- solid surface interactions. Diffuse reflectance spectra were obtained of benzo(f)quinoline spotted from ethanol, of the hydrochloride of benzo(f)quinoline spotted from ethanol, and of benzo(f)quinoline spotted from 0.1 M HCl ethanol solution on MN silica gel for column chromatography, on an EM silica gel chromatoplate, and on a Brinkmann N-HR silica gel chromatoplate. The reflectance spectra showed that with the neutral benzo(f)quinoline sample, the neutral form was on the surface of the silica gel samples. For the hydrochloride sample, the spectra also indicated that the neutral form was present. Examination of the ultraviolet absorption spectra of the neutral compound and hydrochloride in ethanol showed the hydrochloride spectrum to be very similar to the spectrum for the neutral compound. This indicated that the hydrochloride was in equilibrium with a relatively large fraction of the neutral counterpart in ethanol solution. All the reflectance spectra for the benzo(f)quinoline samples spotted from 0.1 M HCl solution showed that the protonated form of the compound was adsorbed on silica gel.

In other experiments, EM silica gel chromatoplates were pretreated with hydrochloride acid by soaking the chromatoplates for 10 sec in an acidic water solution. The acid treated chromatoplates were dried 0.5 h at 110°C prior to use as RTP supports. Enhanced RTP was observed from neutral benzo(f)quinoline spotted from an ethanol solution onto the acid-pretreated chromatoplates. Because the acid studies indicated a change in the silica gel chromatoplates with acid treatment, infrared spectroscopy was used to examine silica gel samples after ethanol or acid treatment. Figure 7.3A shows the infrared spectra of column chromatography MN silica gel. The band at about 1870 cm^{-1} is an overtone band of silica gel, and the band at 1630 cm^{-1} has been assigned to a water deformation band (21). As shown in Figure 7.3A, there is essentially no difference in the infrared spectra of the acid-treated silica sample compared to the ethanol-treated silica sample. Figure 7.3B gives the infrared spectra of EM silica gel samples scraped from an aluminum backed chromatoplate. The ethanol-treated silica gel sample gave broad bands at 1560 cm^{-1} and 1720 cm^{-1} in addition to the 1870 cm^{-1} band and an ill-defined band at 1630 cm^{-1}. With acid treatment, the 1560 cm^{-1} band disappeared and the 1720 cm^{-1} band became more prominent. The changes in the infrared

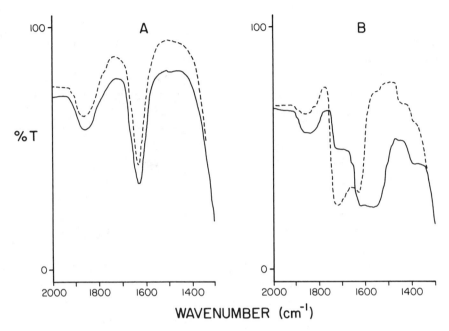

Figure 7.3. Infrared spectra of ethanol treated (——), and 0.1 M HCl ethanol treated (- - -) silica gel sample. (A) MN silica gel for column chromatography, (B) EM silica gel from a chromatoplate. (Reprinted with permission from C.D. Ford and R.J. Hurtubise *Anal. Chem.*, 59, 2144. Copyright 1987 American Chemical Society.)

spectrum with acid treatment were consistent with carboxylate anions being converted to carboxylic acid groups (22). After consulting an EM patent, the patent indicated that the sodium salt of polyacrylic acid was used as a binder in the manufacture of silica gel chromatoplates (23).

The infrared spectra shown in Figure 7.3A indicated that the form of the binder EM used changed upon acid treatment. Also, strong RTP was observed only when the chromatoplate was treated with acid. For these reasons and the fact that the sodium salt of polyacrylic acid was used as a binder, polyacrylic acid was investigated as a possible material for inducing RTP. Polyacrylic acid was mixed with column chromatography MN silica gel 60 to give mixtures containing varying percentages of polyacrylic acid. Benzo(f)quinoline showed weak RTP when adsorbed on only MN silica gel (Table 7.5). A fixed amount of benzo(f)quinoline from a

0.1 M HCl ethanol solution was adsorbed onto several mixtures of MN silica gel-polyacrylic acid containing increasing amounts of polyacrylic acid. The RTP signals increased almost linearly to 10% polyacrylic acid and then decreased above 20% polyacrylic acid. Also, 100% polyacrylic acid gave only weak RTP from benzo(f)-quinoline. Polyacrylic acid-sodium chloride mixtures were also explored with benzo(f)quinoline, and similar RTP results were obtained, except RTP signals were somewhat greater. It is known that the modulus of polymers increases when the polymer is mixed with an inorganic powder of higher modulus (24). Generally, modulus measures the resistance to deformation of materials to external forces (25). The polymer is restricted in its ability to rotate and migrate by the powder in the mixture. With polyacrylic acid-silica gel mixtures and polyacrylic acid-NaCl mixtures, it was postulated that the silica gel and NaCl serve to restrict the general movement of the polymer matrix, thereby giving the matrix greater rigidity and permitting greater RTP signal intensities from adsorbed compounds.

Ford and Hurtubise (14) proposed that the protonated form of benzo(f)quinoline was adsorbed flatly on pure silica gel without binder, and that hydrogen bonding between the protonated form and silanol groups was sufficient to hold the molecule rigidly enough so a weak RTP signal could be observed. When acidic polymers or their salts, such as polyacrylic acid, are used as binders for commercial silica gel chromatoplates, other adsorbate-solid-surface interactions are implicated. The results from acid studies, luminescence, reflectance, infrared, and binder studies showed that enhanced RTP signals were obtained only when the binder was in the acidic form (14). The presence of carboxyl groups dispersed throughout the silica gel provided sites for strong hydrogen bonding interaction with the phosphor. Infrared results for the EM chromatoplates showed that the amount of binder was about 5% by weight (14).

Any mechanism of interaction between protonated benzo(f)-quinoline and EM silica chromatoplates, which has a carboxyl-containing polymer as a binder, must involve interactions between the adsorbed compound and the binder. Most likely, the adsorbate-solid-surface interactions would involve hydrogen bonding between the π-electron system of the adsorbate and the carboxyl groups of the binder. Because the carboxyl groups in the polymer would be more strongly acidic than the silanol groups, the carboxyl group would be expected to form a stronger hydrogen bond with the π-ring system of protonated benzo(f)quinoline. This would favor holding the phosphor more rigidly to the surface. In addition, the NH$^+$ moiety in benzo(f)quinoline could form a hydrogen bond with the carbonyl oxygen of the carboxyl group in the polymer. In other experiments, phenanthrene showed a moderate RTP on EM silica gel chromatoplates when spotted from 0.1 M HCl ethanol solution. The results indicated that the excess acid converted carboxylate groups of the polymer binder to carboxyl groups, which then interacted with the π-electrons of phenanthrene to hold the compound rigidly enough to obtain RTP (14). Other interactions are most likely responsible for RTP from benzo(f)quinoline, and additional research is needed to clarify these aspects.

Ford and Hurtubise (16) proposed that the main interaction causing RTP from terephthalic acid adsorbed on EM silica gel chromatoplates was hydrogen bonding between the surface silanol hydroxyl groups and the hydroxyl and carbonyl groups of terephthalic acid. However, because it was found later that a salt of polyacrylic acid was present in EM chromatoplates, and polyacrylic acid was largely responsible for inducing RTP from benzo(f)quinoline, it was important to reexamine the previously reported mechanism for terephthalic acid. Hurtubise and Smith (19) adsorbed neutral, basic (0.1 M), and acidic (0.1 M) terephthalic acid and coumarin-3-carboxylic acid solutions on Brinkmann Sil-G chromatoplates and Grace column-chromatography silica gel. The Brinkmann Sil-G chromatoplates were known to contain a binder other than polyacrylate (14), and the Grace silica gel contained no binder. An RTP signal would be obtained from phosphors adsorbed on Grace silica gel if the silanol groups were responsible for inducing RTP. Experiments showed that little or no RTP was obtained from terephthalic acid or coumarin-3-carboxylic acid adsorbed on either support under all conditions investigated. The lack of RTP from the two carboxylic acids on the Grace silica gel showed that hydrogen bonding of the carboxyl groups with silanol hydroxyl groups was not the mode of interaction that induced RTP. The results with the Brinkmann Sil-G chromatoplates showed that the combination of binder in the chromatoplates and the silica gel gave little RTP from adsorbed phosphors. However, additional experiments with polyacrylic acid-sodium chloride mixtures yielded relatively strong RTP from terephthalic acid. These experiments showed that the salt of polyacrylic acid in the EM chromatoplates was responsible for inducing RTP from adsorbed terephthalic acid (19). In general, it has been found that silica gel does not give strong RTP from the adsorbed phosphors that have been investigated.

Burrell and Hurtubise (26) investigated extended luminescence calibration curves to study surface interactions in solid-surface luminescence. Calibration curves extended well beyond the normal linear range for solid-surface fluorescence, and solid-surface phosphorescence gave some unique characteristics for benzo(f)quinoline adsorbed on an EM silica chromatoplate under neutral and acidic conditions. The EM silica chromatoplates contained a polyacrylate binder. The solid-surface fluorescence curves leveled off, whereas the solid-surface phosphorescence curves passed through a maximum and then decreased. Figure 7.4 shows the room-temperature fluorescence and RTP extended luminescence calibration curves for the protonated form of benzo(f)quinoline. Both the room-temperature fluorescence and RTP signals were relatively strong and both signals were useful analytically. The room-temperature fluorescence and RTP curves became nonlinear at approximately the same point. However, the room-temperature fluorescence curve leveled off and was parallel to the nanogram axis as indicated in Figure 7.4A. The RTP curve showed a decrease in RTP intensity with increasing amounts of $B(f)QH^+$. As discussed earlier in this section, the RTP of $B(f)QH^+$ results from the protonation of the carboxylate groups of the salt of polyacrylic acid binder in the EM chromatoplates. The carboxyl groups formed can interact strongly with $B(f)QH^+$ and permit RTP to be observed. The RTP curve in Figure 7.4B is important in analytical work because it shows that the overall shape of the RTP curve is very different from the RTF curve in Figure 7.4A. The flat portion of the room-temperature fluorescence curve shows that the room-temperature fluorescence signals in this region are independent of the amount of adsorbate. The room-temperature fluorescence intensity of $B(f)QH^+$ becomes independent of the amount of $B(f)QH^+$ near 1 μg. Data for 100 ng and 1 μg of $B(f)QH^+$ showed essentially the same fluorescence excitation and emission wavelengths. However, the 20-μg sample showed new bands in the excitation spectrum, and the emission spectrum was shifted 16 nm compared to the 100-ng sample. The wavelength results for the $B(f)QH^+$ samples showed that multilayer formation of $B(f)QH^+$ molecules was occurring in the plateau region of the room-temperature fluorescence plot.

As shown in Figure 7.4B, the RTP intensity for $B(f)QH^+$ begins to decrease above 1000 ng. This is essentially the same

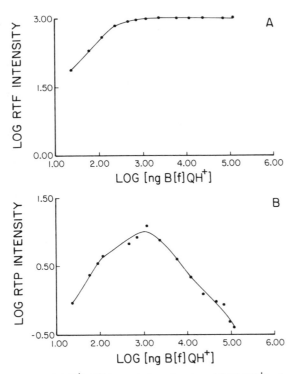

Figure 7.4. (A) log (B(f)QH$^+$ RTF intensity) vs. log (ng of B(f)QH$^+$) adsorbed on an EM chromatoplate. Each data point represents the average of three to five runs. (B) log (B(f)QH$^+$ RTP intensity) vs. log (ng of B(f)QH$^+$) adsorbed on an EM chromatoplate. Each data point represents the average of three to five runs. (Reprinted with permission from G.J. Burrell and R.J. Hurtubise, *Anal. Chem.*, <u>59</u>, 965. Copyright 1987 American Chemical Society.)

point at which the room-temperature fluorescence curve for B(f)QH$^+$ levels off. The decrease in the RTP curve beyond 1000 ng was most likely due to two major effects, namely, inner-filter effect and collisional deactivation. For example, it was discussed earlier in this section that for strong RTP to be observed from the chromatoplate, B(f)QH$^+$ has to interact directly with the carboxyl groups of the binder in the chromatoplate (14). B(f)QH$^+$ molecules in a multilayer would not give RTP because the interactions between molecules would not be strong enough, and the B(f)QH$^+$ molecules would not interact directly with carboxyl groups. The multilayers of phosphor molecules would act as an inner filter and prevent maximum excitation of those $B(f)QH^+$ molecules interacting directly with the carboxyl groups in the chromatoplate. Also, collisional deactivation would most likely occur because several $B(f)QH^+$ molecules can compete for an adsorption site at the higher amounts of $B(f)QH^+$. The collisional deactivation would be favored because the experiments were carried out at room temperature.

To gain additional insights into the interactions responsible for the RTP of $B(f)QH^+$ in Figure 7.4, phosphorescence lifetime values (τ_p) were obtained at several different amounts of $B(f)QH^+$. Samples of 0.1, 1.0, and 5.0 μg of $B(f)QH^+$ gave good linearity for the ln(intensity) vs. time plots. However, 20-μg and 100-μg samples showed curvature for the ln(intensity) vs. time plots, with the 100-μg sample showing the greatest curvature. The curvature of the graphs indicated nonexponential decay of the RTP. The RTP lifetimes were approximately constant from 0.10 to 5.0 μg. At 20 μg and 100 μg, the τ_p values decreased. Also, the RTP intensity was decreasing as a function of the amount of $B(f)QH^+$ near 5 μg (Figure 7.4). It was concluded that multilayers were formed and the inner-filter effect was most likely operative (26). This was explained by using Equation (7.7)

$$\tau_p = \frac{1}{k_p + k_m + k_q[q]} \qquad (7.7)$$

where k_p is the rate constant for phosphorescence, k_m is the rate constant for radiationless transition from the triplet state, k_q is the rate constant for bimolecular quenching such as with oxygen, and $[q]$ is the concentration of the quencher. For τ_p to remain constant, most likely k_p, k_m, and $k_q[q]$ are constant at 0.10, 1.0, and 5 μg. Niday and Seybold (27) proposed that k_m was a measure of the rigidly held mechanism for RTP, and k_q was a measure of the effect of oxygen on RTP or how efficiently the matrix protects the phosphor from oxygen. Assuming the rate constants in Equation (7.7) did not vary at 0.10, 1.0, and 5.0 μg, the decrease in RTP intensity must be due to an inner-filter effect, whereby the layer or layers of molecules above the molecules interacting directly with the carboxyl groups were prevented from being excited as effectively compared to lower amounts of $B(f)QH^+$. The 20-μg and 100-μg samples showed a decrease in lifetime, and Figure 7.4

shows that at these amounts the RTP intensity of B(f)QH$^+$ is very low. The decrease in RTP lifetime for the 20-μg and 100-μg samples is most likely a combination of the inner-filter effect and collisional deactivation, which could occur because of the relatively large amount of adsorbed molecules.

The room-temperature fluorescence and RTP excitation and emission spectra for B(f)QH$^+$ adsorbed from acidic solutions onto EM chromatoplates were also investigated (26). At 1.0 and 20 μg, the excitation spectra had practically the same excitation wavelengths (Figure 7.5). However, the 100-ng sample did not have an excitation band near 376 nm as did the 1.0-μg sample. The room-temperature fluorescence and RTP excitation wavelengths for the

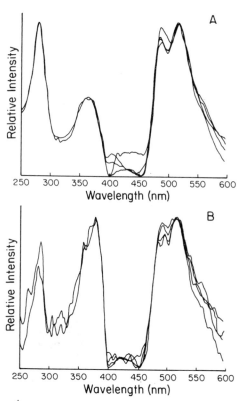

Figure 7.5. (A) B(f)QH$^+$ RTP spectra for 100 ng adsorbed on the EM chromatoplate: λ_{ex}=280, 320, 370, and 390 nm; λ_{em}=516 and 550 nm. Spectra normalized to the strongest emitting species. (B) B(f)QH$^+$ RTP spectra for 20 μg adsorbed on the EM chromatoplate: λ_{ex}=280, 320, 370, and 390 nm; λ_{em}=516 and 550 nm. Spectra normalized to the strongest emitting species. (Reprinted with permission from G.J. Burrell and R.J. Hurtubise, *Anal. Chem.*, 59, 965. Copyright 1987 American Chemical Society.)

100-ng B(f)QH$^+$ samples were the same within experimental error. However, a comparison of the room-temperature fluorescence and RTP excitation wavelengths for the 1.0-μg samples showed that the wavelengths were not the same. This was seen more dramatically for the 20-μg B(f)QH$^+$ room-temperature fluorescence and RTP samples (Figure 7.5). It was concluded that the difference in the excitation wavelengths for these samples was most likely a result of the singlet and triplet emitting molecules being in different environments. The singlet emitting molecules would fluoresce from the silica gel matrix and in the multilayers, whereas the triplet emitting molecules would phosphoresce only from the silica matrix because direct interaction with carboxyl groups was needed for RTP. Interestingly, the RTP emission wavelengths were the same for the 100-ng, 1.0-μg, and 20-μg B(f)QH$^+$ samples, except the band at 485 nm was shifted to 498 nm for the 20-μg sample. These results showed that the B(f)QH$^+$ molecules in the triplet state were interacting with the carboxyl groups in the silica gel matrix because of the similarity of the RTP emission wavelengths. The shapes and wavelength maxima of B(f)QH$^+$ RTP emission spectra were found to be essentially independent of the wavelength of the exciting radiation at all amounts adsorbed on the chromatoplates. In summary, it was determined that fluorescence could occur from molecules adsorbed on the surface and in multilayers of molecules while phosphorescence only occurred from molecules adsorbed in the chromatoplate matrix and not in multilayers of molecules (26).

In other studies, the room-temperature fluorescence and RTP data were obtained for B(f)QH$^+$ and B(h)QH$^+$ from the samples adsorbed on EM silica gel chromatoplates submerged in chloroform/n-hexane solvents (28). These chromatoplates also contained a polyacrylate binder. The luminescence results revealed several of the interactions of the nitrogen heterocycles with the chromatoplates. RTP excitation and emission spectra, RTP intensities, and RTP lifetimes were obtained for 100-ng samples of B(f)QH$^+$(I) and B(h)QH$^+$(II) adsorbed on the chromatoplates under dry conditions and submerged in n-hexane, and in binary solvents of 1%, 5%, 30%, 50%, 75% CHCl$_3$, and pure CHCl$_3$. The phosphorescence lifetimes for B(f)QH$^+$ (1.1 ± 0.02 s) and B(h)QH$^+$ (0.73 ± 0.02 s) remained essentially constant on the dry chromatoplates and the chromatoplates submerged in the solvents. In addition, the corresponding RTP and excitation and emission wavelengths for the two compounds on the dry chromatoplate and submerged in all the solvents showed no change or very small changes. These results indicated

(I)   (II)

that the solvents were not significantly perturbing the interactions between the phosphors and the chromatoplate. However, the RTP intensities of both compounds decreased significantly when exposed to n-hexane, yet analytically useful signals were still observed. Collisional deactivation by adsorbed or solution-phase solvent molecules did not completely eliminate the RTP. With increasing volume percentages of chloroform, further decreases in RTP intensities of the adsorbates were observed. By measuring the solution fluorescence from the n-hexane to 50% $CHCl_3$ (v/v) solvents, after removal of the chromatoplate sections, it was determined that the RTP intensity decreases were not a result of desorption by the adsorbates from the chromatoplate because no fluorescence was detected in these solvents. The solutions which contained less than 50% $CHCl_3$ in n-hexane did not displace either $B(f)QH^+$ or $B(h)QH^+$ from the surface. Chloroform and 75% $CHCl_3$ were the two strongest solvents, and they displaced a few nanograms of the adsorbates within the time span of the experiment, which contributed somewhat to the RTP intensity changes observed. However, the relatively large loss of RTP intensity with n-hexane present was a result of the collisional deactivation of a fraction of the phosphor molecules that were readily exposed to the solvent molecules. The molecules were assumed to be randomly adsorbed on the surface of the silica, and within the pores at various depths. Therefore, their adsorption location determined their susceptibility to collisional deactivation by n-hexane. The RTP data also showed that there are at least two populations of the adsorbed phosphor molecules, namely, the molecules whose RTP was initially quenched by the solvent and those that remained phosphorescent with the other solvents present.

Because of the RTP data for the two nitrogen heterocycles and the general porous nature of the silica gel, two reasons were put forth by Burrell and Hurtubise (28) as to why RTP was observed in the presence of organic solvents. The phosphors were protected by the porous nature of the silica and the strong interaction of the phosphor molecules with the carboxyl groups from the binder in the silica matrix. As considered earlier in this section, the protonated form of B(f)Q interacts with the carboxylate groups of the polyacrylate binder and this permitted a very strong interaction between $B(f)QH^+$ and the surface. $B(h)QH^+$ would interact in a similar fashion. However, because the nitrogen atom is sterically hindered in $B(h)QH^+$, it may not interact as strongly with the matrix as $B(f)QH^+$.

The room-temperature fluorescence changes effected by n-hexane and n-hexane:chloroform solvents on $B(f)QH^+$ and $B(h)QH^+$ were different compared to the changes in RTP considered above. The room-temperature fluorescence from the two nitrogen heterocycles increased substantially when submerged in n-hexane. However, when the adsorbates were submerged in solvents containing chloroform, the room-temperature fluorescence intensity decreased relative to the samples submerged in n-hexane. The room-temperature fluorescence enhancement with n-hexane present indicated that the molecules in the singlet state were affected differently by environmental factors than were the molecules in the triplet state. For example, the molecules in the singlet are adsorbed on silanol groups and carboxyl groups. However, the molecules in the triplet state that are phosphorescent are adsorbed on carboxyl groups (14). Also, the lifetime of the singlet state is about 10^{-8} s. This would favor less collisional deactivation compared to molecules in the triplet state. In addition, the porous structure of the silica gel would prevent direct interaction of a portion of the fluorophors with solvent molecules. In other experiments, the excitation and room-temperature fluorescence emission wavelengths were unaffected by the presence of the solvents. This showed that the protonated forms of $B(f)QH^+$ and $B(h)QH^+$ were adsorbed strongly on the chromatoplates, and the adsorption interactions of the fluorophors were not affected by the solvents in a way that would result in spectral changes.

Another important aspect in the interpretation of the room-temperature fluorescence and RTP data from $B(f)QH^+$ and $B(h)QH^+$ was the differences in the excited singlet-state and excited triplet-state acidities of the adsorbates. The acidities of excited triplet-states of $B(f)QH^+$ and $B(h)QH^+$ (pK_a = 6.4 and 7.1, respectively) were reported to be much greater than the corresponding excited singlet-state acidities of these molecules (pK$_a$ = 10.3 and 9.8, respectively) (29,30). It was assumed that the corresponding pK$_a$'s on silica would be about the same. Because the protonated triplet-state molecules were much more acidic than the protonated singlet-state molecules, the triplet-state molecules would interact more strongly with the surface than the less acidic excited singlet-state molecules. As a result, the molecules emitting the phosphorescence are less affected by the relatively weak solvents than would be the less acidic, less strongly adsorbed fluorophors. By comparing the room-temperature fluorescence and RTP intensities of B(f)QH$^+$ and B(h)QH$^+$ from n-hexane to 50% CHCl$_3$, it was observed that there was a much greater relative decrease in the room-temperature fluorescence intensities compared to the corresponding RTP intensities. For the 75% CHCl$_3$ and the CHCl$_3$ solvents, some of the B(f)QH$^+$ and B(h)QH$^+$ was desorbed. Thus, the effects of these two solvents were more pronounced than those of the other solvents.

Because luminescence was observed from B(f)Q and B(h)Q adsorbed on chromatoplates with various solvents present, the adsorbates' room-temperature fluorescence and RTP intensities were considered as a function of the chromatographic solvent strength of each solvent. Solvent strength in liquid-solid chromatography is defined as the adsorption energy per unit area of the solvent (4). Thus, the solvent strength can be considered as a measure of how adsorbed solvent molecules would affect the luminescence of adsorbed B(f)Q and B(h)Q. The solvent strength is not linearly related to the volume percentage of chloroform in the bulk solvents because of the preferential adsorption of the stronger solvent component (4). The room-temperature fluorescence and RTP intensities of the submerged adsorbates were compared to the solvent strength of n-hexane and n-hexane:CHCl$_3$ binary solvents. The room-temperature fluorescence from neutral and protonated B(f)Q and B(h)Q decreased when adsorbed on chromatoplates that were submerged in binary solvents with increasing CHCl$_3$ content. The decrease in the room-temperature fluorescence from the submerged adsorbates was not attributed to solution effects alone, but partly to adsorbed CHCl$_3$ perturbing the adsorption interactions of the fluorophors with the adsorption sites. Scott and Kucera (31) obtained the isotherm for the adsorption of CHCl$_3$ from n-heptane onto silica gel. They showed that the data fit the Langmuir function. In the work by Burrell and Hurtubise (28), it was shown that the decreases in the room-temperature fluorescence were related to the solvent strengths for n-hexane through 50% CHCl$_3$.

The decrease in fluorescence intensity was approximately proportional to the amount of $CHCl_3$ adsorbed on the silica base on the results of Scott and Kucera (31) for adsorbed $CHCl_3$. The decrease in fluorescence intensity for 75% $CHCl_3$ and $CHCl_3$ was due in part to the small amount of solute being eluted from the surface.

For the case of neutral samples of B(f)Q and B(h)Q adsorbed on silica, it was assumed that the phosphor molecules were adsorbed on silanol groups in the silica. Both of these compounds gave weak RTP signals (28). The RTP intensities of $B(f)QH^+$ and $B(h)QH^+$ were compared to solvent strengths of the binary solvents in which they were submerged. Little RTP change was observed for the adsorbates submerged in the solvents with solvent strengths from 0 (n-hexane) to 0.20 (50% $CHCl_3$). This showed that the adsorbed solvent molecules were not affecting the RTP signals to any great extent. Several aspects could be involved in the small RTP changes observed. For example, as discussed earlier, the protection of the phosphors by the matrix from the solvents and the strong adsorption of the phosphors on the carboxyl groups of the adsorbent makes the RTP from the compounds rather insensitive to solvent competition effects. Also, as considered earlier, the acidities of $B(f)QH^+$ and $B(h)QH^+$ are relatively high in the triplet state. Therefore, the relative high acidities of the triplet states and the resulting interactions with the solid matrix are stronger than the relatively weak solvent interactions. The decrease in the RTP from $B(f)QH^+$ and $B(h)QH^+$ for the samples submerged in the strongest solvents, namely, 75% $CHCl_3$ and $CHCl_3$, showed that both solvent competition for carboxyl sites and the desorption of a small amount of the adsorbate from the surface were important.

In summary, it has been shown that for the phosphors investigated, silica is not a good matrix for inducing RTP. However, if silica gel contains a polyacrylate binder that has been converted to the acid form, then strong RTP can be obtained from adsorbed phosphors. RTP occurs from the phosphors that are adsorbed directly to the surface and not in multilayer of molecules in the silica-polymer matrix. Also, the porous nature of the silica protects the phosphor from quenching interactions as was indicated by the observation of RTP of phosphors adsorbed on silica gel chromatoplates submerged in n-hexane:chloroform solvents. A rather detailed discussion of several phosphor interactions with polyacrylic acid-salt mixtures is given in the next section.

7.3.   Polyacrylic Acid

As considered in the previous section, polyacrylic acid mixed with silica gel induced RTP from a variety of adsorbed organic compounds. Because of this discovery, several aspects of the conditions and the interactions needed for RTP from compounds adsorbed on polyacrylic acid-salt mixtures have been studied (19,20,32-34). Also, Dalterio and Hurtubise (32) investigated several polymers containing polar functional groups as surfaces for obtaining RTP from hydroxyl aromatics. In no case did a polymer alone yield strong RTP from adsorbed compounds. The polymers examined had to be mixed with an inorganic salt for them to be a useful material for obtaining RTP. It was found that polyacrylic acid (secondary standard, mol wt 2,000,000)-salt mixtures induced the strongest RTP from the model compounds investigated. RTP analytical data were reported for nine compounds using 1% polyacrylic acid-sodium bromide mixtures (32). The main repeating group in polyacrylic acid is illustrated below.

$$-CH_2-\underset{\underset{\displaystyle |}{|}}{\overset{\overset{\displaystyle H}{|}}{C}}-COOH$$

Hurtubise and Smith (19) reported the results from two polyacrylic acid-NaCl mixtures for their potential for inducing RTP from terephthalic acid and the dianion of terephthalic acid. One mixture contained unneutralized polyacrylic acid and the other mixture contained neutralized polyacrylic acid. The data indicated several modes of interaction of terephthalic acid and its dianion with polyacrylic acid, and with the sodium salt of polyacrylic acid. Terephthalic acid most likely formed hydrogen bonds with poly-acrylic acid both in the initial "wet" state and final dry state. The dianion of terephthalic acid reacted to some extent with polyacrylic acid in the "wet" state to form terephthalic acid. It should be mentioned that polyacrylic acid itself behaves like an aliphatic carboxylic acid and can be titrated with a solution of sodium hydroxide. It is possible, depending on the extent of the reaction of the dianion with polyacrylic acid, that a mixture of the monoanion, dianion, and the terephthalic acid would remain on the dried surface with various combinations of hydrogen bonds being formed. Terephthalic acid added to the sodium salt of polyacrylic acid could react to form the monoanion and/or dianion in the "wet" state. In the dry state, there could be a mixture of the monoanion, dianion, and terephthalic acid with various combinations of hydrogen bonds. For the case in which the dianion was added to the sodium salt of polyacrylic acid, a relatively high RTP signal was obtained. This was surprising because the dianion would not undergo an acid-base reaction with the sodium salt of polyacrylic acid in the "wet" state and no hydrogen bonds could be formed in the dry solid matrix. The results obtained did not permit a simple interaction mechanism to be presented for all the conditions studied. While hydrogen bonding appeared to be the important interaction holding the compound rigid for neutral terephthalic acid adsorbed on polyacrylic acid-NaCl, it could not explain the RTP results for the dianion adsorbed on the sodium salt of polyacrylic acid-NaCl mixture. Niday and Seybold (27) postulated that various sugars or salts packed into filter paper could inhibit internal molecular movements of the phosphorescent compound and thus enhance RTP. This may be one factor to consider with the terephthalate dianion adsorbed on the sodium salt of the polyacrylic acid-NaCl mixture. However, in other experiments with terephthalic acid, no RTP was observed for the compound adsorbed on sodium acetate. Sodium acetate has been used to induce RTP from certain compounds (Section 7.1). If a simple "matrix packing" mechanism were occurring with sodium acetate, then RTP should have been observed. More work would be needed to elucidate the interactions and conditions for RTP from terephthalic acid.

Ramasamy and Hurtubise (35) investigated the RTP properties of benzo(f)quinoline and other nitrogen heterocycles with polyacrylic acid-salt mixtures under a variety of conditions. Because the polar functional group in polyacrylic acid was the carboxyl functionality, various aliphatic carboxylic acid-sodium chloride mixtures were investigated for their potential to obtain RTP from adsorbed benzo(f)quinoline. Nine 0.5% aliphatic carboxylic acid-sodium chloride mixtures were investigated, but none yielded as strong an RTP signal as benzo(f)quinoline adsorbed on 0.5% polyacrylic acid-sodium chloride mixture from 0.1 M HBr ethanol solution. The 0.5% polyacrylic acid-sodium chloride mixture gave an RTP signal of about four times the signals obtained for the aliphatic carboxylic acid-sodium chloride mixtures. It was postulated that benzo(f)quinoline was entangled in the polymer-salt matrix, and thus, the molecules were held very rigidly, which allowed strong RTP to be observed from the adsorbed compounds. It was also possible that benzo(f)quinoline was buried in the matrix, and collisions with oxygen were minimized. Sodium chloride or some other salt had to be mixed with polyacrylic acid for strong RTP to be observed from benzo(f)quinoline. In addition, an acidic solution of benzo(f)quinoline yielded stronger RTP signals than a neutral solution of the compound because the protonated form of the molecule can interact more strongly with the polymer matrix. Very weak or no RTP was observed for phosphors adsorbed individually on either polyacrylic acid or sodium chloride. A number of other inorganic salts (LiCl, KCl, K_2SO_4, and NaBr) were mixed separately with polyacrylic acid to investigate the effects of the salts on the RTP of the polyacrylic acid-salt mixtures. However, it was found that the polyacrylic acid-sodium chloride mixture induced the strongest RTP signal.

Deanin (24) discussed the effects of inorganic fillers on the thermal and mechanical properties of polymers. In polymers without a filler, the polymer molecule has various degrees of freedom to migrate and rotate. In a polymer mixed with an inorganic salt, some polymer molecules are adjacent to inorganic particles which have practically no mobility. A polymer molecule lying near such a rigid species is restricted in its ability to rotate and migrate. Deanin (24) has emphasized that the most important effect of fillers on thermal properties of polymers is to reduce the coefficient of thermal expansion of the polymer. This means that the mobility and motion of the polymer will be less. With the added rigidity and lower coefficient of thermal expansion for the polymer-salt matrix, the nitrogen heterocycle is an environment with less relative motion. These conditions then favor enhanced RTP.

Samples of polyacrylic acid were reacted with different amounts of NaOH to give 25%, 50%, 75%, and 100% neutralized samples of polyacrylic acid. The RTP intensity of benzo(f)quino-line adsorbed on the 0.5% polyacrylic acid-sodium chloride mixtures that contained partially neutralized polyacrylic acid showed a decrease in RTP intensity with percent neutralization. The RTP signal of adsorbed benzo(f)quinoline dropped by a factor of greater than 10 at 100% neutralization for a 0.5% polyacrylic acid-sodium chloride mixture compared to a similar unneutralized polyacrylic acid-sodium chloride mixture. Because the RTP intensity decreased with the percent neutralization, this showed that some carboxyl groups participated in hydrogen bonding with benzo(f)quinoline to anchor the molecules so RTP could be detected. The results also indicated that specific geometric requirements in the polymer-salt matrix were needed to achieve the optimal environment for strong RTP. Ramasamy and Hurtubise (35) also showed that the solvent used to adsorb the phosphor onto 0.5% polyacrylic acid-sodium surface was important. Eleven solvents were investigated and of the eleven solvents, methanol gave the best results. This solvent yielded an RTP signal from benzo(f)quinoline that was 19.6 times greater than the poorest solvent, chloroform. For methanol-water solvents, a 70% methanol water (v/v) solvent resulted in the largest RTP signal with benzo(f)quinoline. The RTP signal was 3.1 times greater than a sample adsorbed from methanol.

The amount of polyacrylic acid was important for inducing RTP from nitrogen heterocycles (35). When the polyacrylic acid content in polyacrylic acid-NaCl mixtures was varied over a wide range, it was found that relatively large signals were obtained for the nitrogen heterocycles between 0.5 and 1.0% polyacrylic acid. Ethanol was used as a solvent. Beyond 1% polyacrylic acid, the RTP of benzo(f)quinoline decreased nonlinearly, and at 90% polyacrylic acid-NaCl, practically no RTP was observed. Figure 7.6 gives the RTP intensity as a function of percent polyacrylic acid from 0 to 1% polyacrylic acid in polyacrylic acid-NaCl mixtures for five nitrogen heterocycles. As shown in Figure 7.6, for all the compounds, the RTP signals increased with polyacrylic acid content and then the RTP intensity reached a maximum signal and stayed approximately constant over a range of percent polyacrylic acid values. A certain optimal percent polyacrylic acid was needed to obtain maximum RTP signals. The ratios of the number of repeating groups in polyacrylic acid to one phosphor molecule for the nitrogen heterocycles to achieve the maximum RTP signals were calculated. For benzo(f)quinoline, a value of 1.54×10^3 was obtained. From this ratio, it was clear that many of the carboxyl groups did not interact with the phosphor molecules. The large ratio obtained for benzo(f)quinoline indicated that a given molecule would be far from its nearest neighbor. This condition would minimize the interaction of nitrogen heterocycle molecules with each other and permit the polyacrylic acid-NaCl matrix to interact effectively with nitrogen heterocycle molecules. The ratio of the number of repeating groups in polyacrylic acid at the optimal polyacrylic acid concentration to one molecule of dissolved NaCl

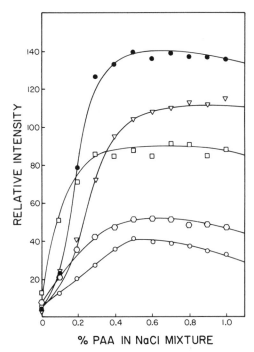

Figure 7.6. Graphs of RTP for nitrogen heterocycle vs. percent PAA in NaCl mixtures. One hundred nanograms of each phosphor was adsorbed from 0.1 M HBr ethanol solutions: (•) benzo[f]quinoline; (∇) 4-azafluorene; (□) phenanthridine;(◯) 13H-dibenzo-(a,i]carbozole; (◦) isoquinoline. (Reprinted with permission from S.M. Ramasamy and R.J. Hurtubise, *Anal. Chem.* 54, 2477. Copyright 1982 American Chemical Society.)

was also calculated. For benzo(f)quinoline, a ratio of 3.8 was obtained. From this information, it was concluded that there are more repeating groups than NaCl molecules at the optimal concentration of polyacrylic acid needed for RTP. The role of NaCl in the RTP of the nitrogen heterocycles is important because strong RTP is not obtained without NaCl. It was postulated that dissolved NaCl initially on the wet surface breaks some of the intra- and/or intermolecular hydrogen bonds of the polyacrylic acid dissolved in ethanol, and this permitted nitrogen heterocycles to form hydrogen bonds with the carboxyl groups. In addition, the NaCl increases the modulus of the polymer-salt matrix. This condition would favor enhancement of RTP.

In other experiments, the RTP of benzo(f)quinoline was obtained as a function of percent polyacrylic acid with methanol as a solvent (35). The solubility of NaCl in methanol is substantially greater than in ethanol. The ratio of repeating groups in polyacrylic acid to one NaCl molecule at the optimal polyacrylic acid concentration for RTP was calculated as 0.28, which was smaller than the ratio obtained for ethanol, namely, 3.8. The ratio of 0.28 indicated that there was an excess of sodium and chloride ions in solution at the optimal polyacrylic acid content. This condition would favor breaking of intermolecular hydrogen bonds of the carboxyl groups in the initial wet state of the solid surface. As previously considered, benzo(f)quinoline gave a greater RTP signal when adsorbed onto 0.5% polyacrylic acid-NaCl with methanol compared to ethanol. With methanol as a solvent and the relatively large amount of NaCl dissolved in methanol, more intra- and/or intermolecular carboxyl hydrogen bonds were broken in polyacrylic acid, and benzo(f)quinoline could interact with a larger number of carboxyl groups, which resulted in a greater RTP intensity with methanol. In the work by Allerhand and Schleyer (36), they reported a very large spectral shift to lower cm^{-1} values for the OH stretching frequency of methanol with halide ions in solution. This was attributed to the anion hydrogen bonding with the OH of the methanol. Their results give support to the concept that Cl^- can interact in solution with the carboxyl groups of polyacrylic acid in addition to OH groups of methanol.

Ramasamy and Hurtubise (20) used reflectance and infrared spectroscopy to study the interactions of benzo(f)quinoline and phenanthrene on polyacrylic acid-salt mixtures. Based on the spectral results from neutral benzo(f)quinoline adsorbed on polyacrylic acid-salt mixtures (0.5% polyacrylic acid-NaCl, 1% polyacrylic acid-NaBr, and 1% polyacrylic acid-KBr), both neutral benzo(f)quinoline and protonated benzo(f)quinoline were adsorbed on the surface. Because the protonated form of benzo(f)quinoline is analytically more important, the protonated form was considered in more detail. It was found that the hydroxyl groups of polyacrylic acid interacted with the π-electrons of protonated benzo(f)quinoline. The NH^+ group formed bonds with the oxygen of either the carbonyl group or the hydroxyl group of polyacrylic acid. Because protonated benzo(f)quinoline formed more bonds and presumably stronger bonds than did neutral benzo(f)quinoline, it was held more rigidly by polyacrylic acid than neutral benzo(f)quinoline. In other experiments, hydrobromic acid solutions of benzo(f)quinoline were adsorbed onto the polyacrylic acid-salt mixtures, thus yielding the hydrobromide of benzo(f)quinoline adsorbed on the surface. An important consideration of the hydrobromide of benzo(f)quinoline was the interactions of the hydrobromide with the surface, particularly bromide ion. The infrared results did not provide direct evidence on bromide ion. However, it was postulated that the positive charge of the NH^+ group could be shared by bromide, carbonyl, and hydroxyl groups.

The hydrocarbon analog of benzo(f)quinoline, phenanthrene, which also yields RTP from polyacrylic acid-salt mixtures, was studied by infrared spectroscopy (20). With phenanthrene, the π-electrons of phenanthrene and the carboxyl groups of polyacrylic acid could interact. The infrared data supported this model, and thus, the carboxyl groups interacted with π-electrons of phenanthrene, which allowed phenanthrene to be anchored to the surface.

The interactions involved in the RTP of hydroxyl aromatics adsorbed on polyacrylic acid-salt mixtures were discussed by Dalterio and Hurtubise (33). It was pointed out earlier in this section that sodium chloride is important for inducing RTP from compounds adsorbed on polyacrylic acid. For 4-phenylphenol, a 0.5% polyacrylic acid-sodium chloride mixture gave an RTP signal 2.4 times greater than a similar sample on 0.5% polyacrylic acid-sodium bromide. The NaBr mixtures yielded the lowest RTP for the salts examined, which showed that the heavy-atom effect was unimportant with polyacrylic acid-sodium bromide mixtures. The polyacrylic acid in several polyacrylic acid-salt mixtures was converted partially or completely to sodium polyacrylate by reaction with NaOH solutions to investigate the change in RTP of three model compounds as a function of percent neutralization of polyacrylic acid. All the compounds yielded the largest RTP signals on unneutralized 0.5% polyacrylic acid-sodium chloride. Interestingly, relatively large RTP intensities were obtained with 4-phenylphenol and 4,4'-biphenol adsorbed on 75% neutralized polyacrylic acid. Also, 4,4'-biphenol gave a relatively strong RTP signal adsorbed on a mixture of 100% neutralized polyacrylic acid. No RTP was observed from 2-naphthol on any of the neutralized polyacrylic acid samples. It was concluded that the polyacrylic acid polymeric chains achieved certain conformations in the presence of NaCl that were favorable for obtaining RTP from 4-phenylphenol and 4,4'-biphenol when the $COO^-/COOH$ ration was 3/1. The results from the neutralization studies indicated that for the model compounds examined, the polymer chain conformation was more important for inducing RTP than the number of carboxyl groups present in the polymer chain.

In other experiments, it was shown that a relatively strong RTP signal could be obtained from 4-phenylphenol adsorbed on 0.8% polyacrylic acid-NaBr mixture that contained a considerable amount of moisture in the solid matrix (33). The RTP signal with moisture present was an unusual occurrence because all previously investigated adsorbents showed optimal RTP with little or no moisture present (37-39). Schulman and Parker (40) showed that with filter paper in the presence of moisture, increased quenching of RTP was favored by permitting the transport of O_2 into the sample matrix. For enhanced RTP to be observed with water present, the H_2O molecules incorporated in the polyacrylic acid-salt matrix must change the matrix structure in a fashion that diminishes oxygen quenching of RTP. Additional studies are needed to develop an understanding of this interesting effect of water on the RTP from 4-phenylphenol adsorbed on polyacrylic acid-salt mixtures.

Solution fluorescence polarization experiments showed that model hydroxyl aromatic compounds associated with polyacrylic acid in ethanol solutions (33). With the addition of either NaCl or NaBr to ethanol polyacrylic acid solution, the fluorescence polarization increased, with the NaBr solution giving the larger fluorescence polarization. The polarization data showed that halide ions in solution interacted with polyacrylic acid molecules by breaking intra- and intermolecular hydrogen bonds between carboxyl groups in the polyacrylic acid polymer. This allowed a number of carboxyl groups to interact with the fluorescent molecules and thus yielded larger fluorescence polarization values. However, there was an inverse relationship between the extent of fluorescence polarization with the salt solutions and the magnitude of the RTP signal. In other experiments, the measurement of phosphorescence polarization of 4-phenylphenol adsorbed on polyacrylic acid-salt mixtures was attempted. However, extensive depolarization of RTP occurred and no useful data were obtained (33). In general, the area of luminescence polarization as applied to solid-surface luminescence is unexplored and is in need of further investigation.

Diffuse reflectance, fluorescence, phosphorescence, and infrared spectrometry were utilized by Dalterio and Hurtubise (34) to study several of the interactions of hydroxyl aromatics and aromatic hydrocarbons in polyacrylic acid-salt mixtures. With ultraviolet diffuse reflectance spectrometry, the ground state of the adsorbed phosphors was investigated. The lowest excited singlet state of the adsorbed molecules was studied by fluorescence spectrometry, and the lowest excited triplet state was studied by phosphorescence spectrometry. These three spectral techniques were used mainly to study interactions of the phosphors with solid matrices. However, infrared spectrometry was used to investigate interactions of the solid supports with phosphors. With ultraviolet diffuse reflectance spectrometry, the model compounds were adsorbed onto pure NaBr, which served as a reference surface. No RTP was observed from the compounds adsorbed on NaBr. The longest wavelength diffuse reflectance bands and wavelength shifts relative to the compounds adsorbed on NaBr were compared for several hydroxyl aromatics and aromatic hydrocarbons adsorbed on 1% polyacrylic acid-NaBr. It has been found for phenol and some substituted phenols that the longest wavelength absorption band, due to a π-π* transition, would shift to the red when the phenolic hydroxyl group acted as a proton donor in hydrogen bond formation and to the blue when the phenolic hydroxyl group acted as proton acceptor (41). Red shifts in the diffuse reflectance spectra relative to the pure NaBr surface were acquired for 4-phenylphenol and 2-phenylphenol adsorbed on 1% polyacrylic acid-NaBr. The red shift indicated that the hydroxyl groups of the phenols were hydrogen bonded to the matrix by a predominately proton donating mechanism. Red shifts were also obtained for biphenyl and naphthalene adsorbed on polyacrylic acid-NaBr. Because these compounds contained no hydroxyl groups, the red shifts were most likely caused by intermolecular π-electron hydrogen bonds (OH-π) between the aromatic hydrocarbons and the carboxyl and hydroxyl containing surfaces. A small blue shift was obtained for 2-naphthol adsorbed on 1% polyacrylic acid-NaBr. The important feature of this particular spectral data was the lack of a relatively large spectral shift. From the blue spectral shift for 2-naphthol, it was concluded that it was behaving either as a proton acceptor or as both a proton donor and proton acceptor (42,43). The different hydrogen bonding interactions occurring with the phenylphenols and 2-naphthol were also partly ascribed to different steric and geometric factors of the phosphor fitting into the solid matrix.

Low-temperature and room-temperature fluorescence and phosphorescence spectra for a number of polycyclic aromatic hydrocarbons and hydroxyl aromatics adsorbed on 1% polyacrylic acid-NaBr were obtained to investigate the interactions of model compounds in excited singlet states and triplet states. The fluorescence λ_{max} values of the model compounds were obtained at low temperature in ethanol glass and at room temperature in ethanol. Generally, the room-temperature fluorescence λ_{max} values for each model compound in ethanol were red shifted with respect to the low-temperature λ_{max} values. In most cases, the room-temperature fluorescence for the model compounds absorbed on 1% polyacrylic acid-NaBr showed no spectral shift or small shifts compared to the low-temperature fluorescence data. The small spectral shifts indicated that the excited molecules did not reorient themselves extensively from the Franck-Condon excited singlet state.

Table 7.6 gives the phosphorescence λ_{max} values for several model compounds at liquid nitrogen temperature in ethanol glass and at room temperature adsorbed on 1% polyacrylic acid-NaBr. Table 7.6 also gives the LTP λ_{max} values of the anions of the hydroxyl aromatics. The LTP λ_{max} of the anions were red shifted by 11-16 nm compared to the LTP λ_{max} values of the neutral hydroxyl aromatics. The RTP λ_{max} value of 4-phenylphenol and 2-naphthol adsorbed on 1% polyacrylic acid-NaBr from neutral ethanol solutions were not red shifted as much as the LTP λ_{max} values of the respective anion (Table 7.6). This was evidence that the triplet emitting species for these compounds on the solid surfaces were the neutral molecules. Surprisingly, RTP λ_{max} of 2-phenylphenol on 1% polyacrylic acid-NaBr was red shifted a greater amount than the LTP λ_{max} of the 2-phenylphenol anion. Because of the experimental conditions, 2-phenylphenol would not ionize in the triplet state on the solid surface. The large shift was most likely related to the reorientation of 2-phenylphenol on the surface and the Franck-Condon states. Little solvent or phosphor reorientation would occur at low temperature. However, at room temperature, there would be reorientation to an equilibrium triplet state, and then phosphorescence would occur from the equilibrium triplet state to the Franck-Condon singlet ground state. With 2-phenylphenol, steric crowding of the phenyl ring would also be a factor, and this probably favored reorientation of the phenyl ring. The RTP λ_{max} shifts in Table 7.6 showed similar trends with respect to the RTF and diffuse reflectance λ_{max} for the model compounds. The RTP λ_{max} shifts for the phenylphenols, biphenyl, and naphthalene were red shifted with respect to the LTP λ_{max} values. The red shifts for the phenylphenols indicated increased hydrogen bonding as proton donors in the triplet state. However, with 2-phenylphenol apparently substantial reorientation of the molecule occurred on the surface. The RTP λ_{max} value of

Table 7.6. Phosphorescence λ_{max} Values for Model Compounds at Low and Room Temperature[a]

	LTP[b]	LTP[b] (0.1 M OH⁻)	$\Delta\lambda$[c]	RTP[d] 1% PAA-NaBr	$\Delta\lambda$[e]	RTP[d] filter paper	$\Delta\lambda$[e]
				λ_{max}, nm			
Biphenyl	464			472	+8	475	+11
4-Phenylphenol	477	488	+11	482	+5	483	+6
2-Phenylphenol	463	479	+16	495	+32	508	+45
Naphthalene	506			511	+5	514	+8
2-Naphthol	526	541	+15	520	-6	493	-4
9-Anthracenemethanol	494			497	+3	493	-1

[a] Phosphorescence λ_{max} where taken as the most intense band of the corrected phosphorescence spectra. Average of duplicate runs. Overall reproducibility $\pm$ 1 nm.

[b] Sample concentrations were between 10 and 50 µg/mL in 100% ethanol. Solutions were frozen at ~77 K by liquid N_2.

[c] $\Delta\lambda = \lambda_{anion}$, LTP - $\lambda_{neutral}$, LTP.

[d] 200 ng of phosphor was adsorbed from ethanol for all RTP spectra.

[e] $\Delta\lambda = \lambda_{RTP} - \lambda_{neutral}$, LTP.

Reprinted with permission from R. A. Dalterio and R. J. Hurtubise, *Anal. Chem.*, 56, 336. Copyright 1984 American Chemical Society.

2-naphthol was blue shifted compared to its LTP λ_{max} value, indicating that 2-naphthol most likely acted as a proton donor and proton acceptor.

Infrared spectra were obtained for polyacrylic acid in KBr, model compounds in KBr, and for polyacrylic acid with model hydroxyl aromatics or polycyclic aromatic hydrocarbons in KBr (34). Both the carboxyl OH stretching frequencies and the carbonyl stretching frequencies of polyacrylic acid were investigated. When hydroxyl aromatics were present with polyacrylic acid, the carboxyl OH stretching band could still be examined because it was broader and more intense than the phenolic OH band. The maximum frequency positions of the broad bands were determined by taking the first derivative of the spectra to increase the accuracy of determining band maxima. The shifts in the infrared band of the polyacrylic acid OH stretching vibration indicated either an increase or decrease in hydroxyl association depending on which phosphor was adsorbed. The OH stretching band of polyacrylic acid with 4-phenylphenol present was shifted to a lower wavenumber by 24 cm^{-1}, compared to polyacrylic acid alone. The shift to lower wavenumber indicated a net increase in the hydrogen bonding association of the polyacrylic acid hydroxyl groups in the presence of 4-phenylphenol. With 2-naphthol adsorbed on polyacrylic acid-salt mixtures, the polyacrylic acid hydroxyl stretching band shifted to a higher wavenumber by 59 cm^{-1}. This suggested a net decrease in the hydrogen bonding association of the polyacrylic acid hydroxyl groups. Dalterio and Hurtubise (34) suggested that the interaction of 2-naphthol with the polymer caused more polyacrylic acid inter- and intramolecular hydrogen bonds to be disrupted than were formed. For biphenyl and naphthalene mixed with polyacrylic acid, the polyacrylic acid OH stretching band was shifted to lower wavenumbers by 51 and 78 cm^{-1}, respectively, with respect to the hydroxyl stretching frequency of polyacrylic acid alone. A net increase in polyacrylic acid hydroxyl association was implied with biphenyl or naphthalene present with the polymer. Also, shifts to smaller wavenumbers were observed for the polyacrylic acid carbonyl stretching frequency when any of the four model compounds were present with polyacrylic acid. The shifts ranged from 3 to 6 cm^{-1}, and this implied a net increase in the hydrogen bonding association of the polyacrylic acid carbonyl groups.

In other experiments, a polyacrylic acid-KBr disk was placed in the reference beam of an infrared spectrophotometer. In the sample beam of the spectrophotometer, a KBr disk was placed that contained polyacrylic acid and either 4-phenylphenol or 2-naphthol. With this experimental setup, the absorption bands of polyacrylic acid were canceled (34). The polyacrylic acid hydroxyl stretching band was essentially eliminated, and the phenolic OH stretching band could be observed for the model compounds. The hydroxyl stretching bands of 4-phenylphenol and 2-naphthol were shifted to smaller wavenumbers by 12 and 17 cm^{-1}, respectively, when mixed with polyacrylic acid, compared to the hydroxyl stretching bands of these compounds alone in the KBr. For 4-phenylphenol and 2-naphthol, the shifts of the phenolic OH stretching bands to lower wavenumbers indicated increased hydrogen bonding for these compounds with polyacrylic acid in KBr disks. The infrared shifts for 4-phenylphenol and 2-naphthol were generally correlated with the reflectance and luminescence data discussed earlier, namely, increased hydrogen bonding for these compounds. For 4-phenylphenol, the 12 cm^{-1} red shift implied increased hydrogen donation, although the results did not eliminate the possibility that 4-phenylphenol behaved as both a proton donor and acceptor. The larger red shift for 2-naphthol suggested the compound could be acting as both a proton donor and proton acceptor. For example, it has been reported that the OH frequency for phenol is shifted to a lower frequency when the OH group acts simultaneously as a donor and an acceptor compared to the situation with the OH group acting individually as either a proton donor or a proton acceptor (44).

In summary, the interactions that cause RTP from compounds adsorbed on polyacrylic acid/salt matrices are somewhat complex. With the solution used to adsorb the phosphor on the polymer/salt matrix, a portion of the salt and a portion of the polymer from the solid matrix dissolve in the solution as it is placed on the matrix. The dissolved salt breaks intermolecular hydrogen bonds between the polymer molecules. This permits the dissolved phosphor to interact with the polymer molecules more effectively. After the solvent is evaporated, a complex matrix of polymer, salt, and phosphor is formed. The phosphor interacts strongly with the polymer via hydrogen bonding, and the salt packs the matrix, with the result that the modulus of the matrix is increased. Depending on the phosphor, it can behave as a proton donor and/or a proton acceptor through hydrogen bonding with the polymer. Also, there can be a net increase in hydrogen bonding character or a net decrease in hydrogen bonding character of the polymer depending on which phosphor is adsorbed. The previous summary of interactions is not applicable in all aspects to adsorbed phosphors because it was found that the disodium salt of terephthalic acid adsorbed on the sodium salt of polyacrylic acid/salt mixture gave a strong RTP signal. With this system, hydrogen bonds could not be formed. Thus, most likely, additional factors can be involved for producing RTP from adsorbed phosphors on polyacrylic acid/salt matrices.

7.4.   Cyclodextrin-Salt Matrices

It has been shown that 80% α-cyclodextrin-NaCl mixtures (45,46), 1% α-cyclodextrin-NaCl mixture (47) and β-cyclodextrin-NaCl mixtures (48), are analytically useful in obtaining room-temperature fluorescence and RTP from a variety of compounds. However, the interactions of adsorbed compounds have been studied more extensively for α-cyclodextrin-NaCl mixtures than for β-cyclodextrin-NaCl mixtures. It is well known that cyclodextrins have the unique ability to form inclusion complexes with a large number of organic and inorganic compounds (49). To develop a better understanding of phosphor-α-cyclodextrin interactions in salt matrices, Bello and Hurtubise (50) prepared the α-cyclodextrin inclusion complexes of benzo(f)quinoline, phenanthrene, 4-phenyl-phenol, and p-aminobenzoic acid. Inspection of the solid, prepared, inclusion complexes under an ultraviolet hand-lamp showed that the four complexes gave moderately strong RTP signals. There are basically two types of crystal structures for α-cyclodextrin complexes. These are cage-like and channel-like structures (51). However, the results from Bello and Hurtubise (50) did not allow an exact statement about the nature of the structures of the prepared inclusion complexes or the compounds adsorbed on the 80% α-cyclodextrin-NaCl mixture. Nevertheless, several insights were obtained by using luminescence and diffuse reflectance spectrom-etry to study the interactions of adsorbed compounds (50).

The room-temperature fluorescence and RTP spectra of the prepared inclusion complexes were compared to the corresponding spectra of the model compounds adsorbed on 80% α-cyclodextrin-NaCl mixtures. The room-temperature fluorescence emission spectra and RTP emission spectra of p-aminobenzoic acid, 4-phenylphenol, and phenanthrene were essentially the same as the respective emission spectra of the three model compounds adsorbed on the 80% α-cyclodextrin mixtures. Very good correlations were obtained with the various bands by comparing the appropriate spectra with the prepared complex and the compounds adsorbed on the 80% α-cyclodextrin-NaCl mixture. The similarity in the spectral shape and band positions of the luminescence spectra of p-aminobenzoic acid, 4-phenylphenol, and phenanthrene indicated that the luminescent compounds were included in a similar fashion with the 80% cyclodextrin-NaCl mixture and the prepared inclusion complexes (50).

The excitation spectrum for benzo(f)quinoline adsorbed on 80% α-cyclodextrin-NaCl was similar to the excitation spectrum for the prepared inclusion complex (Figure 7.7). However, Figure 7.7 shows that the respective room-temperature fluorescence and RTP emission spectra of the prepared inclusion complexes and samples adsorbed on 80% α-cyclodextrin are very different. The luminescence spectra of the prepared inclusion complex shows broad bands, and the room-temperature fluorescence spectrum of the inclusion complex was shifted substantially to longer wavelengths. To clarify these differences, a comparison of the spectra of the prepared inclusion complex of benzo(f)quinoline to that of the solution fluorescence spectrum of the protonated form of benzo(f)quinoline was acquired. The fluorescence emission band maximum for the prepared inclusion complex was blue shifted by 6 nm compared to that of the solution spectrum. However, the overall shape of both spectra were the same. Because of the similar spectral characteristics in the room-temperature fluorescence spectrum of the prepared inclusion complex to the solution fluorescence emission spectrum of the protonated benzo(f)quinoline, it was concluded that a large fraction of the emitting species from the prepared inclusion complex was in the protonated form (50).

By comparing the excitation spectrum of the prepared inclusion complex and the excitation spectrum of the protonated benzo(f)quinoline in solution, it was found that the two excitation spectra were very different from one another. The excitation spectrum of the prepared inclusion complex was of the neutral form of benzo(f)quinoline. However, the excitation spectrum from the acid solution gave protonated benzo(f)quinoline. Because of the previous results and the results in Figure 7.7, it was shown that with the prepared inclusion complex a majority of the benzo(f)quinoline was undergoing excited state protonation. Also, most likely, the nitrogen atom in benzo(f)quinoline was oriented towards one of the rims of the α-cyclodextrin cavity where the nitrogen atom could interact with the hydroxyl groups at the rim of α-cyclodextrin. The

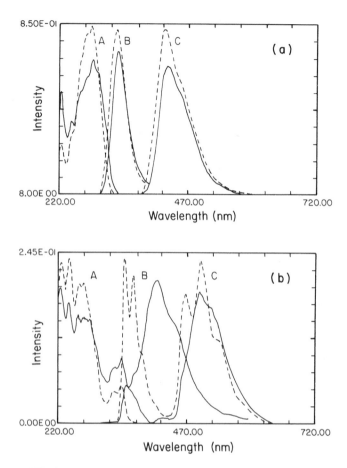

Figure 7.7. (a) Excitation (A), RTF emission (B), and RTP emission (C) spectra of PABA adsorbed on a 80% α-cyclodextrin-NaCl mixture (- - -) and the prepared inclusion complex of PABA (——). (b) Excitation (A), RTF emission (B), and RTP emission (C) spectra of B(f)Q adsorbed on an 80% α-cyclodextrin-NaCl mixture (- - -) and the prepared inclusion complex of B(f)Q (——). (Reprinted with permission from J.M. Bello and R.J. Hurtubise, *Anal. Chem.*, 59, 2395. Copyright 1987 American Chemical Society.)

luminescence spectral results for benzo(f)quinoline adsorbed on 80% α-cyclodextrin-NaCl showed that benzo(f)quinoline did not give excited state protonation on the 80% α-cyclodextrin mixture. Also, Figure 7.7 shows that the fluorescence and phosphorescence emission spectra of benzo(f)quinoline adsorbed on 80% α-cyclo-dextrin-NaCl are different than the corresponding spectra from the prepared inclusion complex. Thus, it was concluded that with benzo(f)quinoline on the 80% α-cyclodextrin mixture, the pyridinic nitrogen was most likely inside the α-cyclodextrin cavity where it could not interact with the hydroxyl groups of the α-cyclodextrin.

The diffuse reflectance spectra of the four model compounds, which were discussed earlier in this section, adsorbed on 80% α-cyclodextrin-NaCl and the diffuse reflectance spectra of the prepared inclusion complexes were reported (50). It was found that the diffuse reflectance spectra of the prepared inclusion complexes of benzo(f)quinoline, phenanthrene, and p-aminobenzoic acid were essentially the same as the spectra of the compounds on the 80% mixture. The similarity of the diffuse reflectance spectra of compounds adsorbed on the 80% α-cyclodextrin-NaCl mixture with the spectra of the respective prepared inclusion complexes showed that these model compounds were interacting in approximately the same fashion as in the ground state with the 80% α-cyclodextrin mixture and in the prepared inclusion complexes.

In other experiments, the amount of the four model compounds that did not interact with the α-cyclodextrin in the 80% α-cyclodextrin-NaCl mixture was determined by an extraction method and fluorescence spectroscopy (50). Only 0.7-2% of 4000 ng of a given analyte could be removed from 80% α-cyclodextrin mixture. These results indicated that the four model compounds were interacting strongly with the α-cyclodextrin to form inclusion complexes.

The luminescence, reflectance, and extraction results obtained by Bello and Hurtubise (50) did not permit a definitive statement on the exact structures of the prepared inclusion complexes or the compounds adsorbed on the 80% α-cyclodextrin-NaCl. Detailed X-ray analysis would be needed to obtain the appropriate structural information. Nevertheless, the luminescence and reflectance spectral results and extraction results showed that inclusion complexes were formed for the four model compounds. p-Aminobenzoic acid, 4-phenylphenol, and phenanthrene gave similar luminescence spectral results for the prepared inclusion complexes and with the 80% α-cyclodextrin-NaCl mixture. Benzo(f)quinoline behaved in a different fashion and underwent excited state protonation with the prepared inclusion complexes but not with the 80% α-cyclodextrin mixture. The results of this work are important analytically because they define the type of components on the 80% α-cyclodextrin mixture.

Bello and Hurtubise (52,53) further pursued the interactions of benzo(f)quinoline, phenanthrene, p-aminobenzoic acid, and

4-phenylphenol by obtaining a variety of luminescence properties of the compounds adsorbed on α-cyclodextrin-NaCl mixtures. The room-temperature fluorescence and RTP intensities of the four model compounds were obtained from mixtures of α-cyclodextrin-NaCl ranging from 0.0005% to 80% α-cyclodextrin. The results showed the importance of the initial wet chemistry in the sample preparation procedures for the observation of luminescence signals for compounds adsorbed on α-cyclodextrin-NaCl mixtures. The data indicated that to obtain the optimum room-temperature fluorescence and RTP signals, the solvent used to adsorb the analytes had to be saturated with α-cyclodextrin. This indicated that the formation of an inclusion complex in the adsorbing solvent prior to solvent evaporation was important for the observation of RTP. With a saturated solution of α-cyclodextrin, the reaction between the cyclodextrin and the model compounds would favor inclusion complex formation in solution to give maximum product yield. They also concluded that with α-cyclodextrin-NaCl mixtures, the packing of the solid matrix was not very important for enhancing the RTP signal when the adsorbing solutions were beyond the saturation point of α-cyclodextrin. This conclusion is in contrast to the results obtained for p-aminobenzoic acid adsorbed on sodium acetate-NaCl mixtures discussed in Section 7.1. For p-aminobenzoic acid adsorbed on sodium acetate-NaCl mixtures, it was found that the maximum RTP signal was obtained on pure sodium acetate, and matrix packing of the sodium acetate was very important to achieve the optimal RTP signal.

Fluorescence quantum yield values (ϕ_f), phosphorescence quantum yield values (ϕ_p), and phosphorescence lifetime values were obtained for benzo(f)quinoline, phenanthrene, p-aminobenzoic acid, and 4-phenylphenol adsorbed on various mixtures of α-cyclodextrin-NaCl at 20°C and -180°C (52,53). Tables 7.7 and 7.8 show the quantum yield values and phosphorescence lifetime values, respectively. It was found that the fluorescence quantum yield values for a given compound on the 80% α-cyclodextrin-NaCl and on the 0.5% α-cyclodextrin mixture at 23°C were essentially the same. This is indicated in Table 7.7 for phenanthrene and benzo(f)quinoline. For 0.05% α-cyclodextrin-NaCl mixtures, both the ϕ_f and ϕ_p values were less at 23°C compared to the other α-cyclodextrin-NaCl mixtures (Table 7.7). The difference in the room-temperature fluorescence and RTP quantum yields of the compounds is related to the different amounts of α-cyclodextrin initially dissolved in the adsorbing solvent (methanol). The 0.05%

Table 7.7. RTP and RTF Quantum Yields for B(f)Q and Phenanthrene Adsorbed on α-Cyclodextrin-Sodium Chloride Mixtures at 23 °C and -180°C[a,b]

	23°C		-180°C	
	ϕ_f	ϕ_p	ϕ_f	ϕ_p
80% α-Cyclodextrin				
Phenanthrene	0.29 ± 0.01	0.13 ± 0.02	0.29 ± 0.03	0.26 ± 0.04
B(f)Q	0.41 ± 0.01	0.09 ± 0.02	0.40 ± 0.03	0.22 ± 0.04
0.5% α-Cyclodextrin				
Phenanthrene	0.30 ± 0.05	0.10 ± 0.03		
B(f)Q	0.38 ± 0.05	0.10 ± 0.03		
0.05% α-Cyclodextrin				
Phenanthrene	0.16 ± 0.03	0.04 ± 0.01	0.26 ± 0.04	0.11 ± 0.02
B(f)Q	0.22 ± 0.03	0.05 ± 0.01	0.36 ± 0.04	0.12 ± 0.02

[a] Results are the average of at least two determinations.
[b] 95% Confidence level based on pooled standard deviation values.

Reprinted with permission from J.M. Bello and R.J. Hurtubise, *Anal. Chem.*, 60, 1291. Copyright 1988 American Chemical Society.

Table 7.8. Phosphorescence Lifetimes of B(f)Q and Phenanthrene Adsorbed on α-Cyclodextrin-Sodium Chloride Mixtures[a]

	80% α-CD		0.05% α-CD	
	$\tau_p(23°C)$,s	$\tau_p(-180°C)$,s	$\tau_p(23°C)$,s	$\tau_p(-180°C)$,s
Phenanthrene	1.84 ± 0.11	3.11 ± 0.14	1.94 ± 0.10	2.96 ± 0.16
B(f)Q	0.87 ± 0.04	2.48 ± 0.16	0.95 ± 0.05	2.41 ± 0.02

[a]Reproducibility based on the 95% confidence level.
Reprinted with permission from J. M. Bello and R.J. Hurtubise, *Anal. Chem.*, 60, 1291. Copyright 1988 American Chemical Society.

α-cyclodextrin mixture did not yield a saturated α-cyclodextrin solution, whereas saturated α-cyclodextrin solutions were obtained with the 0.5% and 80% mixtures. Therefore, in the 0.05% mixture, fewer α-cyclodextrin molecules could interact with the analyte, and the analyte would not be included as effectively in the 0.05% mixture as in the other two α-cyclodextrin mixtures. Interestingly, the phosphorescence lifetimes did not change much at 23°C in going from 80% α-cyclodextrin to 0.05% α-cyclodextrin (Table 7.8). Also, for the 80% α-cyclodextrin mixture, the ϕ_f values did not change from 23°C to -180°C (Table 7.7). This is important analytically because it is not necessary to lower the temperature to obtain the maximum ϕ_f value for the four compounds.

Bello and Hurtubise (52,53) calculated values for the triplet quantum yield (ϕ_t), the rate constant for phosphorescence (k_p), and the rate constant for radiationless transition from the triplet state (k_m) for benzo(f)quinoline, phenanthrene, p-aminobenzoic acid, and 4-phenylphenol adsorbed on α-cyclodextrin-NaCl mixtures using the approaches described in Section 7.1. For a given compound on a particular α-cyclodextrin-NaCl mixture, the k_p values were essentially the same at 23°C and -180°C. At room temperature, the k_m values of the compounds were relatively high on the α-cyclodextrin-NaCl mixtures. The relatively large k_m values at room temperature indicated that the nonradiative transition from the triplet state made an important contribution to the loss of phosphorescence emission from the compounds adsorbed on α-cyclodextrin-NaCl mixtures. At -180°C, the k_m values were considerably smaller than at 23°C. However, only p-aminobenzoic acid gave a k_m value of zero at -180°C.

As with p-aminobenzoic acid adsorbed on sodium acetate (Section 7.1), the percentages of radiative and nonradiative transitions were calculated for the four model compounds adsorbed on α-cyclodextrin-NaCl mixtures, and energy diagrams were constructed. For phenanthrene, p-aminobenzoic acid, and 4-phenyl-phenol at room temperature, there was a substantial amount of absorbed energy lost through the nonradiative transition from the singlet state to the ground state for the compounds adsorbed on 0.05% α-cyclodextrin-NaCl. For phenanthrene and p-aminobenzoic acid adsorbed on 80% α-cyclodextrin-NaCl at room temperature, there was no nonradiative loss of energy from the singlet state. However, for 4-phenylphenol, the radiative loss was reduced to 40% on the 80% α-cyclodextrin-NaCl mixture. Interestingly, benzo(f)-quinoline gave no nonradiative loss of energy from the singlet state on the 80% and 0.05% α-cyclodextrin-NaCl mixtures. For the four model compounds investigated on α-cyclodextrin-NaCl mixtures, by lowering the temperature from 23°C to -180°C, the increase in percentage of phosphorescence ranged from 1.9 to 7.0-fold. 4-Phenylphenol showed the largest increase in percentage of phosphorescence at -180°C, most likely because of the flexible nature of its chemical structure, which would not permit it to be held as rigidly as the other model compounds. Several other conclusions were made from the energy diagrams, and they proved to be very valuable in assessing how the energy absorbed by the adsorbed compounds was distributed within the compounds (52,53).

In summary, there were several important factors revealed related to the interactions of adsorbed compounds on α-cyclo-dextrin-NaCl mixtures. It was necessary to have a saturated solution of α-cyclodextrin prior to adsorbing the luminescent component on the surface so an inclusion complex could be formed in solution. The NaCl served the function of breaking intermolec-ular hydrogen bonds between α-cyclodextrin molecules in solution and increasing the modulus of the dry α-cyclodextrin-NaCl matrix. However, it was found that matrix packing with α-cyclodextrin-NaCl mixtures was not as important compared to the sodium acetate matrix. In the α-cyclodextrin-NaCl matrix, it was possible for two or more α-cyclodextrin molecules to interact with one phosphor molecule in the solid state. This partly explains why rather large molecules such as triphenylene gave a strong phosphorescence when adsorbed on 80% α-cyclodextrin-NaCl (45).

7.5.   Filter Paper

Filter paper is the most widely used solid surface for obtaining RTP from adsorbed organic compounds. Relatively little is understood about the interactions that are responsible for RTP with filter paper. However, several advances have been made in elucidating interaction and conditions needed for solid-surface RTP. The first workers to investigate, in some detail, the phosphor interactions with filter paper were Schulman and Parker (40). They considered the effects of moisture, oxygen, and the nature of the support-phosphor interaction using two model compounds. Earlier, Schulman and Walling (54,55) suggested that surface adsorption of phosphorescent compounds to the support inhibited collisional deactivation of the triplet state and minimized oxygen quenching when the sample was dried. Schulman and Parker (40) also proposed that hydrogen bonding of ionic organic molecules to hydroxyl groups on the solid surface was the primary mechanism of providing the rigid sample matrix for RTP. They showed that by silanization of the hydroxyl groups in filter paper the RTP of sodium 1-naphthoate was reduced by 90%. In addition, they proposed that moisture acts to disrupt hydrogen bonding and aids in the transport of O_2 into the filter paper. Their results showed that both moisture and oxygen could independently quench RTP. This was arrived at from relative intensity data for humidified argon and oxygen. Some of their data are given in Table 7.9 and Figure 7.8 for samples adsorbed on Whatman No. 1 filter paper. In Figure 7.8, NaBPCA refers to sodium 4-biphenylcarboxylate. In the absence of oxygen, moisture acts by itself as a powerful quencher (Figure 7.8). At low humidity, a somewhat moderate quenching effect was noted. However, at high humidity, quenching was quite dramatic (Figure 7.8). Schulman and Parker (40) and Schulman and Walling (54,55) concluded that moisture competes with surface hydroxyl groups for hydrogen bonding to the phosphor molecules and ties up hydroxyl groups so the phosphor is not held rigidly. Quenching by triplet ground-state oxygen occurred in the absence of moisture, but the extent of oxygen quenching was facilitated notably by the presence of moisture. Schulman and Parker (40) concluded that moisture must be regarded as the most important contributor to quenching of RTP because it can transport oxygen into the sample matrix and allow collisional deactivation to occur. Alkaline solutions of only two model compounds, sodium 4-biphenylcarboxylate and sodium 1-naphthoate, were investigated by Schulman and Parker (40). There is a need for more investigations to study the effects of moisture and of oxygen on the RTP of other compounds adsorbed on filter paper under neutral, acidic, and basic conditions.

In our research laboratory, recent work has shown that the RTP of protonated benzo(f)quinoline adsorbed on filter paper showed both static and dynamic quenching as a function of percent relative humidity. The static quenching contribution was about 4.5-fold larger than the dynamic quenching component. The results indicated that the filter paper matrix was contributing to the quenching of RTP because the matrix was not as rigid with adsorbed moisture compared to the dry filter paper. With adsorbed moisture, the hydrogen bonding network in filter paper was not as stable. This condition permitted the static quenching of the adsorbed phosphor to occur, most likely as a result of enhanced

Table 7.9. Relative Intensities of Sodium 4-Biphenylcarboxylate Samples in Ar and O_2 as a Function of Relative Humidity

$\%H$ [a]	I_{Ar} [b]	Rsd in Ar [c]	I_{O_2} [d]	Rsd in O_2 [c]	Q_{H_2O} [e]	Q_{O_2} [f]
0	100	1.2	70.9	1.5	0	29.1
3.2	98.1	1.1	70.3	1.3	1.9	27.8
8.5	91.1	2.4	61.3	1.9	8.4	30.3
18.8	57.6	1.2	25.5	3.8	42.4	32.1
37.1	12.8	2.6	2.5	14	87.2	10.3
58.3	2.6	7.1	0.4	2.9	97.4	2.2
80.5	0.7	4.3	0.1	17	99.3	0.6
100	0.3	11	0.0		99.7	0.3

[a] Percent relative humidity of gas at 298 K.
[b] Intensity in argon relative to 0% humidity in argon (I^o_{Ar}).
[c] Calculated from triplicate runs and given in percent.
[d] Intensity in oxygen relative to (I^o_{Ar}).
[e] $I^o_{Ar} - I_{Ar}$.
[f] $I_{Ar} - I_{O_2}$.
Reprinted with permission from E.M. Schulman and R.T. Parker, *J. Phys. Chem.*, 81, 1932. Copyright 1977 American Chemical Society.

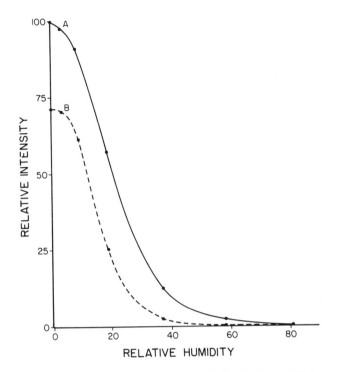

Figure 7.8. Relative intensities of NaBPCA samples in AR (curve (A)=I_{Ar},——) and O_2 (curve (B) = I_{O_2}, - - -) as a function of humidity. (Reprinted with permission from E.M. Schulman and R.T. Parker, *J. Phys. Chem.*, 81, 1932. Copyright 1977 American Chemical Society.)

vibrational modes in the filter paper matrix.

Wellons et al. (56) compared the phosphorescence signals of several compounds adsorbed from alkaline solutions onto filter paper at room temperature and at 77 K in a solid matrix. The data obtained indicated that the degree of rigidity with which the molecules were held on filter paper compared to a solid solution of 77 K. The authors commented that the molecules that had the most ionic sites showed the greatest rigidity on the surface. Vanillin and 2,4-dithiopyridimine are double charged species in strongly alkaline solution, and these two compounds gave very strong RTP signals on filter paper. Most likely ionic interactions were occurring on the filter paper. However, with two uncharged compounds, sulfaquanidine and 5-acetyluracil, relatively strong RTP was observed. With these compounds, probably, hydrogen bonding was occurring with the filter paper surface.

Bower and Winefordner (57) investigated the effect of sample environment on the RTP of several polycyclic aromatic hydrocarbons. They found that the heavy-atom effect gave significant enhancement of RTP for the polycyclic aromatic hydrocarbons. It was postulated that silver ion formed π-complexes with the π-electron cloud of the aromatic hydrocarbons. They considered the silver ions to be bonded to the molecule and to the functional groups on filter paper. This bonding between the phosphor molecules and the support would give the necessary rigidity for the observation of RTP. In other work, Aaron et al. (58) studied the effects of ion-exchange filter paper and of heavy atoms on the RTP of several indoles. Anion-exchange filter paper (Whatman DE-81) gave the largest RTP signals with iodide present. For the particular case of 5-fluorindole with thallium(I) ions present, the use of cation-exchange CM23 carboxymethylcellulose resin on S&S 903 filter paper enhanced the RTP signal about six times compared to Whatman DE-81 filter paper. The authors commented that the smaller RTP signal on the DE-81 filter paper might be due to the perturbation of hydrogen bonding in DE-81 filter paper by thallium(I) ions. This would result in a decrease in the RTP signal. In addition, they theorized that thallium(I) ions may favor the adsorption of planar 5-fluoroindole molecules at the surface of the CM resin on S&S 903 filter paper by a charge-transfer complex which was stabilized by hydrogen bonding with the carboxymethylcellulose groups of the paper.

Andino et al. (59) employed X-ray photoelectron spectrometry to investigate the surface processes involved in the RTP of 3,5-diiodotyrosine, 5-hydroxytryptophan, carbaryl, and bis(8-quinolinate)platinum(II) adsorbed on Whatman No. 1 filter paper with a heavy atom present. The X-ray photoelectron spectral results gave no evidence of a strong chemical interaction between the heavy atom and phosphor adsorbed on the surface of the paper. By using grazing angle experiments to examine surface layers of different thickness, the phosphorescent molecules and heavy-atom probe appeared to be evenly distributed throughout the sampling depth. The authors emphasized that their initial results on the relative amounts of phosphor molecules compared to that of the heavy atoms on the surface and the magnitude of penetration of the molecules into the bulk of the filter paper need to be confirmed by other molecular probes. In general, the ideal probe has yet to be discovered.

Su and Winefordner (60) presented a model of the heavy-atom-analyte-substrate interaction in room-temperature phospho-rimetry based on the earlier X-ray photoelectron spectral results (59). They concluded that the heavy atom, iodide ion, was shared by glucopyranosyl groups in filter paper. The electron cloud intensity of the iodide ion repelled the electron clouds surrounding oxygen atoms of the filter paper, which were bonded to specific carbons in the filter paper. The net effect of these interactions resulted in 'pushing' the electron distribution closer to the specific carbon atoms bonded to the oxygen atoms in filter paper. The electron cloud shift would then strengthen the bonding between the carbon and oxygen atoms. The authors commented that this phenomenon explained two earlier observations (59): (a) the iodide ion did not interact strongly with the phosphor; (b) the bond energies of specific carbons in filter paper increased when iodide ion was adsorbed on the filter paper. Also, their results supported the concepts of hydrogen bonding or ionic interaction between the phosphor and oxygen atoms in filter paper.

White and Seybold (61) investigated the effect of added alkali halide salts on the room-temperature fluorescence and phosphorescence of 2-naphthalenesulfonate adsorbed on filter paper. They reported the normal external heavy-atom effect on the luminescence. The dependence of fluorescence quenching on perturber concentration was described by a modified version of the Perrin equation. The modified Perrin equation gave an excellent fit for all the halogen quenching data. The Stern-Volmer and normal Perrin models sufficiently represented the quenching data for the lighter halogens but not iodide results. For the phosphorescence data, it was shown that external heavy atoms increased the radiative triplet decay constant more than the competing triplet state nonradiative constant. In other work, Meyers and Seybold (62) reported the effects of external heavy atoms and other factors on the room-temperature fluorescence and phosphorescence of tryptophan, tyrosine, and various derivatives of the previous two compounds. Several compounds showed considerable RTP enhancement with sodium iodide. The RTP of tryptophan on filter paper was increased 455-fold, its methyl ester 340-fold, and that of indole 370-fold by the addition of sodium iodide to the surface. Niday and Seybold (27) investigated matrix effects on the lifetime of RTP. They obtained the phosphorescence half-life of 2-naphthalene-sulfonate on filter paper with various compounds adsorbed on the filter paper. Some of the compounds adsorbed to filter paper were NaF, NH_4Cl, H_3BO_3, glycine, glucose, and sucrose. In all cases, with added compound, an increase in the phosphorescent lifetime was observed. However, in no situation was the phosphorescent lifetime as long as that observed at 77 K in a rigid mixture of ethyl ether, isopentane, and ethyl alcohol (EPA) in a ratio of 5:5:2. The authors emphasized that one explanation for RTP is that the matrix holds the adsorbed compound rigid and thereby prevents vibrational motions necessary for nonradiative decay from the triplet state. They also concluded that packing the matrix with salts and sugars further inhibited internal motion of the phosphorescent compound. It was considered that the added compounds "plug up" the channels and interstices of the matrix, decreasing oxygen permeability and protecting the phosphorescent molecules from quenching by oxygen.

McAleese and Dunlap (63) postulated a matrix isolation mechanism for RTP from cellulose paper samples based on the swelling property of cellulose. Cellulose undergoes substantial swelling in the presence of various strongly polar solvents. The solvents benzene, acetone, propanol, ethanol, methanol, and water were used in their research, and the magnitudes of RTP intensities from several phosphors were compared in reference to the solvent used to adsorb the compound on the filter paper. Swelling of filter paper would favor entry of phosphor molecules into the submicroscopic pores in the filter paper. After drying the matrix, the molecules could become trapped between cellulose chains, and this would provide the necessary rigidity for phosphorescence to be observed. The authors contended that phosphor immobilization with filter paper could not be explained by a simple hydrogen bonding mechanism.

Dalterio and Hurtubise (32) reported the effects of sodium halide salts (NaCl, NaBr, NaI) on the RTP intensity of 4-phenylphenol adsorbed on filter paper. NaI on filter paper enhanced the RTP of 4-phenylphenol to the greatest extent. The RTP with NaI was not likely the result of only the heavy-atom effect, because NaCl caused about a 10-fold increase in RTP compared to the untreated paper. The NaI caused a 17-fold increase in RTP compared to the untreated paper. A given salt could contribute to matrix packing by which the excited phosphor molecules would be more rigidly held within the matrix and thus less prone to deactivation. For the case of NaBr, it was also shown that the maximum RTP of 4-phenylphenol depended on the amount of NaBr present on the filter paper. In contrast to the previous results, it was found that benzo(f)quinoline samples spotted on filter paper from 0.1 M HBr-ethanol solution containing either NaBr or NaCl did not intensify the RTP signal compared to just 0.1 M HBr ethanol solution. The RTP signals remained about the same for all the samples (35). In related work, Ramasamy and Hurtubise (20) utilized reflectance spectroscopy to determine the forms of benzo(f)quinoline and quinoline adsorbed on filter paper. Comparison of the absorption wavelengths for benzo(f)quinoline adsorbed on filter paper from ethanol with the absorption wavelengths in ethanol indicated that the neutral form of benzo(f)quinoline was the predominant form adsorbed on filter paper. For a 0.1 M acid solution of benzo(f)quinoline adsorbed on filter paper, the data showed that the cation of benzo(f)quinoline was adsorbed on the surface. Similar results were obtained for quinoline.

Dalterio and Hurtubise (34) used a variety of spectral techniques to explore the interactions of hydroxyl aromatics and aromatic hydrocarbons on filter paper. With ultraviolet diffuse reflectance spectroscopy, the spectral shifts of the compounds on filter paper were obtained and compared to the compounds on NaBr. Red shifts were observed for all the compounds adsorbed on filter paper except for 2-naphthol and 1,2-dihydroxynaphthalene which gave blue shifts. The red shifts suggested the hydroxyl group of a hydroxyl aromatic was hydrogen bonded to the filter paper by a predominately proton donating interaction. They attributed the red shifts for the aromatic hydrocarbons to intermolecular π-electron hydrogen bonds (OH---π) between the aromatic hydrocarbons and the hydroxyl group of filter paper. A small blue shift was obtained for 2-naphthol on filter paper. It was speculated that 2-naphthol was acting as both a proton donor and a proton acceptor. Because 1,2-dihydroxynaphthalene contains two adjacent hydroxyl groups which form an intramolecular hydrogen bond, the interpretation of its spectral shift on filter paper was not simple. However, the rather large blue shift observed on filter paper indicated that one or both of the oxygens of the hydroxyl groups of the compound accepted protons from the hydroxyl groups of filter paper.

With phosphorescence spectroscopy, the phosphorescence λ_{max} values were obtained for several model compounds at low temperature (LTP) in ethanol glass and at room temperature adsorbed on filter paper (34). The LTP λ_{max} values for the anions of the hydroxyl aromatics were also obtained. The anions showed red shifts between 11-16 nm with respect to the LTP λ_{max} values of the neutral hydroxyl aromatics. The magnitude of phosphorescence red shifts for 4-phenylphenol and 2-naphthol indicated the triplet emitting species on filter paper were the neutral molecules. However, the RTP λ_{max} value of 2-phenylphenol on filter paper was red-shifted greater than the LTP λ_{max} of the 2-phenylphenol to anion. It was unlikely that 2-phenylphenol ionized in the triplet state on filter paper because of the conditions of the experiment. The large red shift appeared to be due to the reorientation of 2-phenylphenol on the surface and the Franck-Condon states. For example, at low temperature, very little solvent or phosphor reorientation normally occurs. However, at room temperature, there would be reorientation to an equilibrium triplet state, and then, phosphorescence would take place from the equilibrium triplet state to the Franck-Condon singlet ground state (34). Most likely, the steric crowding of the phenyl ring in 2-phenylphenol favors reorientation of the phenyl ring. The RTP λ_{max} for the phenylphenols, biphenyl, and naphthalene were red shifted compared to the respective LTP λ_{max} values. For the phenylphenols, the red shifts suggested increased hydrogen bonding as proton donors in the triplet state. As discussed, 2-phenylphenol apparently undergoes significant reorientation on the surface which causes a large red shift. The RTP λ_{max} of 2-naphthol was blue shifted compared to its LTP λ_{max} value, indicating that 2-naphthol acted as a proton acceptor in the triplet state or behaved as both a proton donor or acceptor.

Multiple internal reflectance infrared spectroscopy was used mainly to study the OH stretching vibration of filter paper with adsorbed model hydroxyl aromatics and salts (34). As considered above, by adsorbing inorganic salts onto filter paper with certain phosphors, the RTP of the phosphor could be substantially increased. This effect appears to be different than the heavy-atom effect. Table 7.10 shows the hydroxyl stretching frequencies of filter paper alone, which adsorbed NaCl and NaBr, and adsorbed model compounds with the salts. In Table 7.10, it can be seen that with NaCl and NaBr adsorbed onto filter paper, the cellulose hydroxyl stretching band shifts to lower wavenumbers by 9 and 12 cm^{-1}, respectively, with respect to filter paper with no adsorbed salt. This indicated increased hydrogen bonding association of the filter paper with the salts present. Allerhand and Sehleyer (36) showed by infrared spectroscopy that chloride and bromide ions can form hydrogen bonds with the hydroxyl groups of methanol. With filter paper, the ions from the salts most likely fill in spaces in the matrix and form hydrogen bonds to cellulose OH groups. The overall

Table 7.10. Multiple Internal Reflectance Infrared Hydroxyl Stretching Band of Filter Paper with Adsorbed Salts and Model Compounds.[a,b]

	No phosphor	4-Phenyl-phenol	Δcm^{-1} [c]	Biphenyl	Δcm^{-1}	2-Naphthol	Δcm^{-1}	Naphtha-lene	Δcm^{-1}
				Wavenumber, cm^{-1}					
Filter paper no salt	3282	3304	+22	3251	-31	3258	-24	3266	-16
Filter paper with NaCl[d]	3273	3329	+56	3226	-47	3257	-16	3228	-45
Filter paper with NaBr[e]	3270	3328	+58	3263	-7	3259	-11	3262	-8

[a]A sample of 5 mg of model compound adsorbed on each of two 14 x 49 mm filter paper sheets.
[b]Average of duplicate runs. Overall reproducibility of band maxima ± 3 cm^{-1}
[c] Δcm^{-1} (phosphor on filter paper)-cm^{-1} (no phosphor).
[d]A sample of 2.3 mg of NaCl adsorbed on each filter paper sheet.
[e]A sample of 9.7 mg of NaBr adsorbed on each filter paper sheet.
Reprinted with permission from R.A. Dalterio and R.J. Hurtubise, *Anal. Chem.*, 56, 336. Copyright 1984 American Chemical Society.

effect would be a more tightly packed matrix which would restrict moisture and oxygen penetration and favor RTP. Packing the filter paper matrix could also increase the modulus of the matrix and thus hold the phosphor more rigidly within the matrix. With model compounds added to filter paper, additional shifts in the OH stretching bands were observed. However, the direction of the shifts was different (Table 7.10). For adsorbed 4-phenylphenol, the OH band of filter paper shifted to larger wavenumbers by 22, 56, and 58 cm^{-1} on filter paper with no salt, with NaCl and with NaBr, respectively (Table 7.10). The adsorption of 4-phenylphenol onto filter paper most likely caused the disruption of some of the hydrogen bonds in the filter paper matrix. This type of shift could occur by replacing stronger OH hydrogen bonds in the filter paper with weaker hydrogen bonds in the filter paper by the formation of weaker bonds to the phosphor. However, reflectance and luminescence data showed that 4-phenylphenol acted as a proton donor. Thus, 4-phenylphenol also participated in hydrogen donation to the hydroxyl groups of filter paper. For the adsorption of 2-naphthol, naphthalene, or biphenyl onto filter paper, the filter paper OH band shifted between 7 and 47 cm^{-1} to smaller wavenumbers in all instances. This indicated a net increase in hydrogen bonding association of filter paper OH groups, presumably by proton donation by the OH groups of the filter paper to the phosphors. The different effects on the OH stretching band of the filter paper were probably influenced by steric factors related to the phosphors fitting into and interacting with the filter paper matrix in different ways. It was found that phenolic OH stretching bands of 4-phenylphenol and 2-naphthol did not interfere with the observation of the filter paper OH band, because there was not enough compound on the surface to give an observable band from the adsorbed compound. The infrared results showed that a variety of hydrogen bonding interactions were taking place between the filter paper and phosphors. One type of hydrogen bond interaction did not predominate in all the cases investigated. It was discovered that the filter paper and the phosphors behaved as proton donors, proton acceptors, or simultaneously as proton donors and acceptors.

Vo-Dinh et al. (64) studied the phosphorescence line narrowing of coronene and phenanthrene adsorbed on cellulose substrate. It is well known that a normal feature of RTP emission spectra from compounds adsorbed on surfaces is that broad bandwidths are obtained from the RTP spectra. However, little is understood about the band broadening mechanisms and the interactions of the phosphors with solid substrates. In their study, phosphorescence line narrowing spectroscopy was used to resolve the vibrational structure of phenanthrene and coronene adsorbed on filter paper. This technique involved using a narrow-band light source, usually a laser, to excite selectively a subset of guest molecules that had a specific electronic transition energy. The phosphorescence line narrowing spectra that resulted gave extremely sharp structural features with spectral line widths of less than 1 cm^{-1}. The 457.9 nm and 514.5 nm lines of an argon ion laser were used to directly excite the phosphor to the triplet state from the singlet ground state. The transition probability was enhanced by the external heavy-atom effect of thallium acetate impregnated on the filter paper substrate. Also, measurements were carried out from 4 to 300 K to study the effects of temperature on the spectral bandwidths of phosphorescence and to obtain information on the coupling between the guest's electrons with the intermolecular vibrations of the host matrix (electron-phonon coupling). Differences were observed between the interactions of phenanthrene and coronene with the paper substrate. The strength of electron-phonon coupling was weak for coronene, but moderately strong for phenanthrene. For coronene, they concluded that the coronene molecules adsorbed on filter paper undergo no significant energy-randomizing process such as geometrical reorientation on the paper or phonon-assisted energy transfer during the phosphorescence lifetime. They also found that the phosphorescence line narrowing spectra obtained with the paper substrate and butyl bromide glass were similar. Generally, their results are of considerable interest because filter paper provides a substrate which permits fundamental studies of phosphorescence and other photophysical processes from low temperature to room temperature.

Suter et al. (65) considered several aspects of the hydrogen-bonding properties of Whatman No. 4 filter paper using benzo(a)-phenazine, 1,4-diazatriphenylene, and 'Michlers ketone' as phosphors. The spectral results from the phosphors adsorbed on filter paper at room temperature and at 77 K were compared to spectral results obtained from the phosphors dissolved in rigid solvents at 77 K. The technique of total luminescence spectrometry was employed in their work. This approach involved measuring the luminescence intensity as a function of both the excitation and emission frequencies. The excitation and emission frequencies were projected onto a two-dimensional plane, yielding an isocontour two-dimensional spectrogram. By comparing the several two-dimensional spectrograms obtained, they concluded that the Whatman No. 4 paper had a high proton donor capability. Their results also showed that the RTP of benzo(a)phenazine and 1,4-diazatriphenylene was due to hydrogen-bonded chromophores. However, the results for 'Michlers ketone' were not as clear.

The effects of external heavy-atom perturbers on the room-temperature phosphorescence spectrum of dibenzo(f,h)quinoxaline adsorbed on Whatman No. 4 filter paper has been studied (66). The heavy atom modified the phosphorescence spectral features. The authors determined that the spectral modifications were related to the enhancement of bands belonging to totally symmetric vibrations of the phosphor. Time-resolved phosphorimetry experiments showed that inhomogeneous interactions between phosphor and heavy atom could lead to substantial differences in the phosphorescence spectra of a single chemical species. This result is important in RTP work because it is helpful in avoiding misinterpretation of phosphorescence spectra in time-resolved phosphorescence measurements.

Experimental values of fluorescence quantum yield, phosphorescence quantum yield, and phosphorescence lifetime were obtained at temperatures from 23°C to -180°C for 4-phenylphenol adsorbed on filter paper. From the experimental values, rate constants for phosphorescence and radiationless transition from the triplet state were calculated along with triplet formation efficiency (67). The general approach used to calculate the luminescence parameters was described in Section 7.1. Table 7.11 gives the ϕ_f and ϕ_p values from 23°C to -180°C for 4-phenylphenol adsorbed on filter paper. The data show a 54.5% increase in ϕ_f from 23°C to -180°C. The ϕ_p values in Table 7.11 showed a very dramatic increase from 23°C to -180°C, namely, 0.020 to 0.53. The very large increase in the ϕ_p value on cooling to -180°C implied that 4-phenylphenol was not as strongly held on filter paper at room temperature compared to low temperature. This is most likely due to the phenyl ring of 4-phenylphenol not strongly interacting with the surface. It was found that the rate constant for phosphorescence did not vary with temperature, which showed that the rate constant was mainly a function of the molecular structure of 4-phenylphenol. However, the rate constant for the radiationless transition from the triplet state decreased with temperature. This indicated that this rate constant depended on how rigidly the 4-phenylphenol was held in the filter paper matrix. A similar conclusion was reached for p-aminobenzoic acid adsorbed on pure sodium acetate and on

Table 7.11. Fluorescence and Phosphorescence Quantum Yields from 23° to -180° C for 4-Phenylphenol Adsorbed on Filter Paper

Temperature, C°	Fluorescence quantum yield, ϕ_f [a]	Phosphorescence quantum yield, ϕ_p [a]
23	0.33	0.020
0	0.47	0.072
-40	0.46	0.12
-80	0.45	0.25
-120	0.50	0.36
-180	0.51	0.53

[a]Average of duplicate runs. The pooled standard deviations for the fluorescence and phosphorescence quantum yields were 0.04 and 0.01, respectively.

Reprinted with permission from S.M. Ramasamy and R.J. Hurtubise, *Talanta*, 1989, 39, 315.

mixtures of sodium acetate with sodium chloride (Section 7.1). It was not possible to obtain all of the rate constants for the various radiative and nonradiative processes for 4-phenylphenol with the experimental conditions employed (67). However, by using ϕ_f, ϕ_t, and ϕ_p values, it was possible to construct energy diagrams (Section 7.1). Figure 7.9 shows the diagrams obtained for 4-phenylphenol adsorbed on filter paper. From these diagrams, it is noted that the percentage internal conversion decreases with temperature and becomes zero at -180°C. It was found earlier that the percentage internal conversion was zero at -180°C for 4-phenylphenol adsorbed on 80% α-cyclodextrin-NaCl and 0.05% α-cyclodextrin-NaCl (52). Figure 7.9 shows that the extent of intersystem crossing from the S_1 state to the T_1 state increased significantly for 4-phenylphenol as the temperature was lowered. Interestingly, the intersystem crossing from the triplet state to the ground state did not change much with temperature. At -180°C there was no intersystem crossing to the ground state (Figure 7.9). As indicated in Figure 7.9, at room temperature, a large fraction of the energy absorbed is lost by internal conversion. This most likely results from the relatively weak adsorption of 4-phenylphenol on filter paper because the only polar group in the molecule is the hydroxyl group. In addition, the nonrigid structure of 4-phenylphenol can result in several vibrational modes in the ground state and excited singlet state, which could provide channels for nonradiative loss of absorbed

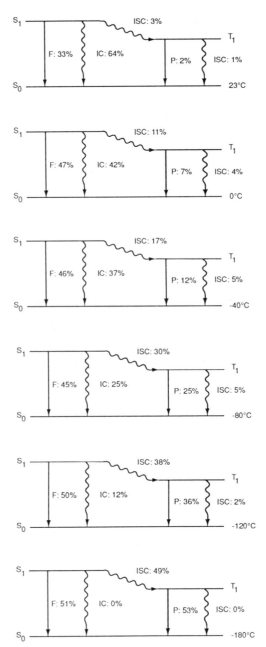

Figure 7.9. Energy diagrams for 4-phenylphenol adsorbed on filter paper at different temperatures. (S_0) ground state; (S_1) singlet state; (T_1) triplet state; (F) fluorescence; (IC) internal conversion; (ISC) intersystem crossing; (P) phosphorescence. (Reprinted with permission from S.M. Ramasamy and R.J. Hurtubise, *Talanta*, 1989, *36*, 315.)

energy. The results in Figure 7.9 support the concept that a rigid analyte-substrate structure is responsible for the RTP from 4-phenylphenol. This was mainly deduced from the large decrease in internal conversion with lower temperature and the increase in intersystem crossing with lower temperature.

The effects of temperature on the solid-surface luminescence properties of the protonated form of benzo(f)quinoline $(B(f)QH^+)$ adsorbed on filter paper were also investigated (68). Similar data and calculated rate constants were obtained for $B(f)QH^+$ as were obtained for 4-phenylphenol. There was a 9.2-fold increase in ϕ_p from 23°C to -180°C for $B(f)QH^+$ adsorbed on filter paper. The increase in ϕ_p indicated that one factor which contributed to the decrease in ϕ_p at room temperature was the rigidity with which the phosphor was held. Ramasamy and Hurtubise (68) showed that graphs of ln ϕ_p versus 1/T for $B(f)QH^+$ and 4-phenylphenol adsorbed on filter paper were similar to a graph of ln(Young's modulus) versus 1/T. Figure 7.10 shows the plots of lnϕ_p versus 1/T for $B(f)QH^+$ and 4-phenylphenol. The tangent lines in Figure 7.10 intersect at -23°C and -46°C for 4-phenylphenol and $B(f)QH^+$, respectively. Most likely the intersection points approximate the temperature at which a phase transition occurs in filter paper. In

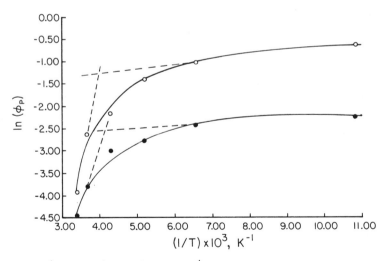

Figure 7.10. Plots of ln ϕ_p vs. 1/T for $B(f)Q^+$ (•) and 4-phenylphenol (○) adsorbed on filter paper. (Data for 4-phenylphenol taken from S.M. Ramasamy and R.J. Hurtubise *Talanta*, 1989, 36, 315. Data for B(f)Q taken from S.M. Ramasamy and R.J. Hurtubise. *Appl. Spectrosc.* 1989, 43, 616. Reprinted with permission.)

polymer research, for example, Rutherford and Soutar (69) and Somersall et al. (70) obtained similar graphs of phosphorescence intensity of various functional groups in polymers versus $1/T$ and related the breaks in the curves to phase transitions occurring in the polymers. Ramasamy and Hurtubise (68) replotted data published by Haughton and Sellen (71) on stress relaxation studied for regenerated cellulose. The stress measurements by Haughton and Sellen were made in the temperature range of 20°C to -185°C. By replotting their data in terms of Young's modulus vs. $1/T$, a graph very similar to the ones in Figure 7.10 was obtained. This correlation suggested that the phosphorescence quantum yield of compounds adsorbed on filter paper was related to Young's modulus. Young's modulus is the ratio of stress to strain as a material is deformed under dynamic load, and it is a measure of the softness or stiffness of a material (72). Ramasamy and Hurtubise (68) made the conclusion that the rigidity of the matrix through the hydrogen-bonding network of the cellulose is an important factor in determining a high ϕ_p for a phosphor adsorbed on filter paper. In reference to Figure 7.10 and the discussion related to modulus, as the temperature increased for the filter paper, the number of hydrogen bonds in the cellulose network decreased, and thus, the matrix became less rigid (73). The results obtained by Ramasamy and Hurtubise (68) indicated that if Young's modulus for filter paper could be maximized at room temperature, then the maximum phosphorescence yield could be obtained for the phosphors at room temperature.

Ramasamy and Hurtubise (68) calculated the triplet formation efficiencies (ϕ_t) for B(f)QH$^+$ adsorbed on filter paper as a function of temperature. It was found that ϕ_t increased significantly as temperature decreased for B(f)QH$^+$. Also, the ϕ_t values for 4-phenylphenol adsorbed on filter paper showed the same general trend with temperature as did B(f)QH$^+$ adsorbed on filter paper. They also constructed energy diagrams for B(f)QH$^+$ as a function of temperature similar to the one in Figure 7.9 for 4-phenylphenol. The diagrams revealed that internal conversion went to 0% at -180°C. However, the percentage of intersystem crossing from the triplet to the ground state increased as the temperature was lowered. This was in contrast to the percentage of intersystem crossing from the triplet to the ground state as temperature was lowered for 4-phenylphenol. The percentage of intersystem crossing from the triplet state for 4-phenylphenol was zero percent at -180°C (Figure 7.9). These different effects for the two phosphors can be partly explained by Equation (7.3). This equation shows that the quantum yield of nonradiative triplet to ground-state transitions (ϕ_{ts}) is a function of the product of three parameters, namely, ϕ_t, k_m. amd τ_p. The three previous terms change with temperature for $B(f)QH^+$ and 4-phenylphenol adsorbed on filter paper. Thus, ϕ_{ts} would be dependent on the relative magnitude of the three terms, ϕ_t, k_m, and τ_p as temperature is varied. Another important conclusion reached by Ramasamy and Hurtubise (68) by comparing k_m and k_p values was that the environmental conditions were much more important for 4-phenylphenol than for $B(f)QH^+$. This is reasonable given the flexible nature of the structure of 4-phenylphenol compared to the rigid structure of $B(f)QH^+$.

In Section 7.1, the calculation of preexponential factors (k_1) and activation energy terms (E_a) was discussed. Table 7.12 compares the E_a and k_1 values for several compounds adsorbed on filter paper and on other surfaces. There is an approximate correlation between the magnitude of the k_1 and E_a values and how strongly a phosphor is bonded to the surface. This has been discussed by Plauschinat et al. (11) and Oelkrug et al. (12). The exact meanings of k_1 and E_a are not known (5). Table 7.12 does

Table 7.12. Preexponential Factors (k_1) and Activation Energy Terms (E_a) for Phosphors Adsorbed on Various Surfaces[a]

Phosphor	Surface	k_1, s^{-1}	E_a, cm^{-1}	Reference
$B(f)QH^+$	Filter paper	3.0	428	68
4-Phenylphenol	Filter paper	2.0	371	67
p-Aminobenzoic acid[b]	Sodium acetate	2.1	392	7
1-Azaphenanthrene	Alumina[c]	1.1	215	11
Naphthalene	Alumina[c]	15	800	12
Phenanthrene	Alumina[d]	2×10^4	1800	11

[a] E_a and k_1 values obtained from plots of $\ln(\tau_p^{-1} - \tau_{p0}^{-1})$ vs. $1/T$.
[b] Anion of p-aminobenzoic acid adsorbed on sodium acetate. E_a and k_1 are average values for short and long decaying components.
[c] Alumina activated at 600°C.
[d] Alumina activated at 100°C. Sample was physisorbed.
Reprinted with permission from S.M. Ramasamy and R.J. Hurtubise, *Appl. Spectrosc.* 1989, <u>43</u>, 616.

show that for physisorbed phenanthrene, k_1 and E_a are relatively large. Its k_1 value is about four orders of magnitude larger than the k_1 values for the other compounds in the table. Thus, k_1 may be a more sensitive indicator than E_a for indicating how strongly a phosphor is adsorbed on a solid matrix.

In summary, several of the factors that enhance or diminish the RTP of adsorbed compounds on filter paper have been elucidated. However, the systems investigated are rather complex and additional research is needed to give a detailed interaction model. If moisture is present on the filter paper, it would "soften" the filter paper matrix and allow oxygen to more readily penetrate the matrix and quench a portion of the phosphor molecules. In addition, adsorbed water would weaken the hydrogen-bonding network in filter paper, which would in turn lower the modulus of filter paper and cause an effective quenching of some of the phosphor molecules. Also, the hydrogen bonding of the phosphor molecules to the hydroxyl groups in the filter is very important for most phosphors. The solvent employed to adsorb the phosphor on filter paper is also important. Polar solvents can swell the filter paper, which would favor entry of phosphor molecules into the submicroscopic pores in the filter paper. Also, salts and other materials like sugars can pack the filter paper matrix, which would favor an increase in the overall modulus of the system. Heavy atom salts induce an increase in intersystem crossing, and this increases the probability of RTP. The structure of the phosphor is also important. Generally, phosphors with rigid structures and polar functional groups give relatively strong RTP, whereas phosphors with polar functional groups but more flexible structures do not give as high an RTP yield. Internal conversion of the luminescent molecule from the singlet state to the ground state and intersystem crossing of the molecule from the singlet state to the triplet state are very important factors in determining the RTP quantum yield. Also, the rate constant for nonradiative transition from the triplet state to the ground state is a very good indicator of the effect of environmental conditions on RTP. In short, how rigidly the phosphor molecules are held in the filter paper matrix, the rigidity of the filter paper matrix, and how effectively the filter paper matrix and other substances added to the filter paper protect the phosphor from various quenchers, such as oxygen and moisture, are the major considerations in yielding high RTP yields for phosphors adsorbed on filter paper.

7.6.   Properties, Interactions, and Experimental Conditions of Solid Surfaces

The previous sections in this chapter summarized several of the results obtained for a variety of solid matrices that have been used to obtain RTP from adsorbed organic compounds. Because of the complexity of the interactions involved for the RTP of adsorbed phosphors, one key interaction or one simple interaction model has not emerged. However, based on earlier research, several important experimental conditions, properties of solid surfaces, and interactions of the phosphors with the solid surfaces have been elucidated. In this section, a general discussion of these aspects will be considered for the solid surfaces, based on the research results presented in the earlier sections of this chapter. Below are itemized several properties, interactions, and experimental conditions that are important for obtaining strong RTP from phosphors adsorbed on solid surfaces. In Table 7.13, a listing of the solid surfaces is given with appropriate citations and assessment of the properties, interactions, and experimental conditions itemized below. It should be realized that a quantitative assessment of all of the items listed below cannot be given at this point in the development of an interaction model for RTP.

(a)   Solution chemistry: This involves any solution interactions and chemistry that occur in the sample preparation step.

(b)   Solution solid-surface interactions: This involves any physical or chemical interactions that occur between the solution used in the sample preparation step and the solid surface.

(c)   Evaporation of the solvent and of moisture: After the sample is deposited on the surface, the surface is dried. This serves to remove the solvent and moisture from the solid surface. In some experiments, the solid surface is dried prior to adding the solution to the surface.

(d)   An inert atmosphere of a dry gas: This condition gives a high RTP signal, but it is not needed in all experimental situations to give an analytically useful RTP signal.

(e)   Hydrogen bonding and/or ionic interactions: This refers to hydrogen bonding and/or ionic interactions in the solid state between the phosphor and solid-matrix.

(f)   Matrix packing: This involves the addition of salts or other materials such as glucose to the solid material. This technique protects the phosphor from atmospheric constituents and makes the solid matrix more rigid.

(g) Matrix isolation: This phenomenon has been proposed for filter paper. After certain polar solvents are deposited on filter paper, the filter paper swells. The swelling of filter paper would favor admission of the phosphor into the complex fibrous network of the filter paper. After the solid matrix is dried, the phosphor molecules become trapped between cellulose chains.

(h) Crystal packing: This is the situation whereby a crystalline solid matrix can form a crystalline network that yields a rigid solid matrix.

(i) Inclusion complex formation: Cyclodextrins can include guest molecules. In the solid state, this results in augmented RTP.

(j) Increased modulus: Modulus is a measure of the softness or stiffness of a material. Generally an increase in the modulus of a material favors enhanced RTP. Any of the items listed above that would increase the rigidity of the solid matrix would increase the modulus of the solid matrix.

The information in Table 7.13 illustrates again that several factors are important in obtaining high phosphorescence yields from adsorbed compounds. Also, the importance of the items listed in Table 7.13 is not the same for each solid matrix. Items (c) and (e) are generally the most important in attaining a high phosphorescence yield, although item (d) is also very important.

Regardless of all the interactions involved with a phosphor and a solid surface, one frequently wants to know what solid surface should be used for a trace organic chemical analysis. Chapter 8 in this book gives several examples of the use of different solid surfaces in trace chemical analysis. The monograph by Vo-Dinh (74) should be consulted for a detailed comparison of the use of sodium acetate, silica gel chromatoplates, polyacrylic acid-salt mixtures, and filter paper in analysis by solid-surface techniques. By far, filter paper is the most widely used solid surface in organic analysis; however, cyclodextrin-salt mixtures have properties that give well defined spectra and high fluorescence and phosphorescence quantum yields from adsorbed compounds (45-48,50,52,53).

Table 7.13. Comparison of Solid Matrices for Phosphorescence[a]

Sodium Acetate

(a) Solution interactions and / or chemistry	very important
(b) Solution solid-surface interactions	important
(c) Evaporation of the solvent and moisture	extremely important
(d) An inert atmosphere of dry gas	very important
(e) Hydrogen bonding and / or ionic interactions	extremely important
(f) Matrix packing	very important
(h) Crystal packing	extremely important
(j) Increased modulus	important

Silica Gel Chromatoplate with a Polyacrylate Binder

(a) Solution interactions and / or chemistry	very important
(b) Solution solid-surface interactions	important
(c) Evaporation of the solvent and moisture	extremely important
(d) An inert atmosphere of dry gas	very important
(e) Hydrogen bonding and / or ionic interactions	extremely important
(f) Matrix packing	important
(j) Increased modulus	important

Polyacrylic Acid-Salt Mixtures

(a) Solution interactions and / or chemistry	very important
(b) solution solid-surface interactions	important
(c) Evaporation of the solvent and moisture	extremely important
(d) An inert atmosphere of dry gas	very important
(e) Hydrogen bonding and / or ionic interactions	extremely important
(f) Matrix packing	extremely important
(j) Increased modulus	very important

Cyclodextrin-Salt Mixtures

(a) Solution interactions and / or chemistry	very important
(b) Solution solid-surface interactions	important
(c) Evaporation of the solvent and moisture	extremely important
(d) An inert atmospher of dry gas	extremely important
(e) Hydrogen bonding and / or ionic interactions	very important

Table 7.13. continued

(f) Matrix packing	extremely important
(h) Crystal packing	extremely important
(i) Inclusion complex formation	extremely important
(j) Increased modulus	important

Filter Paper

(a) Solution interactions and / or chemistry	very important
(b) Solution solid-surface interactions	important
(c) Evaporation of the solvent and moisture	extremely important
(d) An inert atmosphere of dry gas	very important
(e) Hydrogen bonding and / or ionic interactions	extremely important
(f) Matrix packing	very important
(g) Matrix isolation	very important
(j) Increased modulus	very important

[a] See text for the discussion of (a) - (j).

References

1. von Wandruszka, R.M.A.; Hurtubise, R.J. *Anal. Chem.* 1977, 49, 2164.
2. von Wandruszka, R.M.A.; Hurtubise, R.J. *Anal. Chem.* 1976, 48, 1784.
3. de Lima, C.G.; de M. Nicola, E.M. *Anal. Chem.* 1978, 50, 1658.
4. Snyder, L.R. *Principles of Adsorption Chromatography*; Marcel Dekker: New York, 1968; p 199.
5. Honnen, W.; Krablichier, G.; Uhl, S.; Oelkrug, D. *J. Phys. Chem.* 1983, 87, 4872.
6. Senthilnathan, V.P.; Hurtubise, R.J. *Anal. Chem.* 1985, 57, 1227.
7. Ramasamy, S.M.; Hurtubise, R.J. *Anal.Chem.* 1987, 59, 432.
8. Turro, N.J. *Modern Molecular Photochemistry*; Benjamin/-Cummings: Menlo Park, CA, 1978; pp 176-183.
9. Parker, C.A. *Photoluminescence of Solutions*; Elsevier: New York, 1968; pp 88-89, pp 304-305.
10. Kessler, R.W.; Uhl, S.; Honnen, W.; Oelkrug, D. *J. Lumin.*

1981, <u>24/25</u>, 551.

11. Plauschinat, M.; Honnen, W.; Krablichier, G.; Uhl, S.; Oelkrug, D. *J. Mol. Struct.* 1984, <u>115</u>, 351.

12. Oelkrug, D.; Plauschinat, M.; Kessler, R.W. *J. Lumin.* 1979, <u>18/19</u>, 434.

13. Ramasamy, S.M.; Hurtubise, R.J. *Anal. Chem.* 1987, <u>59</u>, 2144.

14. Ford, C.D.; Hurtubise, R.J. *Anal. Chem.* 1980, <u>52</u>, 656.

15. Ford, C.D.; Hurtubise, R.J. *Anal. Chem.* 1979, <u>51</u>, 659.

16. Ford, C.D.; Hurtubise, R.J. *Anal. Chem.* 1978, <u>50</u>, 610.

17. Hurtubise, R.J. *Talanta* 1981, <u>28</u>, 145.

18. Ford, C.D.; Hurtubise, R.J. *Anal. Lett.* 1980, <u>13(A6)</u>, 485.

19. Hurtubise, R.J.; Smith, C.A. *Anal. Chim. Acta* 1982, <u>139</u>, 315.

20. Ramasamy, S.M.; Hurtubise, R.J. *Anal. Chim. Acta* 1983, <u>152</u>, 83.

21. Peri, J.P. *J. Phys. Chem.* 1966, <u>70</u>, 2937.

22. Silverstein, R.M.; Bassler, G.C.; Morrill, T.C. *Spectrometric Identification of Organic Compounds*, 3rd ed.; Wiley: New York, 1974; pp 99-102.

23. Bruckner, K.; Halpaap, H.; Rossler, H. U.S. Patent No. 3502217.

24. Deanin, R.D. *Polymer Structure, Properties, and Applications*; Cahners Books: Boston, MA, 1972; pp 384-392.

25. Nielsen, L.E. *Mechanical Properties of Polymers and Composites*; Marcel Dekker: New York, 1974; p 39.

26. Burrell, G.J.; Hurtubise, R.J. *Anal. Chem.* 1987, <u>59</u>, 965.

27. Niday, G.L.; Seybold, P.G. *Anal. Chem.* 1978, <u>50</u>, 1577.

28. Burrell, G.J.; Hurtubise, R.J. *Anal. Chem.* 1988, <u>60</u>, 564.

29. Grabowska, A.; Pakula, B.; Pancir, J. *Photochem. Photobiol.* 1969, <u>10</u>, 415.

30. Favaro, G.; Masetti, F.; Mazzucato, U. *Spectrochim. Acta, Part A* 1971, <u>27A</u>, 915.

31. Scott, R.P.W.; Kucera, P. *J. Chromatogr.* 1978, <u>149</u>, 93.

32. Dalterio, R.A.; Hurtubise, R.J. *Anal.Chem.* 1982, <u>54</u>, 224.

33. Dalterio, R.A.; Hurtubise, R.J. *Anal.Chem.* 1983, <u>55</u>, 1084.

34. Dalterio, R.A.; Hurtubise, R.J. *Anal.Chem.* 1984, <u>56</u>, 336.

35. Ramasamy, S.M.; Hurtubise, R.J. *Anal. Chem.* 1982, <u>54</u>, 2477.

36. Allerhand, A.; von P. Schleyer, P. *J. Am. Chem. Soc.* 1963, <u>85</u>, 1233.

37. Hurtubise, R.J. In *Molecular Luminescence Spectroscopy: Methods and Applications-Part II*; Schulman, S.J., Ed.;

Wiley: New York, 1988; Chapter 1.

38. Parker, R.T.; Freedlander, R.S.; Dunlap, R.B. *Anal. Chim. Acta* 1980, <u>119</u>, 189.

39. Parker, R.T.; Freedlander, R.S.; Dunlap, R.B. *Anal. Chim. Acta* 1980, <u>120</u>, 1.

40. Schulman, E.M.; Parker, R.T. *J. Phys. Chem.* 1977, <u>81</u>, 1932.

41. Schulman, S.G. In *Modern Fluorescence Spectroscopy*; Wehry, E.L., Ed.; Plenum Press: New York, 1976; Vol. 2, pp 245-246.

42. Ito, M. *J. Mol. Spectrosc.* 1960, <u>4</u>, 125.

43. Nemethy, G.; Ray, A. *J. Phys. Chem.* 1973, <u>77</u>, 64.

44. Hall, A.; Wood, J.L. *Spectrochim. Acta*, Part A 1967, <u>23A</u>, 2657.

45. Bello, J.M.; Hurtubise, R.J. *Appl. Spectrosc.* 1986, <u>40</u>, 790.

46. Bello, J.M.; Hurtubise, R.J. *Anal. Lett.* 1986, <u>19</u>, 775.

47. Richmond, M.D.; Hurtubise, R.J. *Appl. Spectrosc.* 1989, <u>43</u>, 810.

48. Richmond, M.D.; Hurtubise, R.J. *Anal. Chem.* 1989, <u>61</u>, 2643.

49. Szejtli, J. *Cyclodextrins and Their Inclusion Complexes*; Akademia Kiado: Budapest, 1982.

50. Bello, J.M.; Hurtubise, R.J. *Anal. Chem.* 1987, <u>59</u>, 2395.

51. Saenger, W. *Isr. J. Chem.* 1985, <u>25</u>, 43.

52. Bello, J.M.; Hurtubise, R.J. *Appl. Spectrosc.* 1988, <u>42</u>, 619.

53. Bello, J.M.; Hurtubise, R.J. *Anal. Chem.* 1988, <u>60</u>, 1291.

54. Schulman, E.M.; Walling, C. *Science* 1972, <u>178</u>, 53.

55. Schulman, E.M.; Walling, C. *J. Phys. Chem.* 1973, <u>77</u>, 902.

56. Wellons, S.L.; Paynter, R.A.; Winefordner, J.D. *Spectrochim. Acta*, Part A, 1974, <u>30</u>, 2133.

57. Lue-Yen Bower, E.; Winefordner, J.D. *Anal. Chim. Acta* 1978, <u>102</u>, 1.

58. Aaron, J.J.; Andino, M.; Winefordner, J.D. *Anal. Chim. Acta* 1984, <u>160</u>, 171.

59. Andino, M.M.; Kosinski, M.A.; Winefordner, J.D. *Anal. Chem.* 1986, <u>58</u>, 1730.

60. Su, S.Y.; Winefordner, J.D. *Microchem. J.* 1987, <u>36</u>, 118.

61. White, W.; Seybold, P.G. *J. Phys. Chem.* 1977, <u>81</u>, 2035.

62. Meyers, M.L.; Seybold, P.G. *Anal. Chem.* 1979, <u>51</u>, 1609.

63. McAleese, D.L.; Dunlap, R.B. *Anal. Chem.* 1984, <u>56</u>, 2244.

64. Vo-Dinh, T.; Suter, G.W.; Kallir, A.J.; Wild, U.P. *J. Phys. Chem.* 1985, <u>89</u>, 3026.

65. Suter, G.W.; Kallir, A.J.; Wild, U.P.; Vo-Dinh, T. *J. Phys. Chem.* 1986, <u>90</u>, 4941.

66. Suter, G.W.; Kallir, A.J.; Wild, U.P.; Vo-Dinh, T. *Anal. Chem.* 1987, 59, 1644.
67. Ramasamy, S.M.; Hurtubise, R.J. *Talanta* 1989, 36, 315.
68. Ramasamy, S.M.; Hurtubise, R.J. *Appl. Spectrosc.* 1989, 43, 616.
69. Rutherford, J.; Soutar, I. *J. Polym. Sci. Polym. Phys. Ed.* 1980, 18, 1021.
70. Somersall, A.C.; Dan, E.; Guillet, J.E. *Macromolecules* 1974, 7, 233.
71. Houghton, P.M.; Sellen, D.B. *J. Phys. D: Appl. Phys.* 1973, 6, 1998.
72. Rodriguez, F. *Principles of Polymer Systems*; McGraw-Hill: New York, 1970; Chapter 8.
73. Nissan, A.H. *Macromolecules* 1976, 9, 840.
74. Vo-Dinh, T. *Room-Temperature Phosphorimetry for Chemical Analysis*; Wiley: New York, 1984.

CHAPTER 8

APPLICATIONS IN SOLID-SURFACE ROOM-TEMPERATURE PHOSPHORESCENCE

8.1. Introduction

In Chapter 6 the experimental conditions, types of luminescence data, instrumentation, and in Chapter 7 interactions involved with solid-surface room-temperature phosphorescence (RTP) were discussed. In this chapter, some of the applications of RTP will be presented. Parker et al. (1,2) reviewed several of the RTP applications through 1979. Also, Hurtubise (3) has considered RTP applications through 1979. He also presented selected examples of the use of RTP through 1984 (4). Vo-Dinh (5) discussed, in detail, RTP applications through 1982 with some examples from 1983. Because of these earlier discussions of the applications of RTP, the treatment in this chapter will consider mainly applications from 1983 and beyond.

As a result of the investigations of the interactions of phosphors with solid surfaces, four general aspects have emerged that should be considered in experiments with solid-surface RTP. First, before adsorbing the sample solution onto the surface, the initial solution chemistry could be important in yielding strong phosphorescence. For example, a phosphor that is a neutral molecule, a cation, or an anion in solution prior to depositing the sample on the surface can determine, in some situations, if strong phosphorescence signals will be obtained from the phosphor in the adsorbed state. Second, after the sample is adsorbed, the wet-surface chemistry could yield important conditions that give strong phosphorescence. As an example, the solid matrix should be at least partially soluble in the adsorbing solvents so any appropriate wet chemistry can occur. Third, the conditions used to dry the surface such as the length and temperature of the drying period are important. Fourth, the properties of the final dried matrix are very important in providing suitable conditions for strong RTP signals. All four aspects could be important in contributing to the phosphorescence of an adsorbed phosphor. However, the properties of the final dried matrix are usually the most important consideration.

8.2. Pesticides, Polychlorinated Dibenzofurans, and Polychlorinated Dibenzo-p-Dioxins

Pesticides and closely related materials are of important environmental concern, and numerous analytical methods have been developed for these substances. Recently, solid-surface RTP has been applied for the detection and determination of several of these compounds. For example, Aaron et al. (6) reported a comparative study of low-temperature and RTP characteristics of 32 pesticides. Low-temperature phosphorimetry was shown to be sensitive with limits of detection ranging between 0.001 and 30 µg/mL. With RTP, the methodology was simple and specific for some of the pesticides with absolute limits of detection between 10 and 50 ng. The researchers concluded that the RTP technique was more suitable than conventional low-temperature phosphorimetry if one is interested in determining a specific phosphorescent pesticide in the presence of other weakly phosphorescent pesticides. Aaron and Winefordner (7) used sodium iodide as an external heavy-atom perturber to induce RTP from several aromatic pesticides. Except for organochlorinated compounds, sodium iodide increased substantially the RTP of the pesticides. The limits of detection of the naphthalene-like pesticides were in the nanogram range. The authors emphasized that the selectivity of the RTP technique permitted one to determine nanogram amounts of naphthalene-like pesticides in the presence of relatively large amounts of organochlorinated pesticides.

Vannelli and Schulman (8) investigated eighteen pesticides by RTP. They tested the RTP response of the pesticides on several samples of Whatman No. 42 filter paper treated with sodium acetate, sodium iodide, lead acetate, thallium fluoride, a combination of sodium acetate/sodium chloride, and sodium hydroxide/sodium iodide. Thirteen of the pesticides responded well enough to obtain calibration curves. The pesticides were diphenylamine, carbaryl, morestan, mobann, coumaphos, coroxon, asulam, 2-aminobenzimidazole, benomyl, alanap, alanap sodium salt, warfarin, and devrinol.   The limits of quantification (RTP response, which was 2 times the blank response) ranged from 0.7 ng for diphenylamine to 10 ng for devrinol.

Su et al. (9) reported the RTP limits of detection, linear dynamic range, and heavy-atom enhancement factors for six pesticides, namely, phenothiazine, 1-naphthol, warfarin, asulam, carbaryl, and naphthaleneacetic acid.   Iodide was used for most of the experiments as a heavy-atom enhancer, and the limits of detection ranged from 0.8 to 3.0 ng.   Various synthetic mixtures of the pesticides were analyzed without prior separation by changing the substrate and heavy atom.   For example, the increase in the RTP of naphthaleneacetic acid on ion-exchange filter paper (DE-81) with Pb^{2+} as the heavy atom was used to determine this pesticide in a six component mixture.

Asafu-Adjaye and Su (10) reported the RTP of ten phosphorescent pesticides and toxic substances.   They were primarily interested in mixture analysis and combined the use of solid-surface room-temperature fluorescence and RTP.   Parameters such as pH of the sample environment, solid substrate, and heavy atoms were used to selectively determine the components in the mixtures.   The compounds they investigated were asulam, 3,4-dimethoxybenzaldehyde, warfarin, phenothiazine, 1-naphthol, naphthalene acetic acid, 1-naphthoic acid, 2-naphthoic acid, naphthaleneacetamide, and carbaryl.   With rather detailed luminescence information on the conditions for maximum luminescence and the conditions for diminished luminescence, they worked out strategies to determine compounds in synthetic mixtures.   Table 8.1 shows the results obtained for an eight component mixture.   Their results show that by combining solid-surface fluorescence and RTP the components in very complex mixtures of compounds could be determined at the nanogram level.

The effects of heavy-atom-containing surfactants on the RTP of carbaryl were reported by de Lima et al. (11).   They also carried out qualitative and semiquantitative analyses of the surface of filter papers treated with heavy-atom inorganic salts ($TlNO_3$ and $AgNO_3$) and the surfactant salts (thallium dodecyl sulfate and silver dodecyl sulfate) by X-ray photoelectron spectroscopy analysis. X-ray photoelectron spectroscopy gave them information on the relative amounts of heavy atoms and luminescence compounds present on the surface of treated filter paper and the extent of penetration of the compounds into the bulk of the filter paper.   For

Table 8.1. Determination of Eight Compounds in Mixture B

Compounds	Amounts present, ng		% Relative error
	Actual	Exptl $\pm$ std dev[a]	
Asulam	50.0	46.4 $\pm$ 6.4	-7.2
Warfarin	49.1	47.2 $\pm$ 4.0	-3.9
3,4-Dimethoxybenzaldehyde	49.3	54.2 $\pm$ 4.9	+9.9
Phenothiazine	51.1	52.7 $\pm$ 2.6	+3.1
Naphthaleneacetamide	48.5	53.8 $\pm$ 1.0	+10.9
Naphthaleneacetic acid	50.5	49.6 $\pm$ 4.6	-1.8
1-Naphthoic acid	50.9	52.2 $\pm$ 6.0	+8.4
2-Naphthoic acid	50.5	52.7 $\pm$ 0.9	+4.4

[a]Amounts present and standard deviation (std dev) calculated from calibration curves and represent average of three data points from three different sample aliquots.

Reprinted with permission from E.B. Asafu-Adjaye and S.Y. Su, *Anal. Chem.*, 58, 539. Copyright 1986 American Chemical Society.

RTP, they reported that the net phosphorescence signals in the presence of the surfactant salts, spotted from 0.02 M solutions, were higher than in the presence of the corresponding nitrate salts by factors of 7.0 and 9.5 for silver and thallium, respectively. The X-ray photoelectron spectroscopy data showed that the dodecyl sulfate salts were better retained on the surface of the paper than their corresponding nitrate salts. This result correlated well with the 7.0-fold and 9.5-fold increase in RTP with the surfactants present on the filter paper compared to the RTP enhancement with $AgNO_3$ and $TlNO_3$ adsorbed on the filter paper.

In later work, Campiglia and de Lima (12) reported the RTP of carbaryl on low-background filter paper. They reduced the background of filter paper by water extraction and then, exposure of the filter paper to ultraviolet radiation. They also investigated the effects of heavy atoms and the drying temperature on the RTP carbaryl signal. The results from the heavy-atom salts, thallium acetate, lead acetate, and sodium iodide showed that sodium iodide yielded the best overall RTP enhancement. A limit of detection of 110 pg was reported using sodium iodide as a heavy-atom enhancer. They also showed that the drying temperature was important for maximum RTP because with some drying temperatures the RTP signal was diminished. In addition, spectral interference with 1-naphthol present was reported in some instances, and a previous separation of 1-naphthol from carbaryl was recommended.

Khasawneh and Winefordner (13) reported the solution room-temperature fluorescence and solution low-temperature phosphorescence and the solid-surface RTP properties of dibenzofuran, several polychlorinated dibenzofurans, and several polychlorinated dibenzo-p-dioxins. The solution fluorescence limits of detection were in the range of 0.020 to 5.0 ng/mL for dibenzofuran, polychlorinated dibenzofurans, and polychlorinated dibenzo-p-dioxins. For low-temperature phosphorescence, the limits of detection for the same set of compounds were found to be between 0.25 and 3.5 ng/mL, whereas with RTP, the limits of detection were between 0.7 ng and 10.5 ng for dibenzofuran and several polychlorinated dibenzofurans. The authors concluded that luminescence spectrometry was a sensitive and selective technique for the determination of polychlorinated dibenzofurans and dibenzo-p-dioxins. Room-temperature fluorescence was more sensitive than low-temperature phosphorescence and RTP, but the phosphorimetric methods were more selective. The polychlorinated dibenzo-p-dioxins could not be determined by RTP under the experimental conditions they used.

8.3. Polycyclic Aromatic Hydrocarbons and Nitrogen Heterocycles

Polycyclic aromatic hydrocarbons (PAH) and nitrogen heterocycles are of considerable environmental concern. Several of these compounds have been implicated in the initiation of certain forms of cancer (14). Thus, it is important to continue to develop sensitive and selective analytical methods for the characterization and determination of these important classes of compounds. Several solid-surface RTP methods have been developed for these compounds. Vo-Dinh (5) has considered earlier analytical RTP methods such as the analysis of PAH in shale oil and coal liquids. In the next two sections, a general review of the RTP methodology and RTP analytical data for PAH and nitrogen heterocycles since the monograph by Vo-Dinh will be presented.

8.3.1. Polycyclic Aromatic Hydrocarbons

Su and Winefordner (15) evaluated various types of filter paper and Whatman DE-81 anion exchange filter paper as solid substrates for RTP. They also studied the heavy-atom effect, and several PAH were used as model compounds. The heavy-atom effect studies showed that every analyte should possess an optimal combination of substrate material, heavy-atom species and concentration, and sample drying procedure. The PAH investigated were biphenyl, phenanthrene, chrysene, and coronene. Limits of detection were in the range of 0.08 ng to 9.0 ng.

Vo-Dinh (16) gave an overview of the principles and applications of synchronous luminescence and RTP for monitoring in air pollution. He emphasized that RTP and synchronous luminescence can be used for screening large numbers of unfractionated air samples to obtain preliminary information on PAH and nitrogen heterocycles. Also, the two luminescence techniques can be used for detailed identification and quantitation of certain PAH and nitrogen heterocycles that have been separated prior to analysis. Many PAH are weakly phosphorescent. However, in solid-surface RTP, pretreatment of the sample or paper substrate with a heavy atom enhances the RTP of the adsorbed PAH and thus improves RTP detection by increasing phosphorescence

Table 8.2. Limits of Detection (LOD) for Several PNA Compounds by Room-Temperature Phosphorescence

Compound	λex^a (nm)	λem^b (nm)	LOD (ng)
Homocyclics	395	698	0.07
Benzo(a)pyrene	335	543	0.001
Benzo(e)pyrene	343	505	0.025
2,3-Benzofluorene	398	626	0.6
Chrysene	330	518	0.03
1,2,3,4-Dibenzanthracene	295	567	0.08
1,2,5,6-Dibenzanthracene	305	555	0.005
Fluoranthene	365	545	0.05
Fluorene	270	428	0.2
Phenanthrene	295	474	0.007
Pyrene	343	595	0.1

[a] λ_{ex} = excitation wavelength.
[b] λ_{em} = emission wavelength

Reprinted with permission from T. Vo-Dinh, in *Identification and Analysis of Organic Pollutants in Air*, L.H. Keith, Ed., Butterworth Publishers, Boston, 1984, Chapter 16.

intensity. Table 8.2 gives the RTP limits of detection for several PAH.

Vo-Dinh et al. (17) considered the field evaluation of a cost-effective screening procedure for PAH pollutants in ambient air samples. Both synchronous luminescence and RTP were used to estimate the amount of PAH in air particulate extracts that were collected at two wood-burning communities. They obtained good agreement between the screening data and the results obtained by detailed gas chromatography/mass spectrometry and high-performance liquid chromatography. The field results showed that the screening process could readily discriminate between samples whose PAH content differed by considerably less than one order of magnitude. A mixture of ten PAH was used as a reference mixture to investigate the performance of both synchronous luminescence and RTP. The ten PAH were anthracene, benzo[a]anthracene, benzo[a]pyrene, benzo[e]pyrene, chrysene, fluoranthene, phenanthrene, perylene, pyrene, and triphenylene. They concluded that the combination of synchronous luminescence and RTP provided a good estimate of the content of the major PAH species in air samples.

Vo-Dinh (18) developed a personnel dosimeter, which was based on molecular diffusion and detection of polyaromatic pollutants by RTP. The dosimeter was a simple, pen-size device that required no sample extraction for analysis. Figure 8.1 gives a schematic diagram of the dosimeter. The dosimeter was designed to detect vapor concentrations of various multi-ring polyaromatic pollutants at the part-per-billion level. After exposure, the dosimeter was inserted into a luminescence spectrometer for direct determination of the integrated vapor concentration of the pollutants. Detailed studies were performed confirming that the sampling rate of the dosimeter was diffusion controlled. Also, studies were performed on the sorption properties and stability of the dosimeter response, and on the effect of air movement on the performance of the dosimeter.

Both RTP and fluorescence (19) and RTP alone (20) have been used to evaluate the permeation of multi-ring polyaromatic compounds from petroleum products, coal products, and

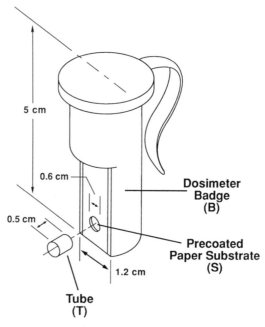

Figure 8.1. Schematic diagram of the PNA dosimeter design. (Reprinted with permission from T. Vo-Dinh, *Environ. Sci. Technol.*, 19, 997. Copyright 1985 American Chemical Society.)

combustion products through protective clothing materials. The protective clothing investigated were rubber gloves made out of various materials. The device used to study the permeation of polyaromatic compounds was a glass vial turned upside down, which contained the petroleum product, coal product or combustion product. A fine mesh wire screen was placed at the opening of the upside down vial. Then a piece of protective clothing, filter paper disc, and wax holding paper were positioned sequentially on top of the fine-mesh wire screen. This sandwich arrangement was mounted in a clamp, which provided a seal between the glove material and the vial. The filter paper acted as a sorbent medium for oil components permeating through the glove material. The filter paper was analyzed directly by RTP for the presence of polyaromatic hydrocarbons. No chemical extraction was necessary for analysis. Measurements were performed after exposure times of 4, 8, 16, and 24 h or until breakthrough was detected. A variety of multi-ring polyaromatic compounds was detected by RTP at the nanogram level. The RTP approach was very simple, cost effective, and required no elaborate equipment.

Vo-Dinh and Uziel (21) reported the RTP spectrum of benzo[a]pyrene-7,t-8,9,10-tetrahydrotetrol (BP-tetrol), which was excited by laser excitation. The BP-tetrol was obtained by acid hydrolysis of the r-7,t-8-dihydroxy-t-9,10-epoxy-7,8,9,10-tetra-hydrobenzo[a]pyrene (BPDE)-DNA adducts. BPDE is considered the ultimate carcinogenic metabolite of benzo[a]pyrene. The BP-tetrol sample was adsorbed on filter paper which was pretreated with a heavy-atom salt, thallium acetate. The detection limit of BP-tetrol by RTP was 15 fmol. Their results demonstrated that RTP would be useful as a simple screening tool for monitoring BPDE-DNA adducts and other benzo[a]pyrene metabolites in biological samples.

Mignardi et al. (22) combined a pulsed source and gated detector with constant energy luminescence spectrometry to demonstrate the analytical power of time-resolution constant-energy synchronous luminescence. Binary mixtures were investigated, and one of the test mixtures was biphenyl and pyrene. The constant-energy difference for the RTP of this mixture was 13, 780 cm^{-1}, and the constant-energy RTP scan easily showed bands for both biphenyl and pyrene. The authors did not use time resolution for this mixture, but illustrated the approach for a mixture of p-aminobenzoic acid and quinine sulfate. They concluded that time-resolved RTP may be useful only for specialized cases, because many phosphors at room temperature have short lifetimes (< 10 ms). However, if two phosphors have lifetime differences of > 5 ms, then it may be possible to combine time-resolution and constant-energy luminescence spectrometry and achieve high selectivity.

Long and Su (23) showed that polyamide high-performance thin-layer chromatographic material was a useful solid substrate for RTP. Limits of detection and linear dynamic ranges for p-amino-benzoic acid were found to be comparable to those found for DE-81 ion-exchange filter paper. They also investigated an electrophoresis media, super sepraphore, as a possible medium for RTP. However, no RTP was observed for either p-aminobenzoic acid or coronene.

Ramos et al. (24) carried out a comparative study of RTP of PAH with organized media, namely, microemulsions and micellar solutions, paper-substrate RTP, and low-temperature phosphorim-etry. They demonstrated the practical use of microemulsion solutions for the RTP of PAH. By the introduction of a new variable, a nonpolar solvent, the technique was more versatile. The authors concluded that on the basis of the analytical figures of merit, phosphorescence of organized media at room temperature appears to be better than low-temperature phosphorimetry and paper-substrate RTP. Nevertheless, the applicability of organized media is limited to those compounds with sufficiently low polarity that prefer the hydrocarbon core of the micelles or microdroplets to the bulk water phase. The observation of RTP from polar molecules, and the very small sample amounts needed are specific advantages of paper-substrate RTP. The effect of the presence of the surfactants sodium dodecyl sulfate, sodium decyl sulfate, Brij 35, and dodecyltrimethylammonium chloride on paper-substrate RTP with PAH and carbazole was reported by Ramos et al. (25). Thallium nitrate was used as a heavy-atom perturber. Enhancements of the sensitivity, ranging from 2 to 9, were found when an anionic surfactant was added or when analytes were spotted from micellar solutions. However, the RTP signal was completely quenched in the presence of the cationic surfactant. Equivalent effects were found for the RTP of the paper background. In the presence of sodium dodecyl sulfate, the limits of detection were in the range of 0.2-3 ng.

Vo-Dinh and Alak (26) reported the enhancement of the RTP of anthracene adsorbed on cyclodextrin treated filter paper. β-Cyclodextrin treated filter paper enhanced the RTP of anthracene greater than α- and γ-cyclodextrin treated filter paper. The cyclodextrin treatment procedure was very simple and improved the analytical usefulness of the RTP method. Alak and Vo-Dinh (27) further reported the selective enhancement of RTP using cyclodextrin-treated filter paper. They compared the RTP enhancement of α-, β-, and γ-cyclodextrin treated filter paper for several PAH. The results showed the RTP enhancement factors for PAH depended on the type of cyclodextrin employed and on the binding strength between the cyclodextrin and the PAH. For example, benzo[ghi]perylene showed a 10-fold enhancement on γ-CD treated filter paper, but only a 2-fold enhancement on β-CD treated filter paper. The limits of detection of the compounds were in the subnanogram range and the linear range of the analytical curves covered approximately 2 orders of magnitude.

Bello and Hurtubise (28) characterized multicomponent mixtures of PAH with α-cyclodextrin-induced solid-surface room-temperature luminescence. The PAH were adsorbed on α-cyclodextrin-NaCl mixtures. With this approach, qualitative analysis of binary mixtures of benzo[a]pyrene and benzo[e]pyrene were easily carried out, even at nanograms levels, by employing room-temperature solid-surface fluorescence and RTP. Also, a technique was developed for the characterization of a nine-component PAH mixture. The approach employed extraction, selective excitation, and detection by room-temperature solid-surface fluorescence, RTP, and solution fluorescence spectroscopy. Nine PAH of various sizes from two-ring naphthalene to ten-ring decacyclene were used as model compounds. Several synthetic mixtures containing different amounts of the nine PAH were analyzed by the approach. The luminescence results showed that seven to nine PAH could be identified in the mixtures at nanogram levels.

8.3.2. Nitrogen Heterocycles

As discussed in the previous section, Vo-Dinh (16) has given an overview of the principles and applications of synchronous luminescence and RTP for air samples to obtain preliminary information on PAH and nitrogen heterocycles. He reported RTP data on acridine, 5,6-benzoquinoline, 7,8-benzoquinoline, carbazole, dibenzocarbazole, and quinoline. He demonstrated how both synchronous luminescence and RTP offered useful tools for polynuclear aromatic analysis in air pollution research.

Senthilnathan et al. (29) reported the effects of adsorbing

polyacrylic acid solutions on the RTP of 1,2-benzocarbazole and 5,6-benzoquinoline on filter paper. The limit of detection for 5,6-benzoquinoline improved 100-fold with polyacrylic acid adsorbed on the filter paper. Without polyacrylic acid, the limit of detection was 4.0 ng; however, with polyacrylic acid, it was 0.04 ng. 1,2-Benzocarbazole did not show an improved limit of detection, but the relative standard deviation of both 5,6-benzoquinoline and 1,2-benzocarbazole improved with poly-acrylic acid solutions spotted on filter paper.

Senthilnathan and Hurtubise (30) combined nitrogen heterocycles and pyrene to form various binary and ternary mixtures of the compounds. The compounds were determined at the nanogram level by using both room-temperature fluorescence and RTP with selective excitation and emission for the compounds adsorbed on silica gel or 30% acetylated cellulose chromatoplates. With fluorescence and phosphorescence calibration curves, it was possible to determine all the compounds in a given mixture without the isolation of the components. For the mixtures, the smallest amount of material that could be determined accurately was about 2.5 ng. Table 8.3 gives the results obtained for a ternary mixture of phenanthridine, 4-azafluorene, and acridine. Senthilnathan and Hurtubise (31) used derivative solid-surface room-temperature fluorescence and phosphorescence for the identification of components in binary and ternary mixtures of nanogram amounts of nitrogen heterocycles. Direct and first- and second-derivative RTP excitation, room-temperature fluorescence and RTP emission spectra were employed for the identification of the compounds in the mixtures by matching spectral wavelengths from mixtures with spectral wavelengths from standards. Filter paper was employed as a solid surface, and the compounds investigated were 5,6-benzo-quinoline, 4-azafluoroene, and phenanthridine. Derivative RTP spectra were readily obtained at the 0.5 ng level.

Nitrogen heterocycles have also been investigated by the room temperature luminescence of the compounds adsorbed on α-cyclodextrin-NaCl mixtures by Bello and Hurtubise (32,33). For 5,6-benzoquinoline adsorbed on 80% α-cyclodextrin-NaCl, the RTP limit of detection was 0.8 ng, the relative standard deviation was 5.0%, and the linear dynamic range was 0-200 ng. Alak and Vo-Dinh (27) enhanced the RTP of several nitrogen heterocycles at the nanogram level on cyclodextrin-treated filter paper. α-, β-, and γ-Cyclodextrins were investigated. Their results showed different RTP enhancement factors for the nitrogen heterocycles depending

Table 8.3. Ternary Mixtures: Phenanthridine-4-Azafluorene-Acridine and 5,6-Benzoquinoline-Acridine-Pyrene

Amount present, ng			Amount present, ng		
Phenanthridine	4-Azafluorene	Acridine	Phenanthridine	4-Azafluorene	Acridine
30	40	15	32	38	15
35	35	10	38	34	11
40	30	5	44	30	5
5,6-Benzoquinoline	Acridine	Pyrene	5,6-Benzoquinoline	Acridine	Pyrene
30	30	30	31	26	30
25	25	25	26	24	26
20	20	20	21	20	20

Reprinted with permission from V.P. Senthilnathan and R. J. Hurtubise, *Anal. Chem.*, 56, 913. Copyright 1984 American Chemical Society.

on the type of cyclodextrin employed. The nitrogen heterocycles investigated were acridine, 7,8-benzoquinoline, 2,6-dibenzoquinoline, dibenzothiophene, 2,6-dimethylquinoline, indole, isoquinoline, phenazine, and quinoline.

Vo-Dinh et al. (34) reported the RTP and fluorescence spectra of the eight benzoquinoline isomers. The isomers could be resolved into linear or angular subgroups on the basis of their fluorescence and RTP spectra by using fixed excitation wavelengths. Second-derivative and synchronous scanning techniques were combined to improve the selectivity of the RTP and fluorescence methods. These luminescence approaches were used to estimate three benzoquinoline isomers in a coal tar fraction. They concluded that the luminescence approaches could be used as screening tools. However, they needed to be combined with appropriate separation methods for good quantitative results for complex samples.

Abbott and Vo-Dinh (35) reported the RTP characteristics of azaarenes in the presence of mercury(II) chloride. The RTP of the polycyclic aromatic hydrocarbons investigated was quenched in the presence of mercury(II) chloride. They combined the use of Hg(II) for selective quenching of polycyclic aromatic hydrocarbons and a heavy-atom perturber for the RTP detection of selected nitrogen-containing compounds in the presence of polycyclic aromatic hydrocarbons. In one example, Hg(II) was used as an enhancing agent for the analysis of isoquinoline content in industrial waste water. They also carried out experiments to determine the external heavy-atom enhancement factor induced by $HgCl_2$ on various nitrogen heterocycles with different pK_a values. For pK_a values from 0 to 5, the RTP enhancement factor increased slowly with increasing pK_a values. However, with pK_a values greater than 5, the RTP enhancement factor was no longer simply dependent on the basicity of the azaarene. This indicated that other factors were involved in enhancing the RTP.

8.4.   Biphenyl and Polychlorinated Biphenyls

Polychlorinated biphenyls (PCB) have unique chemical and physical properties. For example, they are resistant to degradation and reactions with strong acids or bases under vigorous conditions. They have been widely used as heat transfer fluids, flame retardants, hydraulic fluids, and dielectric fluids. The extensive production of these compounds and their improper disposal has resulted in classifying PCB as environmental contaminates. These substances have been listed as carcinogens by EPA (36). Because of the environmental problems associated with PCB, it is important to have sensitive and selective analytical methods for this important class of compounds. Khasawneh et al. have determined both the low-temperature phosphorescence (37) and RTP (38,39) properties of these compounds. The low-temperature phosphorescence properties of these compounds were discussed in Chapter 5.

Khasawneh and Winefordner (38) examined 13 solid surfaces, using 4-chlorobiphenyl as a model compound, to determine the best conditions for maximum RTP intensity. They found that Whatman No. 1 filter paper was the most suitable solid substrate for the RTP of 4-chlorobiphenyl. They reported the RTP characteristics of biphenyl and several PCB congeners. It was found that it was not possible, under the experimental conditions they investigated, to determine one congener in the presence of another. Nevertheless, RTP was shown to be a selective, rapid, inexpensive technique for the determination of individual PCB, with a relative standard deviation between 2 and 10% and approximately 100-pg limit of detection for most compounds.

Khasawneh et al. (39) employed the effects of several heavy atoms on the RTP of biphenyl and several polychlorinated biphenyls on filter paper. The highest RTP signals were obtained with Schleicher and Schull 903 filter paper. They presented RTP data for several polychlorinated congeners and mixtures with and without the presence of different heavy atoms. They were primarily attempting to increase the sensitivity for these compounds by RTP and lower the limits of detection. The RTP lifetimes of these compounds were also obtained, and the possibility of determining one congener in the presence of another was evaluated. RTP data were obtained from the heavy-atom salts $TlNO_3$, KI, NaBr, $Pb(OAc)_2$, $ThCl_4$, and $NaIO_3$. Limits of detection were in the subnanogram to nanogram range. For example, biphenyl gave a limit of detection of 1.0 ng with NaBr, 4-chlorobiphenyl yielded a limit of detection of 0.40 ng with KI, and Aroclor 1221 gave a limit of detection of 0.30 ng with $NaIO_3$ as the heavy-atom salt. They concluded that RTP with a heavy atom present gave good precision, selectivity, sensitivity, and limits of detection for the individual PCB.

8.5.   Pharmaceutical and Clinical Applications

Vo-Dinh (5) has discussed earlier pharmaceutical and clinical applications of RTP. RTP continues to be used in these areas, and several of these applications will be considered in this section.

Glick and Winefordner (40) considered solid-substrate room-temperature luminescence immunoassays. Under appropriate conditions, some compounds exhibit phosphorescence or delayed fluorescence when adsorbed on a solid surface such as filter paper. In their work, they employed eosin (tetrabromofluoroscein) as a label in a separation immunoassay and described the development of a solid-substrate room-temperature luminescence immunoassay. They demonstrated its use with a competitive binding assay for human immunoglobulin G. One of the limitations of compounds adsorbed on filter paper is the RTP background from the filter paper. However, Glick and Winefordner found that this was not a problem in their work. First, the signal was inversely related to the antigen concentration, and at high concentrations where the signal was lowest, the background phosphorescence was not important. Second, the absorption and luminescence properties of eosin used for labeling were such that the background of filter paper was not significant. They showed that the instrumental requirements were simple, the procedure was rapid, and the per assay cost was minimal.

Karnes et al. (41) evaluated Schleicher and Schüll filter paper impregnated with diethylenetriaminopentaacetic acid and Whatman DE-81 anion-exchange paper for quantitation of urinary p-aminobenzoic acid. The two substrates were compared with respect to drying characteristics, pH variation, heavy-atom effect, and other variables. The impregnated Schleicher and Schüll 903 paper was superior in terms of drying characteristics and background interferences. However, the DE-81 paper had the advantages of lower detection limit and wider pH range. They concluded that DE-81 paper was slightly better suited as a substrate. Other RTP data have also been reported for p-aminobenzoic acid on DE-81 paper and Whatman No. 1 filter paper by Scharf et al. (42).

Karnes et al. (43) reported the determination of p-amino-benzoic acid in urine by RTP with application to the bentiromide test for pancreatic function. Orally administered N-benzoyl-L-tyrosyl-p-aminbenzoic acid (bentiromide) is cleaved to benzoyl-tyrosine and p-aminobenzoic acid by the pancreatic enzyme chymotrypsin. They compared the RTP data for urinary p-amino-benzoic acid with similar data obtained by a linear relationship between the Bratton-Marshall colorimetric method (x) and the RTP method (y) of $y = 0.997x + 1.651$. The correlation coefficient for 75 samples in the range 58 to 786 mg/mL was 0.993. The standard deviations of the slope and intercept were 0.002 and 0.300, respectively. They concluded that the RTP approach was technically simple and required no expensive reagents or materials. These characteristics favor routine use, and the same equipment could potentially be employed for other clinical applications.

Long et al. (44) considered the feasibility of surface analysis of homogeneous tablets by RTP. They used propranolol, p-aminobenzoic acid, and acetylsalicylic acid as model compounds and obtained both fluorescence and phosphorescence data. Potassium iodide was used as a heavy-atom salt, and starch or carboxymethylcellulose were employed as excipients. They prepared their own tablets and obtained calibration curves for the model compounds. In addition, two methods of tablet preparation were evaluated, and the effect of the pressure used in manufacturing the tablets on RTP was considered. For the two methods of sample preparation, dry mixing yielded tablets which had observable spots of green phosphorescence, whereas the tablets prepared by evaporation were more uniform. Also, RTP calibration curves had a greater slope for the evaporation method of tablet preparation. Long et al. (44) concluded that the RTP technique could be used for quality-control purposes because the concentrations of tablet ingredients could be correlated with RTP intensities by a reference method such as high-performance liquid chromatography.

Khasawneh et al. (45) reported the phosphorescence spectral characteristics, lifetimes, and limits of detection for 30 pharmaceutical compounds. A particular study was carried out on eight phenothiazine derivatives. Phenothiazine derivatives are used mainly as tranquilizers. Both low-temperature and room-temperature phosphorescence data were reported. RTP analytical figures of merit were presented for chloropromazine, prochlorperazine, promazine, trimeprazine, triflupromazine, trifluoperazine, thioridazine, thioridazine, and bendroflumethiazide. Eight of the nine compounds which showed RTP were phenothiazine derivatives. A comparative study of the phenothiazine derivative spectra showed that the influence of substituents was very small. The limits of detection by RTP were in the range of 3 ng to 104 ng, and the relative standard deviations were from 1.1 to 3.5%.

Garcia Alvarez-Coque et al. (46) considered the effects of anionic, nonionic, and cationic surfactants and thallium(I) nitrate on the paper-substrate RTP of eleven phenothiazine derivatives. They observed enhancement factors within the range of 1.5-10 after the addition of sodium dodecylsulfate, dodecyltrimethylammonium chloride and thallium(I) nitrate. The limits of detection were in the range of 0.2-2 ng, and they also reported results for p-aminobenzoic acid, carbazole, and o-terphenyl. The authors also investigated the influence of moisture on the RTP signals. The RTP intensity of chlorpromazine in the presence of sodium dodecylsulfate under forced drying conditions as a function of time is shown in Figure 8.2. As indicated in Figure 8.2, the RTP intensity increased continuously, but was still increasing at a rate of 0.15% min^{-1} after 1.5 h. They observed somewhat slower rates when the flow of nitrogen was 6 and 8 L min^{-1}. However, no significant improvements were found with a flow rate higher than 10 L min^{-1}. In related work, Sushe et al. (47) reported a solid-surface phosphorescence method for the determination of phenothiazine. The phenothiazine was dissolved in acetone and Chinese chromatographic filter paper was used as the solid matrix. They considered the effects of drying temperature, drying time, spot size, and the

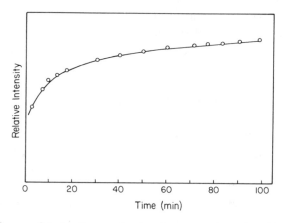

Figure 8.2. Influence of drying time on the phosphorescence intensity. Pieces of Whatman No. 1 paper were spotted with 2 μL of 0.25 M TlNO$_3$ and 5 μL of 25 μg mL^{-1} chlorpromazine hydrochloride in a 80/20 ethanol/water mixture, treated at 110°C for 10 min and introduced into the sample compartment with the nitrogen flow at 10 L min^{-1}. (Reprinted with permission from M.D. Garcia Alvarez-Coque, G.R. Ramos, A.M. O'Reilly, J.D. Winefordner, *Anal. Chim. Acta* 1988, 204, 247.)

concentration of KI on phenothiazine phosphorescence intensity. The analytical calibration curve for phenothiazine was linear from 0.1-150 ng.

Long et al. (48) reported the determination of primary amino active ingredients in pharmaceutical preparations without the need for separation by fluorescamine derivatization and RTP. They investigated tobramycin, phenylpropanolamine hydrochloride, procainamide hydrochloride, and p-aminobenzoic acid. The first and second compounds were not naturally phosphorescent, but the last two compounds were phosphorescent naturally. They reported a rather detailed study of pH, buffer concentration, and general phosphorescence characteristics for the compounds. Also, the determination of phenylpropanolamine hydrochloride in diet capsules was carried out using the derivatization RTP method. They found that a smaller phenylpropanolamine content per capsule was found using whole capsules for analysis compared to analyses in which only the contents of the capsules were analyzed. Most likely the capsule material interfered with the derivatization reaction. They concluded that the derivatization-RTP method was a promising analytical technique for the analysis of nonluminescent and luminescent species.

Asafu-Adjaye et al. (49) reported a method for analyzing drugs on the Toxi-Lab thin-layer chromatographic system by room-temperature luminescence. The Toxi-Lab approach is a thin-layer chromatography method for rapid screening a broad spectrum of drugs in urine and serum. In their work, room-temperature solid-surface phosphorescence and fluorescence were used to detect and quantitate three pairs of drugs after separation on a Toxi-Lab TLC plate. The pairs of drugs investigated were quinine/cimetidine, caffeine/chlordiazepoxide, and phenazopyridine/lidocaine. They found that the effects of the nonluminescent compounds were negligible in each pair for the determination of the luminescent compounds. Their results showed the feasibility of room-temperature solid-substrate luminescence to the Toxi-Lab system of drug analysis.

Gaye and Aaron (50) carried out a comparative study of heavy-atom effects on the RTP of nine biologically important purines on filter paper. The heavy-atom effect yielded a significant RTP enhancement, with the trend being $Tl^+ > Pb^{2+} \geq I^- >> Sm^{3+}$ for most of the purines in Table 8.4. The limits of detection ranged from 400 pg (purine) to 19 ng (theophylline). In other work, Andino et al. (51) studied the luminescence characteristics of

Table 8.4. Comparison of the Effect of Several Ions on the Room-Temperature Phosphorescence Intensity of Purines[a]

Compound	I_p^I / I_p^{Tl}	I_p^{Pb} / I_p^{Tl}	I_p^{Sm} / I_p^{Tl}
Adenine	0.2[b]; 0.4	0.5	0.009
Caffeine	18.3[b]; 12.2	1.4	0.1
Purine	0.3[b]	0.6	--
6-Chloropurine	0.7[b]	1.9	0.1
6-Methylpurine	0.6[b]; 0.2	0.3	0.004
6-Mercapto-purine	0.02	0.2	0.03
Guanine	0.9[b]	--	--
Theophylline	1.25[b]; 5.2	0.3	0.02
Theobromine	3.9[b]	0.6	--

[a]Effect is expressed as the ratio of the phosphorescence intensity in the presence of a heavy ion (I^-, Pb^{2+}, Sm^{3+}) to the intensity obtained in the presence of Tl^+. All solutions used had the same analyte concentration.

[b]In 1 M NaOH.

Reprinted with permission from M.D. Gaye and J.J. Aaron, *Anal. Chim. Acta* 1988, 205, 273.

caffeine and theophylline. The effects of the solvent, pH, the presence of a heavy atom, and the matrix or substrate on the fluorescence and phosphorescence properties of the compounds were considered. They found that for phosphorescence, under the appropriate experimental conditions, both low-temperature and room-temperature phosphorescence could be used as analytical tools for the determination of caffeine and theophylline. In particular, they found that solid-surface RTP was sensitive and simple. The approach did not demand the use of liquid nitrogen and did not have the problems associated with working with frozen matrices.

8.6.   Other Applications and Aspects in RTP

In this section, the applications that did not fit into the earlier sections and some other perspectives related to RTP methodology will be discussed.

Miller (52) has surveyed the applications of solid-surface

luminescence measurements in the fields of cryogenics and RTP, fluorescence immunoassays, and other areas. He discussed the potential of thin-layer phosphorimetry in conjunction with thin-layer chromatography (TLC). It was suggested that combining TLC and RTP would encourage the development of phosphorescent labels used in TLC. He also suggested the use of phosphorescent markers on postage stamps to aid in mail sorting processes.

Aaron et al. (53) reported the effects of ion-exchange filter paper and of heavy atoms on the RTP of twelve indoles. Anion-exchange Whatman DE-81 filter paper gave the largest RTP signals in the presence of iodide. Schleicher and Schuell 903 paper treated with diethylenetriaminepentaacetic acetic acid or supporting carboxymethylcellulose resin was not as useful. It was necessary to use a heavy atom to observe analytically useful RTP signals from indoles. The heavy atom, iodide, induced greater signals than thallium(I). The lifetimes of the indoles were rather short and showed two decaying components. For example, for 5-methyl-indole, the long component gave a lifetime of 5.2 ms, and the short component gave a lifetime of 1.2 ms. The RTP absolute limits of detection were between 0.2 and 14 ng. Analytical data were presented for indole, 5-bromoindole, 3-carbinolindole, 5-cyano-indole, 5-fluoroindole, 3-formylindole, 5-methoxyindole, 2-methyl-indole, 3-methylindole, and 5-methylindole.

Andino et al. (54) reported the RTP results of eight indole-carboxylic acids under different pH conditions and on different ion-exchange filter papers. The RTP excitation and emission wavelengths did not change significantly with pH. Figure 8.3 shows the excitation and emission spectra for 2-indolecarboxylic acid on DE-81 filter paper under different pH conditions. The greatest RTP signals were acquired from neutral solutions adsorbed on DE-81 anion-exchange filter paper. However, alkaline (pH $\sim$ 13) solutions on Schleicher and Schuell 903 filter paper treated with diethylenetriaminepentaacetic acid gave stronger signals than neutral or acidic (pH $\sim$ 1.6) solutions on the same solid surface. The authors found heavy-atom enhancement factors ranging between 4 and 550 using iodide as the heavy atom. The limits of detection were between 100 and 500 pg, which illustrated the usefulness of RTP for the determination of indolecarboxylic acids in trace organic analysis.

Scharf and Winefordner (55) reported the phosphorescence excitation and emission spectra and lifetimes of acetophenone, benzophenone, p-aminobenzophenone, and Michler's ketone

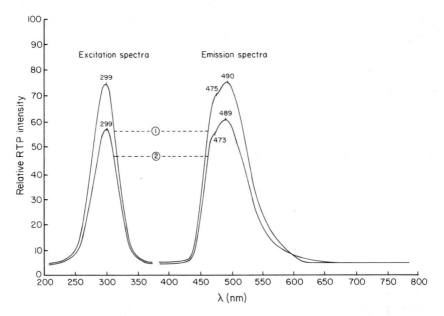

Figure 8.3. Effect of pH on the RTP spectra of 2-indolecarboxylic acid (10^{-3} M) on DE-81 filter paper, in the presence of 1M KI. (1) Excitation and emission spectra of neutral water-ethanol (50:50 v/v) solution. (2) Excitation and emission spectra of basic (0.5M NaOH) water-ethanol (50:50 v/v) solution. (Reprinted with permission from M. Andino, J.J. Aaron, J.D. Winefordner, *Talanta* 1986, 33, 27.)

adsorbed on Whatman No. 1 filter paper at various temperatures and compared the results with the phosphorescence characteristics in different solvent glasses at 77 K. Acetophenone and benzophenone yielded phosphorescence on filter paper only at 208 K. They made various comparisons of the characteristics of the low lying triplet states. For example, in some cases a $^3(n,\pi^*)$ state was observed and in other cases a $^3(CT)$ state was observed. Lowering the temperature appeared to increase the phosphorescence intensity for ketones which phosphoresce in the $^3(n,\pi^*)$ triplet state, but affected it only slightly for analytes which phosphoresce in the $^3(\pi,\pi^*)$ triplet state. They postulated that RTP arose for aromatic ketones and aldehydes with low-lying $^3(\pi,\pi^*)$ or $^3(CT)$ triplet states.

Jones et al. (56) constructed a Becquerel-disc phosphoroscope from a commercially available optical chopper with variable frequency and digital read-out for lifetime measurements in RTP. With a continuous source, RTP lifetimes in the range of 1-1000 ms could be measured with better than 4% relative standard deviation. They concluded that the Becquerel-disc phosphoroscope was ideally suited for the measurement of RTP lifetimes of analytes on solid substrates.

Murray et al. (57) considered lifetime spectra for RTP. They discussed phosphors adsorbed on solid surfaces that gave multiple lifetimes. They described the application of a constrained, regularized least-squares method, available as a computer program package called CONTIN, to the analysis of RTP decay. They concluded that CONTIN was an extremely versatile program for the analysis of phosphorescence lifetime data.

Campiglia et al. (58) discussed the use of an inorganic phosphor as a reference signal in solid-surface RTP. They considered a commercial phosphor based on zinc and cadmium sulfide embedded in a poly(methyl methacrylate) film. Figure 8.4 gives the excitation and emission spectra of the phosphorescent pigment. The utility of the phosphor was demonstrated with the aging of an xenon arc lamp. The inorganic phosphor was very useful for the daily intensity calibration of the instrument, and it proved to be stable. The authors did not recommend the phosphor

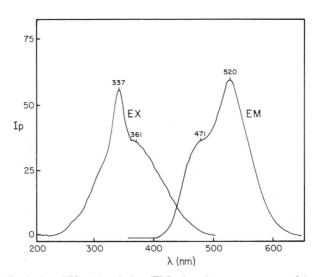

Figure 8.4. Excitation (EX) and emission (EM) phosphorescent spectra of the phosphorescent pigment. (Reprinted with permission from A.D. Campiglia and C.G. de Lima, *Anal. Chem.*, 60, 2165. Copyright 1988 American Chemical Society.)

as a secondary standard because additional experiments were needed to evaluate its RTP behavior over a longer period of time. However, it could be readily employed for the daily calibration of spectrophorimeters.

Ohshima et al. (59) demonstrated a unique application of phosphorescence. Benzene molecules in the lowest-excited triplet state were allowed to collide with biacetyl and benzophenone surfaces, and the phosphorescence from the surface due to energy transfer from the T_1 state of benzene was observed. They discussed the possible application of this method to the detection of metastable species.

Inoue and Ebara (60,61) considered electron-impact-induced phosphorescence of organic molecules and a spectrophosphorimeter for the measurement of the phosphorescence. They emphasized that the measurement of the phosphorescence of organic compounds at low pressures is expected to yield important information about the photophysical nature of molecules. However, only a few results of such measurements have been reported. The spectrophosphorimeter they described consisted of a vacuum chamber with an electron source, a grating monochromator, a photon-counting apparatus, and a microcomputer system that controlled the electron source, the monochromator, and the photon counter, and, at the same time, performed data handling.

8.7.   Concluding Comments

The applications reported earlier by Vo-Dinh (5) and the ones discussed in this section show that solid-surface RTP is applicable to a wide variety of compound classes and samples. Detection limits in the nanogram and subnanogram ranges are easily achieved. In addition, high selectivity can be obtained in many analytical situations. Also, solid-surface RTP has been shown to be applicable to analytical problems where probably no other technique would be as effective. A case in point is the dosimeter badge discussed earlier in the chapter (18). Solid-surface RTP is inexpensive, and it can be readily automated. Also, very small sample sizes are employed, which is particularly advantageous for biological and environmental samples. Because of the many advantages of RTP, it will continue to be used in the analysis of organic compounds at the nanogram and picogram levels. If background signals can be minimized for some of the solid substrates such as filter paper, even smaller amounts of organic compounds will be routinely detected and determined.

References

1. Parker, R.T.; Freedlander, R.S.; Dunlap, R.B. *Anal. Chim. Acta* 1980, 119, 189.
2. Parker, R.T.; Freedlander, R. S.; Dunlap, R.B. *Anal. Chim. Acta* 1980, 120, 1.
3. Hurtubise, R.J. *Solid-Surface Luminescence Analysis*; Marcel Dekker: New York, 1981; Chapter 7.
4. Hurtubise, R. J. In *Molecular Luminescence Spectroscopy: Methods and Applications - Part II*; Schulman, S.G., Ed.; Wiley: New York, 1988; Chapter 1.
5. Vo-Dinh, T. *Room-Temperature Phosphorimetry for Chemical Analysis*; Wiley: New York, 1984; Chapter 9.
6. Aaron, J.J.; Kaleel, E.M.; Winefordner, J.D. *J. Agric. Food Chem.* 1979, 27, 1233.
7. Aaron, J.J.; Winefordner, J.D. *Analusis* 1978, 7, 168.
8. Vannelli, J.J.; Schulman, E.M. *Anal. Chem.* 1984, 56, 1030.
9. Su, S.Y.; Asafu-Adjaye, E.; Ocak, S. *Analyst* 1984, 109, 1019.
10. Asafu-Adjaye, E.B.; Su, S.Y. *Anal. Chem.* 1986, 58, 539.
11. de Lima, C.G.; Andino, M.M.; Winefordner, J.D. *Anal. Chem.* 1986, 58, 2867.
12. Campiglia, A.D.; de Lima, C.G. *Anal. Chem.* 1987, 59, 2822.
13. Khasawneh, I.M.; Winefordner, J.D. *Talanta* 1988, 35, 267.
14. Searle, C.E., Ed.; *Chemical Carcinogens* 2nd ed., Vol. 1; ACS Monograph 182; American Chemical Society: Washington, DC., 1984.
15. Su, S.Y.; Winefordner, J.D. *Can. J. Spectrosc.* 1983, 28, 21.
16. Vo-Dinh, T. In *Identification and Analysis of Organic Pollutants in Air*; Keith, L.H., Ed.; Butterworth: Boston, MA, 1984; Chapter 16.
17. Vo-Dinh, T.; Bruewer, T.J.; Colovos, G. C.; Wagner. T.J.; Jungers, R.H. *Environ. Sci. Technol.* 1984, 18, 477.
18. Vo-Dinh, T. *Environ. Sci. Technol.* 1985, 19, 997.
19. Vo-Dinh, T.; White, D.A. *Am. Ind. Hyg. Assoc. J.* 1987, 48, 400.
20. White, D.A.; Vo-Dinh, T. *Appl. Spectrosc.* 1988, 42, 285.
21. Vo-Dinh, T.; Uziel, M. *Anal. Chem.* 1987, 59, 1093.
22. Mignardi, M.A.; Laserna, J.J.; Winefordner, J.D. *Microchem.*

J., <u>38</u>, 313 (1988).

23. Long, W.J.; Su, S.Y. *Microchem. J.*, <u>37</u>, 59 (1988).
24. Ramos, G. R.; Khasawneh, I. M.; Garcia-Alvarez-Coque, M. C.; Winefordner, J. D. *Talanta* 1988, <u>35</u>, 41.
25. Ramos, G. R.; Garcia-Alvarez-Coque, M. C.; O'Reilly, A. M.; Khasawneh, I. M.; Winefordner, J. D. *Anal. Chem.* 1988, <u>60</u>, 416.
26. Vo-Dinh, T.; Alak, A. *Appl. Spectrosc.* 1987, <u>41</u>, 963.
27. Alak, A.; Vo-Dinh, T. *Anal. Chem.* 1988, <u>60</u>, 596.
28. Bello, J.M.; Hurtubise, R.J. *Anal. Chem.* 1988, <u>60</u>, 1285.
29. Senthilnathan, V.P.; Ramasamy, S.M.; Hurtubise, R.J. *Anal. Chim. Acta* 1984, <u>157</u>, 203.
30. Senthilnathan, V.P.; Hurtubise, R.J. *Anal. Chem.* 1984, <u>56</u>, 913.
31. Senthilnathan, V.P.; Hurtubise, R.J. *Anal. Chim. Acta* 1985, <u>170</u>, 177.
32. Bello, J.; Hurtubise, R.J. *Appl. Spectrosc.* 1986, <u>40</u>, 790.
33. Bello, J.M.; Hurtubise, R.J. *Anal. Lett.* 1986, <u>19</u>, 775.
34. Vo-Dinh, T.; Miller, G.H.; Abbott, D.W.; Moody, R.L.; Ma, C.Y.; Ho, C.-H. *Anal. Chim. Acta* 1985, <u>175</u>, 181.
35. Abbott, D.W.; Vo-Dinh, T. *Anal. Chem.* 1985, <u>57</u>, 41.
36. Windholz, M., Ed.; *The Merck Index*; Merck & Co., Inc.: Rahway, NJ, 1983; p 1091.
37. Khasawneh, I. M.; Winefordner, J. D. *Microchem. J.* 1988, <u>37</u>, 86.
38. Khasawneh, I.M.; Winefordner, J. D. *Microchem. J.* 1988, <u>37</u>, 77.
39. Khasawneh, I.M.; Chamsaz, M.; Winefordner, J.D. *Anal. Lett.* 1988, <u>21</u>, 125.
40. Glick, M.R.; Winefordner, J.D. *Anal. Chem.* 1988, <u>60</u>, 1982.
41. Karnes, H.T.; Schulman, S. G.; Winefordner, J.D. *Anal. Chim. Acta* 1984, <u>164</u>, 257.
42. Scharf, G.; Smith, B.W.; Winefordner, J.D. *Anal. Chem.* 1985, <u>57</u>, 1230.
43. Karnes, H.T.; Bateh, R.P.; Winefordner, J.D.; Schulman, S.G. *Clin. Chem.* 1984, <u>30</u>, 1565.
44. Long, W.J.; Su, S.Y.; Karnes, H.T. *Anal. Chim. Acta* 1988, <u>205</u>, 279.
45. Khasawneh, I.M.; Garcia Alvarez-Coque, M.C.; Ramos, G.R.; Winefordner, J.D. *J. Pharm. Biomed. Anal.* 1989, <u>7</u>, 29.
46. Garcia Alvarez-Coque, M.C.; Ramos, G.R.; O'Reilly, A.M.; Winefordner, J.D. *Anal. Chim. Acta* 1988, <u>204</u>, 247.

47. Sushe, Z.; Changsong, L.; Gongmin, C. *Anal. Chem.* (People's Republic of China) 1987, 15, 980.
48. Long, W.J.; Norin, R.C.; Su, S.Y. *Anal. Chem.* 1985, 57, 2873.
49. Asafu-Adjaye, E.B.; Su, S.Y.; Karnes, H.T. *J. Anal. Toxicol.* 1987, 11, 70.
50. Gaye, M.D.; Aaron, J.J. *Anal. Chim. Acta* 1988, 205, 273.
51. Andino, M.M.; de Lima, C.G.; Winefordner, J.D. *Spectrochim. Acta* 1987, 43A, 427.
52. Miller, J.N. *Pure & Appl. Chem.* 1985, 57, 515.
53. Aaron, J.J.; Andino, M.; Winefordner, J.D. *Anal. Chim. Acta* 1984, 160, 171.
54. Andino, M.; Aaron, J.J.; Winefordner, J.D. *Talanta* 1986, 33, 27.
55. Scharf, G.; Winefordner, J.D. *Talanta* 1986, 33, 17.
56. Jones, B.T.; Smith, B.W.; Berthod, A.; Winefordner, J.D. *Talanta* 1988, 35, 647.
57. Murray, K.A.; Gonzales, E.; Gregory, R. B.; Street, K.W. *Appl. Spectrosc.* 1989, 43, 351.
58. Campiglia, A.D.; de Lima, C.G. *Anal. Chem.* 1988, 60, 2165.
59. Ohshima, S.; Kondow, T.; Kuchitsu, K. *Bull. Chem. Soc. Jpn.* 1985, 58, 1833.
60. Inoue, A.; Ebara, N. *Chem. Phys. Lett.* 1986, 126, 58.
61. Inoue, A.; Ebara, N. *Appl. Spectrosc.* 1986, 40, 410.

CHAPTER 9

SENSITIZED AND QUENCHED PHOSPHORESCENCE IN SOLUTION AT ROOM TEMPERATURE

9.1. Introduction

In Chapter 5, the room-temperature phosphorescence of compounds in fluid solutions was discussed. It was pointed out that analytically useful phosphorescence from compounds in fluid solutions at room temperature from a large number of components has not been achieved. However, two methods based on room-temperature phosphorescence in liquid phases (RTPL) have proven to be important in chemical analysis. These are sensitized and quenched RTPL. Sensitized and quenched RTPL have been used primarily in liquid chromatography and flow injection analysis. In this chapter, the general theory, instrumentation, experimental aspects, and applications associated with these approaches will be considered.

9.2. General Theory for Sensitized RTPL

In sensitized RTPL, the analyte is detected indirectly by the RTPL of an acceptor. After excitation, the analyte acts as an energy donor (D) according to

$$D(T_1) + A(S_0) \rightarrow D(S_0) + A(T_1)$$

where A is the acceptor, and S_0 and T_1 are the electronic ground states and the lowest triplet states, respectively. Sensitized RTPL is considered to have excellent analytical potential when the above reaction is diffusion controlled (1,2). This condition is normally fulfilled if the triplet state of the acceptor is lower in energy than the

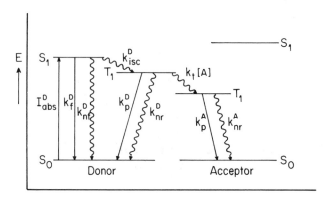

Figure 9.1. Energy diagram for a model system in sensitized phosphorescence: $(I_{abs}{}^D)$ is the rate of light absorption by the donor; $(k_f{}^D)$, $k_{nf}{}^D)$, and $(K_{isc}{}^D)$ are the rate constants in s^{-1} of the intramolecular deactivation of the donor via fluorescence, internal conversion, and intersystem crossing, respectively; $(k_p{}^D)$ and $(k_{nr}{}^A)$ are the overall rate constants of intra- and intermolecular nonradiative deactivation in s^{-1}; $(k_t[A])$ is the apparent rate constant of the energy transfer reaction in s^{-1}. (Reprinted with permission from J.J. Donkerbroek, C. Gooijer, N.H. Velthorst, and R.W. Frei, *Anal. Chem.*, **54**, 891. Copyright 1982 American Chemical Society.)

triplet state of the donor. Under these conditions, the energy transfer reaction competes successfully with other deactivation pathways of $D(T_1)$. The acceptor must have a high phosphorescence quantum yield in liquid solutions but also a low triplet-state energy. Also, the sensitized RTPL method can only be applied at an excitation wavelength for the donor and not the acceptor. Figure 9.1 shows an energy diagram for a model system in sensitized phosphorescence.

The intensity of sensitized RTPL of a donor-acceptor couple is given by the product of the rate of light adsorption by the donor $(I^D{}_{abs})$, its efficiency of intersystem crossing $(\Theta^D{}_{ISC})$, the efficiency of the energy transfer from the donor to the acceptor $(\Theta_t{}^{DA})$, and the phosphorescence of the acceptor $(\Theta_p{}^A)$ (2,3). Also, for uncorrected spectra the photomultiplier output, P(sens), depends on both the excitation wavelength and the emission wavelength. Equation (9.1) relates the photomultiplier output to several factors (2).

$$P(sens) = kI_{0,ex}(2.3\epsilon^D[D]\ell)\Theta_{isc}{}^D\Theta_t{}^{DA}\Theta_p{}^A S_p{}^A \qquad (9.1)$$

In Equation (9.1), k is an instrumental constant independent from λ_{ex} and λ_{em}; I^D_{abs} is approximately equal to $I_{0,ex}(2.3\epsilon^D[D]\ell)$; $I_{0,ex}$ denotes the intensity of the light source at λ_{ex}; ϵ^D is the molar absorptivity of D at that wavelength; [D] is the donor concentration; and ℓ is the optical pathlength. Sp^A is a dimensionless parameter related to the wavelength dependence of the detection system. Equation (9.1) is applicable only if the acceptor does not absorb radiation at λ_{ex}. Additional considerations of the theory will be given later in the chapter.

9.3. Experimental and Instrumental Aspects for Sensitized RTPL

Essential for RTPL detection is the long triplet state lifetime of the acceptor in the solvent under consideration (4). To assume this, the solvent must be deoxygenated. Donkerbroek et al. (1) described a special sample cell for RTPL measurements that permitted efficient removal of oxygen and protection of the solution for several hours from re-entrance of oxygen. The sample cell could be readily used with commercial luminescence instrumentation. For flow injection analysis and liquid chromatography, the instrumentation is somewhat more sophisticated (4,5). A diagram of the experimental set-up used by Gooijer et al. (4) is given in Figure 9.2. Essential to the system in Figure 9.2 is the eluent vessel which is needed to obtain sufficient deoxygenation by purging with nitrogen gas led over a heterogeneous reduction catalyst. In addition, the eluent vessel, pump, injection valve, column, and detector are interconnected by stainless-steel capillaries. The overall system is closed by leading the output capillary back to the eluent vessel. During elution, the valve to waste is opened to avoid contamination of the eluent. For detection devices, commercially available fluorescence detectors have been employed. Gooijer et al. (4) pointed out that with a single detector instrument, fluorescence, sensitized phosphorescence, and quenched phosphorescence detection could be carried out.

9.4. Applications for Sensitized RTPL

Because the rate of energy transfer of donor to acceptor is proportional to the acceptor concentration, there is some flexibility

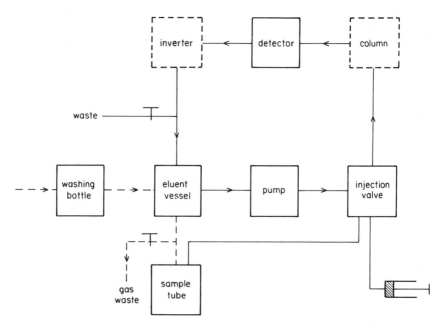

Figure 9.2. Diagram of the dynamic system for flow injection analysis and liquid chromatography with phosphorescence detection. The broken and solid lines represent stainless-steel capillaries for the nitrogen gas and the liquid stream, respectively. (Reprinted with permission from C. Gooijer, N.H. Velthorst, and R.W. Frei, *Trends Anal. Chem.* 1984, 3, 259.)

in choosing optimal experimental conditions (1). The efficiency of the energy transfer process depends on the triplet lifetime of the donor in the absence of the acceptor (τ_0^D) and the rate of the transfer reaction. Generally, even for molecules with short triplet lifetimes, such as 0.1 μs, detection by sensitized RTPL would be theoretically possible (1). To demonstrate the analytical applicability of sensitized RTPL, Donkerbroek et al. (1) investigated the sensitized RTPL of benzophenone in hexane with 1,4-dibromonaphthalene as an acceptor. They showed that direct measurement of RTPL of benzophenone in hexane was not possible. However, they established that sensitized RTPL in hexane could be observed with benzophenone as the donor and 1,4-dibromonaphthalene as the acceptor. The resulting calibration curve gave a regression coefficient of 0.992, and the limit of detection of benzophenone was 4×10^{-7} M. However, the acceptor did not meet all the criteria for a universal acceptor. Its major drawback was the high absorption in regions where most potential donors absorb.

Donkerbroek et al. (2) considered in some detail the sensitized RTPL with 1,4-dibromonaphthalene and biacetyl as acceptors with solvents that are frequently employed in liquid chromatography. They considered three important aspects: (a) the excitation wavelength ranges for the analytes; (b) the minimum triplet-state energies of the analyte necessary to make the energy transfer process diffusion controlled; and (c) the phosphorescence efficiencies of the acceptors at room temperature in various solvents. Based on the absorption spectrum of 1,4-dibromonaphthalene, they concluded that it could be used as an acceptor at donor excitation wavelengths greater than 330 nm. Because of the broad absorption range for biacetyl, the question arose whether it was possible to use a biacetyl concentration which was either low enough to make its excitation wavelength insignificant or high enough to guarantee an acceptable value for the efficiency of energy transfer (θ_t^{DA}). However, they showed that biacetyl had more favorable acceptor properties than 1,4-dibromonaphthalene. This conclusion was based partly on the triplet-state energies of the acceptors. Table 9.1 summarizes the properties of the two acceptors. The most important difference between 1,4-dibromonaphthalene and biacetyl was that dibromonaphthalene could only be applied to detect analyte species absorbing at wavelengths above 330 nm. However, for biacetyl, such a restriction did not exist because of its low molar absorptivity. In earlier work, Donkerbroek et al. (1) obtained a limit of detection of 4.0×10^{-7} M for benzophenone by using sensitized RTPL with 1,4-dibromonaphthalene as an acceptor. Donkerbroek et al. (2) showed in later work that more favorable limits of detection could be obtained with sensitized RTPL with biacetyl as an acceptor. For example, the limit of detection of benzophenone with a biacetyl concentration of 10^{-4} M and an excitation wavelength of 280 nm was 3.2×10^{-8} M in acetonitrile:water (1:1). Limit of detection data were also presented for 4-bromobenzophenone, 4-methylbenzophenone, 2-chlorobenzophenone, 4,4'-dibromobiphenyl, and 4-bromobiphenyl (2). Generally, the limits of detection were in the 10^{-8} M range. It was concluded that sensitized RTPL in liquid solutions was a feasible approach to detect many compounds that do not fluoresce, and the approach can be considered as complementary to fluorescence detection. In principle, sensitized RTPL could be useful for all compounds which show phosphorescence in rigid solutions at 77 K, because the primary condition to be fulfilled is an efficient triplet formation. In

Table 9.1. Br$_2$N and BIAC as Potential Acceptors in Sensitized RTPL

	Br$_2$N	BIAC
Allowed excitation range of donor	>330 nm	>220 nm provided that $\tau_D^o > 10^{-5}$
Allowed T$_1$-state energy of donor	>20.2 x 10^3 cm^{-1} ($\equiv$ 495 nm)	>19.7 x 10^3 cm^{-1} ($\equiv$ 507 nm)
Emission properties of acceptors in [a]		
Hexane	++	+
Acetonitrile	+	+
Acetonitrile/ water (1:1)	+	+
Methanol	+	+
Methanol/water (1:1)	+	+
Dichloromethane	-	-

[a]++, very good; +, good; -, not recommendable.

addition, all compounds are detected at one relatively long emission wavelength of the acceptor. Thus, the effects of stray light and fluorescence background are reduced.

Donkerbroek et al. (6) reported on the sensitized liquid-phase RTP applied to the detection of polychloronaphthalenes. They detected mono-, di-, tri-, and tetra-substituted compounds using biacetyl in azeotropic acetonitrile:water. The detection limits they obtained were in the order of 10^{-8} to 10^{-9} M, except for 1,2,3-trichloro- and 1,2,3,4-tetrachloronaphthalene. The detection limits were comparable to those obtained with fluorescence. Phosphorescence measurements at 77 K showed that, with the exception of 1,2,3-trichloronaphthalene, for the tri- and tetra-substituted compounds the triplet-state energies were lower than for biacetyl. Thus, for these polychloronaphthalenes, the influence of reversed energy transfer on sensitized RTPL detection must be considered. In reverse energy transfer, the acceptor molecule in the triplet excited state interacts with a donor molecule in the ground state to form a donor molecule in the triplet excited state and a ground state acceptor molecule. Reversed energy transfer then provides an additional deactivation pathway for the acceptor molecule in the excited triplet state. Donkerbroek et al. (6) presented a theoretical treatment of the effect of reversed energy transfer reaction on the sensitized RTPL signal. They quantified the importance of reversed energy transfer by deriving Equation (9.2).

$$\frac{P(\text{sens})}{P'(\text{sens})} = \frac{\theta_t^{DA}}{\theta_{t,\text{eff}}^{DA}} = 1 + \frac{k_{-t}\tau_0^A[D]}{1 + k_t\tau_0^D[A]} \tag{9.2}$$

In Equation (9.2), P(sens) is the photomultiplier output with no reversed energy transfer; P'(sens) is the photomultiplier output with reversed energy transfer; θ_t^{DA} is the triplet-triplet energy transfer efficiency; $\theta_{t,\text{eff}}^{DA}$ is the effective energy transfer efficiency; k_{-t} is the rate constant of the energy transfer reaction from A to D; τ_0^A is the effective triplet lifetime of A, [D] is the equilibrium concentration of D; k_t is the rate constant of the energy transfer reaction from D to A; τ_0^D is the effective triplet lifetime of D; and [A] is the equilibrium concentration of A.

Equation (9.2) indicates that reversed energy transfer is not only determined by the rate constants k_t and k_{-t}, but also by the triplet lifetimes τ_0^A and τ_0^D and the concentrations [D] and [A]. If $k_{-t}\tau_0^A[D] \ll 1 + k_t\tau_0^D[A]$, Equation (9.2) shows that the reversed energy transfer process can be neglected. This condition is approximated at very low analyte concentrations. Thus, near the detection limit of the analyte, the reversed energy transfer can be ignored (6). The authors concluded that polychloronaphthalenes could be sensitively detected in acetonitrile:water at room temperature with limits of detection ranging from 10^{-8} to 10^{-9} M by fluorescence and by sensitized RTPL with biacetyl as an acceptor. Exceptions were 1,2,3-trichloronaphthalene and 1,2,3,4-tetrachloro-naphthalene, which showed a rapid radiative decay of the singlet state. Because of their triplet state energies, for three- and four-chlorosubstituted compounds (except for 1,2,3-trichloronaph-thalene), the sensitized RTPL signal was affected by a reversed energy transfer reaction. This influenced the linearity of the signal and not the limits of detection. Their study showed that the sensitivity of the fluorescence and phosphorescence was similar for the polychloronaphthalenes. However, they later showed that the RTPL gave improvements in selectivity (7).

Donkerbroek et al. (7) considered RTPL detection of polychloronaphthalene and polychlorobiphenyls in liquid chromatography. The identification of polychlorobiphenyls was strongly assisted by combining ultraviolet detection and detection based on sensitized RTPL of biacetyl, because the compounds substituted at the α position were not detected by the sensitized RTPL method. They also used quenched RTPL for the polychloronaphthalenes. Quenched RTPL will be discussed in the next section of this chapter. In addition, they combined ultraviolet detection, sensitized RTPL, and quenched RTPL detection for the liquid chromatographic detection of complex mixtures. This was demonstrated for some industrial mixtures of polychloronaphthalenes and for the polychlorobiphenyl mixture Aroclor (7).

Sensitized RTPL was shown to be successfully applicable as a detection method for both continuous-flow and chromatographic systems (8). Limits of detection for a series of halogenated naphthalenes and biphenyls were reported in the subnanogram region. The general experimental set-up for the continuous-flow system is shown in Figure 9.2. As discussed earlier in this chapter, it was important to deoxygenate the eluent. Donkerbroek et al. (8) mentioned that with plug injections without chromatographic separation, the sample solution also had to be deoxygenated to apply sensitized RTPL as a detection method. Also, in their system, the sensitized RTPL peak height in the flow injection experiments depended on the injection volumes. They also performed static experiments with a commercial spectrofluorometer. Limits of detection were reported for twenty halogenated biphenyls and ninety halogenated naphthalenes for both the static and continuous-flow systems. With a few exceptions, in both systems the limits of detection values ranged from 10^{-9} to 10^{-8} M. These results strongly suggested that sensitized RTPL was useful as a detection method in dynamic systems. They also showed that sensitized RTPL was useful in liquid chromatography for the detection of polychlorinated biphenyls and substituted dibenzofurans.

Bauman et al. (9) considered the detection of biacetyl in liquid chromatography using time-resolved sensitized phosphorescence. The determination of biacetyl in beer, wine, and several dairy products is important because of the influence of this compound on the flavor of these food products. The method was based on triplet-triplet energy transfer from a donor, 1,5-naphthalenedisulfonic acid disodium salt, to biacetyl, acting as an acceptor. The determination of biacetyl with high-performance liquid chromatography has been limited because of the low molar absorptivity of biacetyl over the ultraviolet-visible range and its low fluorescence quantum efficiency. Benzophenone, biphenyl, naphthalene, and 1,5-naphthalenedisulfonic acid disodium salt (NDSA) were tested as donors. The best limits of detection for biacetyl were obtained with NDSA. Because of this, its properties as a donor, and the facts that it showed good solubility in polar solvent mixtures, and it showed no retention on a reversed-phase column, NDSA was chosen as a donor molecule. Bauman et al. (9) used a Perkin-Elmer Model LS-2 luminescence detector. This system is provided with a pulsed xenon discharge lamp and a gated detector. The detection limit for biacetyl was reported as 1 ppb. The relative standard deviation was 3.1% at the 10 ppb level and 1.9% at the 100 ppb level. Preliminary experiments showed that biacetyl could be determined in beer samples at the part-per-million level. However, with an improved cleanup procedure, detection limits in the parts-per-billion range should be achieved.

9.5.   General Theory for Quenched RTPL

It is instructive to compare quenched RTPL with sensitized RTPL (Table 9.2). Reaction 3 of the scheme leading to quenched RTPL is the reverse of reaction 3 leading to sensitized RTPL. Therefore, the equilibrium position of the following reaction

$$^3A^* + B \underset{k_{-1}}{\overset{k_1}{\rightleftarrows}} A + {}^3B^*$$

will determine if the analyte can detect more sensitively by sensitized RTPL or by quenched RTPL. As indicated in Table 9.2, for sensitized RTPL a suitable wavelength, λ_{Ex}^A, is selected for the excitation of the analyte, and the phosphorescence of biacetyl is detected. Thus, the analyte produces a positive signal, superimposed on the background signal corresponding to the direct excitation at λ_{Ex}^A. In quenched RTPL, a wavelength, λ_{Ex}^B, is chosen where only biacetyl is excited ($\sim$ 420 nm), where the molar absorptivity of biacetyl is reasonably high ($\sim$ 20 $M^{-1}cm^{-1}$). However, the presence of the analyte leads to a decrease of the

Table 9.2. Reaction Schemes of Sensitized and Quenched RTPL of Biacetyl by Triplet-Triplet Energy Transfer

Sensitized RTPL	Quenched RTPL
(1) Excitation of the analyte (A): $A + h\nu_{ex}^{A} \rightarrow {}^{1}A^{*}$	(1) Excitation of biacetyl (B): $B + h\nu_{ex}^{A} \rightarrow {}^{1}B^{*}$
(2) Intersystem crossing to the triplet state: ${}^{1}A^{*} \rightarrow {}^{3}A^{*}$	(2) Intersystem crossing to the triplet state: ${}^{1}B^{*} \rightarrow {}^{3}B^{*}$
(3) Energy transfer to biacetyl (B): ${}^{3}A^{*} + B \xrightarrow{k_1} A + {}^{3}B^{*}$	(3) Quenching of biacetyl phosphorescence by the analyte: ${}^{3}B^{*} + A \xrightarrow{k_{-1}} B + {}^{3}A^{*}$ $\searrow B + h\nu_{p}^{B}$
(4) Phosphorescence of biacetyl ${}^{3}B^{*} \rightarrow B + h\nu_{p}^{B}$	

Reprinted with permission from J.J. Donkerbroek, N.J.R. van Eikema Hommes, C. Gooijer, N.H. Velthorst, and R.W. Frei, *J. Chromatogr. 1983*, *255*, 581.

direct RTPL intensity, which is indicated by a negative peak in the chromatogram. As shown in Table 9.2, a negative peak can be expected if the deactivation of $^3B^*$ by energy transfer to the analyte is capable of competing with other deactivation processes. This is fulfilled if $k_{-t}[A]$, where k_{-t} is the rate constant and $[A]$ the analyte concentration, and is roughly the same order of magnitude of $1/\tau_0^B$, where τ_0^B is the triplet lifetime of biacetyl in the absence of the analyte and is about 5×10^{-4} s under the experimental conditions applied by Donkerbroek et al. (7).

Donkerbroek et al. (10) derived Equation (9.3), where I(dir)

$$I'(dir)^{-1} = I(dir)^{-1} + \frac{k_Q[Q]}{I^B{}_{abs}\theta^B{}_{isc}k^B{}_p} \qquad (9.3)$$

is the direct RTPL intensity of biacetyl with no quencher present; I'(dir) is the RTPL intensity of biacetyl with a quencher present, k_Q is the bimolecular rate constant of the quenching reaction; $[Q]$ is the concentration of the quencher; $I^B{}_{abs}$ is the rate of light absorption by biacetyl; $\theta^B{}_{isc}$ is the triplet formation efficiency of biacetyl; and $k^B{}_p$ is the rate constant of the phosphorescence process. Equation (9.3) shows that a linear detector based on quenching can be obtained by measuring the reciprocal of the partially quenched biacetyl phosphorescence intensity, that is, by measuring I'(dir). Thus, by plotting I'(dir)$^{-1}$ vs. $[Q]$, a straight line with a slope proportional to k_Q and an intercept equal to the inverted intensity of the unquenched signal would be obtained.

9.6. Applications for Quenched RTPL

The experimental conditions and equipment are essentially the same for quenched RTPL as for sensitized RTPL. These were discussed in Section 9.3.

As discussed in Section 9.4, Donkerbroek et al. (7) detected polychloronaphthalenes in liquid chromatography by ultraviolet and sensitized phosphorescence detection. In the same work, they employed the partial quenching of the direct RTPL of biacetyl by the analyte. They demonstrated the use of the two phosphorescence detection techniques as well as ultraviolet detection with mixtures of Halowaxes, which consist mainly of polychloronaphthalenes containing three or four chlorine atoms per molecule. The authors obtained three chromatograms of a Halowax sample using ultra-violet, sensitized RTPL, and quenched RTPL detection. By using the spectral data from the three detection methods, they were able to characterize fifteen peaks in the chromatograms. Thus, their work showed that by combining ultraviolet detection and phosphorescence detection, new ways were available for the identification and characterization of polychloronated compounds.

Donkerbroek et al. (10) investigated various aspects of quenched RTPL for flow injection analysis and liquid chromatography. They used various compound classes such as chloroanilines, amines, heterocyclic nitrogen compounds, sulfur organics, chlorophenols, other aromatic and aliphatic hydroxyl compounds, and several inorganic ions. As is indicated by Equation (9.3), the sensitivity of the quenched RTPL approach for a particular analyte is proportional to the value of k_Q. Donkerbroek et al. (10) estimated the limit of detection (LOD) for quenched RTPL with Equation (9.4).

$$\text{LOD (in M)} = 10/k_Q \text{ (in M}^{-1}\text{s}^{-1}) \qquad (9.4)$$

Equation (9.4) suggests that the determination of k_Q values for a number of analytes is important as a screening technique to examine the analytical potential of the detection method. Also, it would be expected that analytes with triplet energies significantly lower than the triplet energy of biacetyl would produce diffusion controlled quenching by energy transfer (10). Thus, these analytes would be detectable by the RTPL quenching of biacetyl. However, Donkerbroek et al. (10) investigated mainly compounds with higher triplet energies than biacetyl, which illustrated that, for quenching, other mechanisms could also occur. As an example of detection limits using liquid chromatography, the following were obtained for sulfur compounds: thiourea, 0.5 ng; thiohydantoin, 3.7 ng; ethylene-thiourea, 1.5 ng; 2-mercapto-1-methylimidazole, 1.3 ng. The authors concluded that quenched RTPL has analytical potential for several groups of compounds that may have inherently poor detection properties by other techniques. Also, quenched RTPL can be complementary to sensitized RTPL.

Gooijer et al. (11) reported the detection of platinum(II) complexes in liquid chromatography using quenched biacetyl phosphorescence. The chromatography was performed on a solvent-generated anion-exchange column prepared by coating a C_{18} column with hexadecyltrimethylammonium bromide. Platinum coordination complexes are well known agents with antitumor activity. Cisplatin [CDDP, cis-dichlorodiamineplatinumII)] is used in the treatment of solid tumors. Other platinum compounds that have a high therapeutic index are carboplatin (CBDCA) and iproplatin (CHIP). From chromatograms for CDDP and CBDCA, the detection limits were calculated as 3.0×10^{-7} and 3.3×10^{-7} M, respectively. Experiments on urine samples revealed that CDDP and CBDCA could be observed separately, although interference effects from phosphorescence-quenching compounds in the matrix were not negligible. This problem could possibly be solved by using precolumn sample-handling techniques. Figure 9.3 gives the quenched phosphorescence chromatogram of a urine sample spiked with CDDP and CBDCA. No pretreatment of the sample was employed. In Figure 9.3 only the CDDP peak is observed. However, by increasing the methanol content of the mobile phase, Gooijer et al. (11) were able to observe peaks for both CDDP and CBDCA.

A detection method for the determination of nitrite and

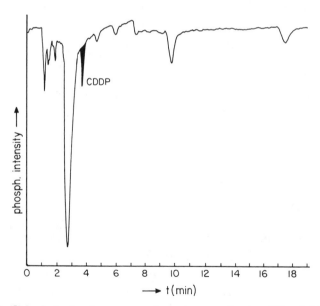

Figure 9.3. Quenched phosphorescence chromatogram of urine (diluted 1:7 in eluent) spiked with CDDP ($1.0 \cdot 10^{-5}$ M) and CBDCA ($1.4 \cdot 10^{-5}$ M). Chromatographic conditions are given in the text; the eluent contained no methanol, the citrate buffer concentration was $1 \cdot 10^{-2}$ M and the pH 5.0. Injection volume: 20 μL. Flow-rate: 1.1 mL min^{-1} (Reprinted with permission from C. Gooijer, A.C. Veltkamp, R.A. Baumann, N.H. Velthorst, R.W. Frei, W.J.F. Van Der Vijgh, *J. Chromatogr.* 1984, 312, 337.)

sulfite, based on the quenched RTP of biacetyl using ion chromatography, was described by Gooijer et al. (12). The detector was
placed directly after a high-performance liquid chromatographic
anion-exchange column. With electronic signal inversion, positive
peaks and calibration graphs that were linear over two orders of
magnitude of concentration were acquired. Sulfite ions were
detected indirectly with the thiosulphate ion, which was produced
by the reaction of sulfite with sulfur. They showed that nitrite could
be determined in process meats without interference from other
components. The determination limits for nitrite were one or two
orders of magnitude better than for other known techniques. A
detection limit of 0.02 nmole was obtained for nitrite by quenched
RTPL. For wine samples, further improvement of the chromatographic procedure was needed for reliable quantitation of the
thiosulphate peak.

Baumann et al. (13) evaluated time-resolved quenched
phosphorescence as a detection method for chromate based on
paired-ion reversed-phase high-performance liquid chromatography.
They used a Perkin-Elmer LS-2 luminescence detector which could
be operated in a pulsed-source time-resolved phosphorescence
mode. The emission signal from the sample was gated, during a
time interval t_g, after the source flash by a delay time t_d. Baumann
et al. (13) derived a modified Stern-Volmer equation that related the
signal intensities in the absence and presence of quenchers, I'_0 and
I', for the time-resolved mode, namely, Equation (9.5).

$$\frac{I'_0}{I'} = (1 + \tau_0 k_q[Q])\exp(t_d k_q[Q]) \qquad (9.5)$$

In Equation (9.5), τ_0 is the triplet lifetime of biacetyl in the absence
of quencher; [Q] is the quencher concentration (M); and the term
k_q is the bimolecular rate constant of the quenching reaction
($M^{-1}s^{-1}$). As shown in Equation (9.5), the intensity ratio is the
product of a linear and an exponential term. The influence of the
exponential term decreases when t_d is shortened. A linear relationship was obtained for I'_0/I' vs. chromate concentration for a t_d value
of 0.01 ms. The authors commented that for routine analysis with
the quenched biacetyl phosphorescence system, a combination t_d =
0.01 and t_g = 1.00 ms could be used. To show the applicability of
their method, tapwater and surface water samples were spiked with
chromate, filtered over a 0.2 μm filter and after deoxygenation

injected onto the column. If time-resolution was applied and the instrumental parameters were chosen carefully, the authors concluded that the sensitivity of the method achievable should be sufficient for the control of drinking water quality at the 50 ppb limit.

Baumann et al. (14) reported immobilized 1-bromonaphthalene for quenched phosphorescence detection in high-performance liquid chromatography. Previously, covalent immobilization of phosphors was not employed in sensitized and quenched phosphorescence detection. Some of the advantages of this approach are: (a) solvent compatibility will be improved; (b) immobilized phosphors will not be consumed; (c) because the phosphor is not a constituent of the eluent, various chemical and electrochemical oxygen removal procedures can be used. One possible disadvantage would be the background scattering caused by the support in the detector cell.   Nevertheless, in contrast to fluorescence, with phosphorescence, the scattering could be efficiently suppressed by the use of temporal resolution. Baumann et al. (14) packed the immobilized phosphor in a pulsed-source time-resolved luminescence detector. Various measuring conditions were optimized, and several immobilized batches were synthesized and compared for their performance. The band broadening of the solid-state system was investigated using nitrite as a test ion. In general, a very narrow chromatographic band was obtained for nitrite ion. Table 9.3 compares the phosphorescence intensity and lifetime, in the absence of quencher, of 1-bromonaphthalene chemically bonded to different carriers.   All the batches tested were very stable. Overnight exposure to continuous ultraviolet radiation did not influence the phosphorescence intensity, and under flow conditions, the batches could be used at least 6 months without losing their characteristics.

9.7.   Concluding Comments

RTPL can be a useful and powerful technique for detection in liquid chromatography and flow injection analysis. The RTPL approach is complementary to fluorescence detection; however, with minor modifications, the same detector can be employed for both RTPL and fluorescence. Also, sensitized and quenched RTPL are complementary to each other. Quenched RTPL, in particular, has considerable potential for use in detector systems. For example,

Table 9.3. Comparison of Phosphorescence Intensity I_p and Lifetime in Absence of Quencher τ° of 1-Bromonaphthalene Immobilized on Different Carriers

Batch no.	Carrier	I_p	τ°, ms
1	Silica Gel 60[a]	62	0.9
2	Silica Gel 60[b]	65	0.9
3	Controlled pore glass[a]	70	1.0
4	Controlled pore glass[b]	38	0.9
5	Silica Gel 100[a]	13	0.8
6	Silica Gel 100[a,c]	9	1.0

[a]Silanization in dry toluene.

[b]Silanization in water-saturated toluene.

[c]Endcapped with trichlorosilane.

Reprinted with permission from R.A. Baumann, C. Gooijer, N.H. Velthorst, R.W. Frei, I. Aichinger, and G. Gubitz, *Anal. Chem.*, 60, 1237. Copyright 1988 American Chemical Society.

the approach could be used for many nonultraviolet-absorbing compounds such as those found in ion chromatography. With instrumental improvements, temporal resolution should be more widely applicable. Also, by exploiting environmental factors, structural features, and the heavy-atom effect, improved sensitivity and selectivity should be achieved.

References

1. Donkerbroek, J. J.; Elzas, J. J.; Gooijer, C.; Frei, R. W.; Velthorst, N. H. *Talanta* 1981, 28, 717.

2. Donkerbroek, J. J.; Gooijer, C.; Velthorst, N. H.; Frei, R. W. *Anal. Chem.* 1982, 54, 891.

3. Birks, J. B. *Organic Molecular Photophysics*; Wiley: New York, 1975; Vol. 2, Chapter 3.

4. Gooijer, C.; Velthorst, N. H.; Frei, R. W. *Trends Anal. Chem.*, 1984, 3, 259.

5. Frei, R. W.; Velthorst, N. H.; Gooijer, C. *Pure Appl. Chem.* 1985, 57, 483.

6. Donkerbroek, J. J.; Veltkamp, A. C.; Praat, A. J. J.; Gooijer, C.; Frei, R. W.; Velthorst, N. H. *Appl. Spectrosc.* 1983, 37,

188.

7. Donkerbroek, J. J.; van Eikema Hommes, N. J. R.; Gooijer, C.; Velthorst, N. H.; Frei, R. W. *J. Chromatogr.* 1983, 255, 581.

8. Donkerbroek, J. J.; van Eikema Hommes, N. J. R.; Gooijer, C.; Velthorst, N. H.; Frei, R. W. *Chromatographia* 1982, 15, 218.

9. Baumann, R. A.; Gooijer, C.; Velthorst, N. H.; Frei, R. W. *Anal. Chem.* 1985, 57, 1815.

10. Donkerbroek, J. J.; Veltkamp, A. C.; Gooijer, C.; Velthorst, N. H.; Frei, R. W. *Anal. Chem.* 1983, 55, 1886.

11. Gooijer, C.; Veltkamp, A. C.; Baumann; R. A.; Velthorst, N. H.; Frei, R. W.; Van Der Vijgh, W.J.F. *J. Chromatogr.* 1984, 312 337.

12. Gooijer, C.; Markies, P. R.; Donkerbroek, J. J.; Velthorst, N. H.; Frei, R. W. *J. Chromatogr.* 1984, 289, 347.

13. Baumann, R. A.; Schreurs, M.; Gooijer, C.; Velthorst, N. H.; Frei, R. W. *Can. J. Chem.* 1987, 65, 965.

14. Baumann, R. A.; Gooijer, C.; Velthorst, N. H.; Frei, R. W.; Aichinger, I.; Gübitz, G. *Anal. Chem.* 1988, 60, 1237.

CHAPTER 10

MICELLE-STABILIZED, CYCLODEXTRIN, AND COLLOIDAL/MICROCRYSTALLINE ROOM-TEMPERATURE PHOSPHORESCENCE

10.1. Introduction

Surface-active agents (surfactants) that form micelles, cyclodextrins, and colloidal/microcrystalline aggregates have been found to produce some very interesting photophysical phenomena and have formed the basis of several new analytical approaches (1). These systems are considered as organized media in which the microscopically organized chemical assemblies are employed to provide a microscopically heterogeneous fluid solution. The organized aggregates have the unique property to organize reactants on a molecular basis, which brings interacting species together with high specificity. In this chapter, micelle-stabilized, cyclodextrin, and colloidal/microcrystalline room-temperature phosphorescence in solution will be considered.

10.2. Micelle-Stabilized Room-Temperature Phosphorescence

10.2.1. Micelles and Surfactants

Several authors have reviewed the photophysical and photo-chemical processes in micellar systems (2-4). In addition, a general review of the use of micelles in analytical chemistry has appeared (5). Micelles consist of surfactants which may be anionic, cationic or nonionic in nature. They are alike in that they each contain a hydrophobic portion and a charged or polar section in each mole-cule. Above a critical micelle concentration (CMC), surfactants assemble in a connivent, semiordered fashion to form submicro-scopic aggregates which are usually spherical or ellipsoidal in shape. Normal micelles, such as those formed from sodium dodecyl sulfate, aggregate in solution with the ionic head groups oriented outward and the hydrophobic portion oriented inwards. This condi-tion forms two regions of widely different polarity. The micelle is more accurately considered as a two-dimensional rather than a one-dimensional amphipathic form onto whose surface hydrophobic and hydrophilic molecules associate with different degrees of binding strength (1,5). The amphipathicity of the surface of the micelle is a property common with the surfaces of proteins and membranes (6). For example, serum albumin has a high affinity for nonpolar molecules such as steroids and also interacts strongly with ions (5). Materials such as sodium diisooctyl sulfosuccinate form inverted micelles in nonaqueous solvents with the charged head groups at the center and the hydrophobic tails outwards. The site of highest polarity is just the opposite for normal versus inverted micelles. It would be expected that inverted micelles would tend to incorporate polar species, which would be repelled out of the nonpolar bulk solvent. With normal micelles, one would expect nonpolar species to be incorporated because they would be repelled out of the polar solvent. The solubilized species is not rigidly held in the micelle, but can move around within or upon the micelle. The analyte is in constant dynamic equilibrium. For dissolved lumiphors, a wide variation of rate constants of association and dissociation would be expected with the micelles, depending on the hydrophobicity and electrostatic nature of the lumiphor. For micelles to influence the photophysical properties of a lumiphor, the luminescent species must have an appreciable residence time in the micelle. Thus, it is important to consider the relationships between the micellar-solute rate constants and the spectroscopic rate constants. Generally, optimum experimental conditions can be obtained by changing the nature and type of surfactant used to form the micellar assembly.

10.2.2. Experimental Conditions

The two fundamental requirements for the observation of micelle-stabilized room-temperature phosphorescence (MS-RTP) for most luminescent components are the presence of a heavy atom and the removal of oxygen. In a typical experiment, the solid analytes are dissolved in sodium lauryl sulfate (NaLS) solution and then diluted with thallium lauryl sulfate/sodium lauryl sulfate (TlLS/NaLS) solution to give a total detergent concentration of 0.10 M with 30/70 ratio of Tl/Na (7). The solution is deoxygenated in the sample cell with high purity nitrogen for 15-20 min. The samples can then be analyzed using conventional fluorescence instrumentation. However, when the phosphorescence and fluorescence bands overlap, time discrimination must be used so that phosphorescence can be measured in the absence of fluorescence.

10.2.3. Theory and Applications

Cline Love et al. (8) were the first to show the analytical usefulness of MS-RTP. Analytical data were presented for naphthalene, pyrene, and biphenyl in Tl/Na and Ag/Na mixed counterion lauryl sulfate micelles. The average precision of the measurements was about 6%, and the sensitivities were comparable with other phosphorescence techniques. Sample temperature was found to have a moderate effect on RTP intensity, and the ratio of heavy metal/sodium in the micelle had a dramatic effect on the relative luminescence intensity up to $\sim$ 10-20% heavy atom. Because of the presence of a heavy atom, the fluorescence signal was greatly diminished and the RTP signal enhanced. Cline Love et al. (8) did not observe RTP in the absence of heavy atoms. Also, no RTP was observed below the critical micelle concentration. Several RTP analytical curves and limits of detection were given. For example, for naphthalene in NaLS/TlLS and NaLS/AgLS micelles, the linear dynamic range for both systems were over two decades and the limits of detection were 7×10^{-7} and 9×10^{-7} M, respectively. One of the most significant advantages of MS-RTP is the ability to work with fluid solutions at room temperature, which eases sample handling and sample preparation. The main disadvantage of the approach is the time required to degas the samples to prevent oxygen quenching.

Skrilec and Cline Love (9) reported the MS-RTP character-istics of functionally substituted aromatic molecules in aqueous Tl/Na lauryl sulfate micellar solution. Data were reported for ketones, aldehydes, alcohols, carboxylic acids, phenols, and amines. In several cases, the limits of detection compared favorably with low-temperature (77 K) phosphorimetry results. They observed some selectivity in that electron-donating substituents produced larger MS-RTP intensities compared to electron-withdrawing

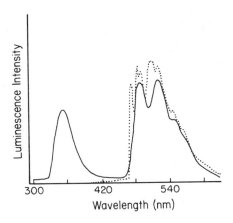

Figure 10.1. Luminescence spectra of 2-naphthol. (——) 2-Naphthol in TlLS/NaLS; residual fluorescence below 450 nm, phosphorescence above 450; phosphorescence recorded at 5X sensitivity for fluorescence signal. (- - -) 2-Naphthol in ethanol, 77 K. (Reprinted with permission from M. Skriliec and L.J. Cline Love, *Anal. Chem.*, 52, 1559. Copyright 1980 American Chemical Society.)

groups substituted on the same luminescent component. Also, some selectivity was observed for positional isomers. Triplet-state lifetimes were compared for micelle-stabilized phosphorescence at room temperature and at low-temperature in ethanol. For the naphthalene series of compounds, the MS-RTP lifetimes ranged from 0.28 to 0.69 ms, whereas the low-temperature lifetimes were from 0.44 to 2.4 s. Figure 10.1 shows the luminescence spectra for 2-naphthol in TlLS/NaLS and at 77 K in ethanol. MS-RTP data were given for seventeen compounds. The compounds were categorized into naphthalene series, pyrene series, and biphenyl series. The limits of detection were in the range of 6.0×10^{-10} M to 27×10^{-8} M.

The influence of analyte heavy-atom micelle dynamics on RTP lifetimes and spectra were reported by Cline Love et al. (10). They were interested in the potential analytical usefulness of MS-RTP lifetimes and the information provided about the molecular dynamics of MS-RTP compared to low-temperature phosphorimetry. Also, they described an instrument capable of measuring lifetimes down to 10 μs. A kinetic model for the partitioning of an excited triplet-state molecule between the aqueous phase and the micelle, with general equations showing the effects of kinetic properties of the system on the lifetime for relatively long-lived species, was developed by Almgren et al. (11). Cline Love et al. (10) considered the implications of the model for the potential analytical utility of MS-RTP. In particular, they pointed out that the dynamic property of fluid systems adds a useful parameter that permits, in some cases, differentiation of species based on triplet-state lifetimes at room temperature, where similar differentiation at 77 K or on solid substrate is not possible.

In the absence of secondary processes, radiative deactivation of the triplet state follows a single exponential decay, and the concentration of excited triplet-state species in the micelles, at any point in time, can be described by a first-order kinetic model. Cline Love et al. (10) showed that the phosphorescence lifetime is described by Equation (10.1).

$$\frac{1}{\tau} = k_- + k_{MP} + k_{qIN}[Q]_{IN} - \frac{k_- k_+[M]}{k_p + k_{qEX}[Q]_{EX} + k_+[M]} \qquad (10.1)$$

Equation (10.1) shows that the observed lifetime depends on the following: (a) rate constants for exit from micelle (k_-), deactivation within the micelle (k_{MP}), and deactivation outside the micelle (k_p); (b) products of concentration of internal quencher and its quenching rate ($k_{qIN}[Q]_{IN}$) and external quencher and its quenching rate ($k_{qEX}[Q]_{EX}$); and (c) reentry rate and concentration of micelles ($k_+[M]$). It should be pointed out that radiationless deactivation rates of the analyte in each location, such as internal conversion, self-quenching, and effects due to the presence of heavy atoms, are included, along with the radiative rates, in k_{MP} and k_p. Cline Love et al. (10) discussed several assumptions and simplifications for Equation (10.1). For example, if the concentration of internal and/or external quenchers is negligible, Equation (10.1) reduces to Equation (10.2).

$$\frac{1}{\tau} = k_- + k_{MP} - \frac{k_- k_+[M]}{k_p + k_+[M]} \qquad (10.2)$$

By comparing typical values for some of the terms in Equation (10.2) and considering the dynamics of the system, some of the rate constants may be substantially larger than others. Almgren et al. (11) showed that k_+ was much larger than k_- for all situations investigated. Also, it is assumed that k_+ is much larger than k_p because k_+ has been estimated to be about 10^9 M^{-1} s^{-1} (10,11). If k_p

were the same magnitude as k_+, that would indicate that the triplet-state lifetime in the external phase would be in the nanosecond range, which is not likely based on the present knowledge of the triplet state. Several other theoretical conclusions were considered, based on Equations (10.1) and (10.2) (10). Single-component MS-RTP lifetimes and two-component MS-RTP lifetimes were also discussed. For example, from single-component lifetime data and solubility data from pyrene, biphenyl, and naphthalene, it was suggested that a convenient method of predicting the approximate ordering of MS-RTP lifetimes based on their solubility characteristics alone was feasible when no other specific interactions were present. Also, by comparing lifetime data in solution at 25°C and 77 K, the dynamic nature of the micellar environment was accentuated. The ordering of MS-RTP lifetimes was different from that at 77 K. This result illustrated the effect of the micelle fluid environment upon the analytes lifetimes. In addition, solvent reorientation, equilibration of the initial excited Franck-Condon state, and several solvent/excited-solute reactions are minimized at 77 K, but would be possible in fluid solution.

Cline Love et al. (10) considered the feasibility of measuring lifetimes and spectra for two-component probe mixtures. Table 10.1 gives the results obtained. The deviation from single-component values in the two-component lifetime determinations was discussed in some detail. For example, the concentrations of the species, overlap of fluorescence and absorbance bands, appropriate excitation wavelengths, and sensitivity of the detector system were considered. They also studied the average

Table 10.1. Triplet-State Lifetime of a Two-Component Mixture by MS-RTP

Component	Mixture[a] lifetimes τ, ms	Single-component lifetimes τ, ms	%Error
Pyrene	1.04 (0.984)	0.93	10.6
Naphthalene	0.62 (0.998)	0.45	27.4

[a]Concentration ratio 1:2 molar pyrene to naphthalene in Tl/NaLS micelle; correlation coefficients in parentheses. The average relative standard deviation of regression coefficient was 8.9% for pyrene and 2.0% for naphthalene.

Reprinted with permission from L.J. Cline Love, J.D. Habarta, and M. Skrilec, *Anal. Chem.*, 53, 437. Copyright 1981 American Chemical Society.

microenvironment for various analytes in heavy-atom micelles by studying their MS-RTP, fluorescence, and absorbance characteristics in different polarity solvents and micelle systems.

A background correction method for interfering radiation present in MS-RTP was reported by Cline Love and Skrilec (12). They developed a simple spectral correction technique for interfering scatter and fluorescence, even for cases where the energy overlap with the phosphorescence band of interest was substantial. When the interfering signal was due to fluorescence and scatter and the signal of interest was due to MS-RTP, then the unwanted signal could be subtracted from the total luminescence signal to obtain only the triplet-state emission signal by taking advantage of the presence of oxygen in the sample solution. Oxygen is an effective triplet-state quencher. The standard and sample solutions were deaerated first, and the MS-RTP intensity of each sample was measured. Then the solutions were saturated with air, and the luminescence emission was measured again using the same instrumental settings. Standard and sample response curves were constructed by plotting relative MS-RTP intensities vs. the analyte concentration. Using the response curves, the background was corrected by using a subtraction method. However, the subtraction method did not correct for the background noise level. The method was applied to a synthetic sample of phenanthrene/2-naphthaldehyde, a coal-derived hydrogen-donor recycle solvent, and an automatic computerized spectrum correction for N-(2-chloroethyl)-carbazole.

The first report of MS-RTP from heterocyclic species dissolved in micellar media was published by Skrilec and Cline Love (13). The MS-RTP energies and triplet-state lifetimes of carbazole, N-(2-cyanoethyl)carbazole, N-(2-iodoethyl)carbazole, and N-(2-chloroethyl)carbazole were observed in mixed-counterion thallium/sodium lauryl sulfate aqueous micellar solution. The lifetimes of the compounds in 77 K ethanol solution and the fluorescence energy maxima of these and other derivatives dissolved in sodium lauryl sulfate, water, and hexane, respectively, were reported. The results they obtained were rationalized as a combination of substituent effects, microenvironment polarity, and heavy-atom effect. Relatively few phosphorescent heterocyclic compounds have yielded MS-RTP. However, apparently there appears to be no intrinsic reason why they cannot be made to give MS-RTP under more favorable experimental conditions. For the carbazole series investigated by Skrilec and Cline Love (13), the most important factor for obtaining MS-RTP appeared to be the relative solubility of the species in the micellar aggregate as influenced by the substituted species polarity and the efficiency of the micelle in organizing external heavy atoms and chromophores for efficient spin-orbit coupling. It should be possible by prudent choice of heavy atoms and surfactant to induce MS-RTP in other heterocyclic aromatic compounds. In later work, Woods and Cline Love (14) reported the MS-RTP properties of phenazine, acridine, and anthracene. They proposed a complex between silver ion and pyridine nitrogen for the nitrogen compounds by which silver ion enhanced the phosphorescence by both the external heavy-atom effect and donor-acceptor complexation. Thallium ion formed a weaker complex and was considerably less effective in inducing RTP in the pyridinic heterocycles. Apparent prototropic equilibria for acridine in micellar solution indicated substantial differences between the bulk phase pH and apparent pH in the region of the micellar assembly where the acridine resided.

Liquid chromatographic phosphorescence detection with micellar chromatography and postcolumn reaction modes were investigated by Weinberger et al. (15). A micellar mobile phase consisting of 0.15 M sodium/thallium lauryl sulfate (70/30) was used for both the separation step and for detection because all the reagents needed for micellization and spin-orbit coupling enhancement were present. In addition, a reversed-phase separation could be carried out and then the micellar reagents introduced postcolumn. The micellar approach was less sensitive than classical fluorescence detection; however, dramatic improvement in selectivity was achieved for the analysis of a mixture of β-naphthol, biphenyl, and phenanthrene by careful selection of excitation and emission wavelengths. The limit of detection was as low as 5 ng in some cases, and linear dynamic ranges covered 3 orders of magnitude. The postcolumn approach was about one order of magnitude less sensitive because of the harmful effects of methanol on the micelle. The precision of both techniques was less than 2% relative standard deviation. The authors emphasized that the intrinsic precision and resolving power of high-performance liquid chromatography facilitated the routine observation of MS-RTP for three reasons: (a) an in situ purification is performed on each solute in the sample, and thus, the detector is presented with a pure solute without the presence of quenchers; (b) with the purification process, it is unnecessary to degas the individual samples, and oxygen is not retained on the column and elutes with the solvent front; (c) the precise flow of the solvent delivery system permits each solute to pass through the detector flow cell underconsistent conditions. This produces a solute illumination time that is constant from sample to sample. Weinberger et al. (15) considered several other aspects of MS-RTP for detection and quantitation in high- performance liquid chromatography. For example, differentiation of fluorescence signals from phosphorescence signals, effect of surfactant concentration, effect of thallium, effect of methanol, conditions for the postcolumn reaction mode, optimization of excitation and emission wavelength filters, sensitivity vs. selectivity, and calibration curves were discussed. Figure 10.2 shows the effect of thallium/sodium ratio in mixed Na/Tl lauryl sulfate micelles on MS-RTP intensity. The ratio had a substantial effect on the response of the system. The rate of increase of response with increasing thallium in Figure 10.2 appears to be exponential for all the solutes studied. Table 10.2 gives the estimated limits of detection for β-naphthol, biphenyl, and phenanthrene for micellar chromatography and the postcolumn reaction approach. As mentioned previously, the postcolumn chromatography approach was less sensitive due to the effect of methanol on the micelle.

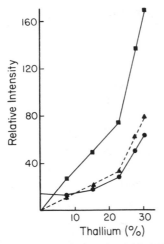

Figure 10.2. Effect of thallium/sodium ratio in mixed Na / Tl lauryl sulfate micelles on MSRTP intensity for micellar chromatography: (•) β-naphthol, (▲) phenanthrene, and (■) nasphthalene; column, Supelco CN; mobile phase, 0.15 M total lauryl sulfate. (Reprinted with permission from R. Weinberger, P. Yarmchuk, and L.J. Cline Love, *Anal. Chem.* 54, 1552. Copyright 1982 American Chemical Society.)

Table 10.2. Estimated Limit of Detection

| | Limit of detection,[a] ng | | |
Compound	Micellar chromatography[b]	Postcolumn reaction[c]	Postcolumn reaction[d]
β-Naphthol	8	22	120
Biphenyl	20	30	60
Phenanthrene	3	9	33

[a]Excitation wavelength, 280 nm for β-naphthol, 260 nm for biphenyl and phenanthrene; emission filter, 550 nm cutoff; the LOD equals 3σ/peak height x ng injected, where σ equals the baseline noise.

[b]Mobile phase, 0.15 M SLS / TILS (70/30); flow rate, 1.5 mL min^{-1}; column, Supelco CN.

[c]Mobile phase, 50% methanol; flow rate, 1 mL min^{-1}; PCR reagent, 0.15 M SLS / TILS (70/30); flow rate, 5 mL min^{-1}; column, Supelco CN.

[d]Mobile phase, 75% methanol; flow rate, 1.5 mL min^{-1}; PCR regent, 0.15 M SLS / TILS (70/30); flow rate, 3 mL min^{-1}; column, Supelco C8.

Reprinted with permission from R. Weinberger, P. Yarmchuk, and L.J. Cline Love, *Anal. Chem.*, 54, 1552. Copyright 1982 American Chemical Society.

However, it was more selective because of increased chromatographic efficiency. The primary advantage of MS-RTP was less spectral interference and more selectivity by detection of signals in the red region of the spectrum, which is less crowded with emitting species than the lower wavelength region. The high-performance liquid chromatography/MS-RTP methodology has important implications for fundamental spectroscopic studies of luminescence, for the study of the nature of micellar aggregates, and for enhanced selectivity in chromatographic analysis. In related work, Arunyanart and Cline Love (16) developed a model for micellar effects on liquid chromatography capacity factors and for the determination of micelle-solute equilibrium constants.

The experimental requirements for synchronous wavelength scanning MS-RTP and the factors affecting peak resolution were discussed and compared with those for synchronous wavelength scanning fluorescence by Femia and Cline Love (17). They identified the compounds in a four-component mixture, and they discussed the criteria to minimize triplet-state energy transfer. It was also shown that considerable improvement in resolution of the synchronous peaks was obtained by second-derivative spectra. In fluorescence synchronous scanning for mixtures of fluorescent compounds which are not spectrally resolved, vibronic congestion in the wavelength region of interest will result in poorly resolved synchronous peaks. For phosphorescent compounds which differ in structure by the number or placement of aromatic rings, wavelength offsets can be used which are unique to each family. For example, biphenyl has a wavelength offset of 180 nm, whereas pyrene has a wavelength offset of 268 nm. This improves the selectivity of MS-RTP over synchronous fluorescence. For example, a mixture can be scanned at a specific offset which is entirely unique to one of the compounds or a family of compounds in the mixture. By employing different phosphorescence offsets, it is possible to selectively identify compounds in a mixture by differences in their ground state and lowest triplet-state spacings. Femia and Cline Love (17) gave representative MS-RTP spectra for a four-component mixture of pyrene, fluorene, triphenylene, and phen-anthrene. The concentrations of the components were in the range of 2.5×10^{-6} to 7.7×10^{-6} M.

Garcia and Sanz-Medel (18) described a technique that employed Na_2SO_3 as an oxygen scavenger in micellar media for MS-RTP. Oxygen is a well known quencher of phosphorescence. Degassing by inert gas purging is the most common method used to minimize oxygen in spite of the fact that the purging is time-consuming and, in micellar media, problematic because of the amount of foam formed. Garcia and Sanz-Medel used naphthalene and other aromatic hydrocarbons to monitor the effectiveness and show the convenience of the proposed method of deoxygenation for MS-RTP analysis. The method was based on Reaction 1.

$$2SO_3^{2-} + O_2 \rightleftharpoons 2SO_4^{2-} \qquad (1)$$

They investigated the reaction time for oxygen consumption by following the increase in phosphorescence intensity of the micellar-solubilized naphthalene. They found that the reaction-time profiles were significantly different, depending on the actual value of the total surfactant concentration used. In general, a steady-state RTP signal was reached in about 15 min. The authors also investigated the effects of Na_2SO_3 concentration and temperature on the RTP signals. They recommended that a Na_2SO_3 concentration of 10^{-2} M and a temperature of 15°C be used. To evaluate the analytical performance of the MS-RTP method with sulfite deoxygenation, the determination of naphthalene was selected as the reference to compare with the N_2 purging RTP procedure. Generally, the analytical figures of merit compared favorably with the N_2 purging method.

Nugara and King (19) carried out a series of experiments demonstrating that the problems arising from precipitation of Tl^+-alkyl sulfate salts at high Tl^+ concentrations in MS-RTP solutions can be circumvented by using mixed surfactant systems that include short chain alkyl sulfate. Also, in situations where the sulfite ion is used as a chemical deoxygenating agent, light absorption by a Tl^+-SO_3^{2-} complex can drastically attenuate the exciting radiation, in particular at wavelengths below 275 nm. This offsets any gains obtained from increasing the Tl^+ content of MS-RTP solutions through the use of mixed surfactants. They determined the stoichiometry, formation constant, and molar absorptivity of the Tl^+-SO_3^{2-} complex by ultraviolet spectrometry. Fluorescence and thallium-induced phosphorescence data were reported for naphthalene, phenanthrene, and pyrene solubilized in a mixed micelle MS-RTP system containing Na_2SO_3 to show the advantages of long wavelength excitation in such MS-RTP systems.

Sanz-Medel et al. (20) reported the MS-RTP of metal chelates and its application to niobium determination. They performed a detailed examination of 8-hydroxyquinoline and some of its derivatives as potential complexing reagents for the RTP determination of Nb(V) in micelles of different charge-type surfactants. Their paper was the first one dealing with the application of MS-RTP for the determination of a metal ion. From a rather extensive investigation, they found that only those complexes with a reagent bearing a sulfonic group displayed MS-RTP in a cationic micellar media using Br_3CH as an external heavy-atom material. The order of phosphorescence emission intensity was Nb(V)-ferron > Nb(V)-8-hydroxy-5-quinolinesulfonic acid > Nb(V)-8-hydroxy-7-quinolinesulfonic acid. Because of the previous results, the authors studied in detail the phosphorescence properties of Nb(V)-ferron complex. The total luminescence spectrum of the niobium complex in cetyltrimethylammonium bromide solution (CTAB) micelles with Br_3CH as a heavy-atom material and Na_2SO_2 as an oxygen scavenger is shown in Figure 10.3, along with time-resolved spectra. The three curves in the lower portion of the figure correspond to different delay times and show that as the delay time is increased, the broad fluorescence band at 480 nm decreases while the phosphorescence peak at 573 nm remains. Analytical performance characteristics were obtained, and the calibration curves were linear, the detection limit was 4 ppb,

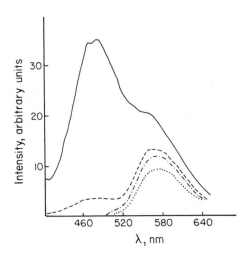

Figure 10.3. Photoluminescence spectra of niobium (V)-ferron complex at pH 5.9 in CTAB micelles for different "delay" times: (——) 0.01 ms, (- - -) 0.02 ms, (-•-) 0.03 ms, (•••) 0.04 ms; $t_g = 2$ ms; scale, X1; [Nb(V)] = 1.07×10^{-5} M, [ferron] = 3.25×10^{-5} M, [CTAB] = 5×10^{-2} M, [Na$_2$SO$_3$] = 1×10^{-2} M, 0.2% (v/v) Br$_3$CH; λ_{exc} 363 nm. (Reprinted with permission from A. Sanz-Medel, P.L.M. Garcia, and M.E.D. Garcia, *Anal. Chem.* <u>59</u>, 774. Copyright 1987 American Chemical Society.)

and the standard deviation was ± 2%. Below is a general reaction scheme for Nb(V).

$$Nb(V) + ferron \xrightarrow{pH\ 6} \begin{array}{l} [Nb(V)\text{-ferron}] \\ \text{binary soluble} \\ \text{complex weakly} \\ \text{fluorescent} \end{array} \xrightarrow[\substack{\text{CTAB} \\ \text{micelle}}]{}$$

$$\begin{array}{l} [Nb(V)\text{-ferron}] \\ \text{micellar complex} \\ \text{nonfluorescent} \end{array} \xrightarrow[- O_2(Na_2SO_3 + h\upsilon)]{+ Br_3CH} \begin{array}{l} [Nb(V)\text{-ferron}] \\ \text{micellar complex} \\ \text{PHOSPHORESCENT} \end{array}$$

The room-temperature liquid phosphorimetry of aluminum-ferron chelate in micellar media for the determination of aluminum has been reported (21). Aluminum ion reacted with 7-iodo-8-quino-linol-5-sulphonic acid (ferron) in cetyltrimethylammonium bromide

(CTAB) micelles to give a highly phosphorescent complex at room temperature. Fernandez de la Campa et al. (21) considered several experimental aspects and phosphorescence characteristics of the complex. For example, spectral characteristics of the complex, influence of surfactant concentration, oxygen exclusion, and heavy-atom effect were discussed. They explored the influence of common anions and about thirty cations on the phosphorescence intensity of the aluminum complex. Titanium(IV), V(V), Nb(V), Ta(V), Fe(II), Fe(III), Co(II), and Ni(II) interfered at all levels tested. Nitrate, sulphate, phosphate, fluoride, iodide, silicate, borate, tartrate, oxalate, and EDTA did not interfere with 100-fold excess relative to aluminum at 1 μg mL^{-1}. Citrate interfered a $\geq$ 100-fold excess. The calibration graph they obtained under the optimized conditions had a linear range up to 500 ng mL^{-1}. The detection limit was found to be 5.4 ng mL^{-1}, the relative standard was 4.5%, and the triplet lifetime of the complex was 0.18 ms. Tap waters from different urban areas and hemodialysis fluids were analyzed for aluminum by the RTP method. The results from the RTP method were compared with the results obtained by graphite-furnace atomic absorption spectrometry (GF/AAS), and satisfactory agreement was obtained. For example, for one tap water sample, the GF/AAS method gave 20.1 $\pm$ 1.4 ng mL^{-1}, whereas the RTP method gave 19.5 $\pm$ 0.8 ng mL^{-1}. For a hemodialysis fluid sample, the GF/AAS approach gave 181.2 $\pm$ 2.8 ng/mL^{-1} and the RTP method gave 187.4 $\pm$ 4.6 ng mL^{-1}.

Garcia et al. (22) investigated the characteristics of the decay of the triplet state of metal chelates in micellar media employing the RTP of Al-ferron complex in cetyltrimethylammonium bromide (CTAB) micelles. With the decay measurements, they discriminated between metal complex species and evaluated the stoichiometry of phosphorescent metal chelates. An extensive study of the effect of foreign ions on the photochemical characteristics of RTP of the aluminum complex was performed. Also, results were compared with results from Nb(V)-ferron phosphorescent complex in CTAB micelles, and likely reaction mechanisms in micellar media were outlined. In one set of experiments, the RTP signal was constant with increasing micelle concentration, which suggested that micelles might act by protecting the triplet excited state from aqueous quenchers and by separating the phosphorescent metal complex molecules from each other. This would minimize radiationless pathways such as self-quenching or triplet-triplet annihilation. The authors considered this as a "dilution effect" because the micellar aggregates served to compartmentalize and separate the metal complex molecules from potential quenchers present in the bulk solution and from themselves. Because no systematic study of the effects of added external metal ions upon the MS-RTP of metal chelates was previously reported, Garcia et al. (22) made a preliminary survey of the influence of about thirty metal ions on the phosphorescence and triplet-lifetime parameters of the aluminum-ferron complex in CTAB micellar media. The metal ions were categorized into four main groups depending upon the nature of their effect upon the AlL_3^{3-} phosphorescence properties, where L^{2-} represents the deprotonated ferron dye. In other work, they found that the phosphorescence of AlL_3^{-3} was observed in the presence of cationic CTAB micellar medium but not in the anionic sodium dodecylsulfate (NaLS) micelles. The previous finding was probably due to the anionic AlL_3^{3-} being electrostatically repelled from the anionically charged NaLS micelles, and thus, binding was not very effective. Their study has contributed to basic knowledge of micelle-stabilized RTP of various inorganic ions (22). In other work, Sanz-Medel et al. (23) discussed metal chelate fluorescence enhancement in micellar media. They also considered mechanisms of surfactant action.

O'Reilly and Winefordner (24) reported one of the first analytical uses of vesicles. However, Riehl et al. (25) used dioctadecyldimethylammonium bromide (DODAB) vesicles in chemiluminescence studies. There are several similarities between micelles and surfactant vesicles; however, there are also many differences. Surfactant vesicles are much larger and have more static aggregates. Once formed, vesicle aggregates cannot be destroyed by dilution, but micelles can be destroyed. The majority of synthetic surfactant vesicle systems demonstrate a temperature-dependent phase transition in contrast to micelle systems (26). Vesicles are prepared by sonication from surfactants such as DODAB. They are single bilayer spherical aggregates with diameters of 500-1000 Å and bilayer thickness of approximately 50 Å. Molecular motions of the individual surfactants in the vesicles involve rotations, kink formation, lateral diffusion on the vesicle plane, and transfer from one interface of the bilayer to another (24). In their work, O'Reilly and Winefordner (24) incorporated magnetite particles in dihexadecylphosphate (DHP) vesicles to influence the luminescence properties of different molecules. Carbazole was used as a model compound for RTP, and sodium sulfite was used as an oxygen scavenger. They compared the luminescence properties of carbazole with and without magnetite particles present. The RTP calibration curves for carbazole were constructed at 297/440 nm and at 330/440 nm in the absence and presence of magnetite, and the linear dynamic range was found to be over 2.5-orders of magnitude. At 297/440 nm, the limits of detection were 30 ng mL^{-1} and 9 ng mL^{-1} in the absence and presence of magnetite, respectively. At 330/440 nm, the limit of detection was 30 ng mL^{-1} both in the presence and absence of magnetite. Luminescence data with and without magnetite were also presented for 1-pyrenecarboxaldehyde, 1-pyrenebutyric acid, and 2-aminonaphthalene, and 7,8-benzoflavone. Generally, the precision of the method ranged from about 1% to 40%. However, additional experimentation may improve the precision. Vesicles provide a realistic model for membranes which are used in biological studies, and thus, luminescence in vesicles may prove to be very important in many aspects of biological investigations (24).

10.3. Cyclodextrin Solution-Phase Phosphorescence

10.3.1. Cyclodextrins

Various aspects of solid-surface luminescence with cyclo-dextrin/salt mixtures were discussed in Chapters 6 and 7. In this part of the chapter, the RTP of organic compounds complexed with cyclodextrins in solution will be considered. There are three types of cyclodextrins that are commercially available. These are α-, and β-, and γ-cyclodextrins. Cyclodextrins are macrocyclic carbo-hydrate molecules composed of α-(1,4) linkages of D(+)-gluco-pyranose units arranged in a torus. Each type of cyclodextrin differs from the other by the number of glucose units. The α-, β-, and γ-cyclodextrins are composed of 6, 7, and 8 glucose units, respectively, and they have the shape of a cone. The interior of a cyclodextrin molecule is hollow, with a specific volume, and is highly hydrophobic. Cyclodextrins form inclusion complexes with both organic and inorganic compounds. The cyclodextrins can selectively interact with a lumiphor based on the dimensions of the lumiphor. Cyclodextrins are water soluble, and chemically and structurally stable. Their structural stability is in contrast to micellar aggregates whose monomer surfactant units are in dynamic equilibrium, constantly leaving and returning to the micelle (1). Water has been extensively used as the solvent of choice for cyclodextrins. However, Patonay et al. (27) and Nelson et al. (28) have considered the influence of aliphatic alcohols on cyclodextrin inclusion complexes. The driving forces for complexation involve several factors, and these factors are not completely understood. Numerous aspects of the chemistry and applications of cyclodextrins have been discussed extensively in the literature (29-31).

10.3.2. Experimental Conditions

The experimental conditions are straightforward, but the exact conditions would depend on the type of compound to be investigated. For example, Turro, et al. (32), in investigating 1-bromo-4-naphthoyl derivatives, filtered samples through Millipore filters and then degassed the samples with nitrogen for 30 min. Scypinski and Cline Love (33) used a somewhat different procedure for polynuclear aromatic hydrocarbons. For the preparation of the inclusion complex, an aliquot of the compound of interest was added to a volumetric flask, and the solvent was evaporated gently on a hot plate. An aliquot of 1,2-dibromoethane was then added, followed by dilution with 0.01 M aqueous cyclodextrin solution. The 1,2-dibromoethane served as a heavy-atom species. The solutions were shaken vigorously by hand. As the complex formed, some precipitation due to inclusion of excess heavy-atom compound by cyclodextrin caused a cloudiness in solution. This did not cause any difficulty in acquiring reproducible, good quality phosphorescence spectra. The prepared sample was placed in a standard fluorescence cell equipped with a Teflon stopper and deaerated for 15 min with ultrahigh-purity nitrogen passed through an indicating oxygen trap. The sample preparation time was about 20 min.

10.3.3. Theory and Applications

Turro et al. (34) investigated the kinetics of inclusion of halonaphthalenes with β-cyclodextrin by time-correlated phosphorescence. The RTP of 1-bromonaphthalene and 1-chloronaphthalene was readily observed in nitrogen-purged aqueous solutions of β-cyclodextrin. Both the phosphorescence intensity and lifetime were increased by the addition of acetonitrile. It was reported that the quenching of halonaphthalene phosphorescence in aqueous solution was significantly inhibited by the addition of β-cyclodextrin as a result of a guest-host complex. The authors reported the rate constants for the formation and dissociation of the

1-bromonaphthalene/β-cyclodextrin complex from an analysis of the dependence of phosphorescence lifetime on nitrite concentration. Their results contributed to the use of RTP as a probe of molecular dynamics in the microsecond to millisecond time domain.

Turro et al. (32) reported that the 4-bromo-1-naphthoyl group is easily quenched by molecular oxygen in homogeneous solvents. Amazingly, when this lumophore was complexed with γ-cyclodextrin in aqueous solution at room temperature, RTP was observed even under 1 atm of oxygen. Phosphorescence decay data showed two types of probe/cyclodextrin complexes were formed with lifetimes of 600 μs and 3.5 ms. Oxygen completely quenched the fast decaying component, but only partially quenched the slow decaying component.

The analysis of static and dynamic host-guest associations of detergent with cyclodextrins by luminescence methodology was investigated by Turro et al. (35). They explored host-guest type associations between β- and γ-cyclodextrins and a series of cationic phosphorescence detergent probes ([n-(4-bromonaphthoyl)alkyl]-trimethylammonium bromide, alkyl=methyl (n=1, BNK-1$^+$), pentyl (n=5, BNK-5$^+$), and decyl (n=10, BNK-10$^+$)). No evidences for association with α-cyclodextrin was found. The association equilibrium constant (K), the association rate constant (k_f), and the dissociation rate constant (k_b) were obtained from a dynamic phosphorescence decay method, with $Co(NH_3)_6^{3+}$ as an aqueous-phase phosphorescence quencher. The k_f values were about 10^7 M^{-1} s^{-1} and were independent of the detergent structure. It was found though that the magnitude of k_b and K varied over several orders of magnitude as a function of the detergent and host structure. They discussed their results in terms of the importance of the dehydration of cyclodextrin and/or probe as a rate-determining step of inclusion. The strength of complexation increased in the order BNK-1$^+$ < BNK-5$^+$ < BNK-10$^+$ for γ-cyclodextrin. This order was consistent with an important role of hydrophobic interactions in the course of inclusion.

Casal et al. (36) reported the phosphorescence of lyophilized complexes between cyclodextrins and β-arylpropiophenones. β-Arylpropiophenones do not phosphoresce in homogeneous solution as a result of excited-state quenching by the β-aryl group. They reported the RTP emission of β-phenylpropiophenone and some of its derivatives in aqueous solution after complex formation with α-, β-, and γ-cyclodextrin. However, after lyophization to dryness, the RTP was enhanced several-fold. They found a dramatic dependence of emission intensity with the size of the guest substrate, and oxygen was an inefficient quencher. Because of the insensitivity toward oxygen, measurements could be carried out under air or pure oxygen. Casal et al. (36) also prepared lyophilized samples of β-phenylpropiophenone with several carbohydrates and observed RTP from the resulting samples. They suggested a co-crystallization mechanism was responsible for the RTP signals.

Scypinski and Cline Love (33) reported the first account of RTP from non-heavy-atom-containing lumiphors in cyclodextrins. 1,2-Dibromoethane was used as an external heavy atom. The resulting phosphorescence was intense, spectrally well resolved, and partially insensitive to quenching by dissolved oxygen. Only molecules that physically entered the cyclodextrin cavity were phosphorescent. This provided substantial selectivity based on lumiphor size. α-, β-, γ-Cyclodextrin were investigated, and α-cyclodextrin was the least successful in inducing RTP, most likely due to its small interior dimensions. Phosphorescence properties were presented for eighteen polynuclear aromatic hydrocarbons. Fluorescence emission was dominant without the presence of a heavy atom. The authors screened several heavy-atom species and found that 1,2-dibromoethane had the desired properties to obtain RTP in fluid solution. Below is illustrated a proposed four-step complexation equilibria between cyclodextrin(CD), lumiphor(L), and heavy atoms(HA) for the formation of the phosphorescent trimolecular complex (33).

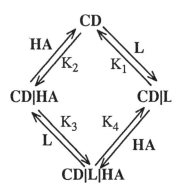

For phosphorescence to occur, K_3 and/or K_4 must be large. Concerning analytical figures of merit, the shapes of the calibration curves were similar to those obtained for micelle-stabilized RTP and were linear over 4 decades. Typical precision was less than

10% relative standard deviation, with limits of detection in the 10^{-11} to 10^{-13} M range. Also, the RTP spectral features with cyclodextrins showed better resolution of the individual vibronic bands compared to micelle-stabilized RTP.

Scypinski and Cline Love (37) reported the cyclodextrin-induced RTP of two- and three-ring nitrogen heterocycles and bridged biphenyls. As discussed in the previous paragraph, the RTP was made possible by the proposed formation of a trimolecular inclusion complex composed of lumiphor, cyclodextrin, and the external heavy-atom species 1,2-dibromoethane. Excitation and emission characteristics and percent phosphorescence unquenched by oxygen for sixteen nitrogen heterocycles in β-cyclodextrin with 1,2-dibromoethane present were given. Also, the excitation, unquenched fluorescence, and phosphorescence wavelengths of biphenyl and four bridged derivatives in β-cyclodextrin with the external heavy-atom species were discussed. Figure 10.4 shows the excitation, fluorescence emission, and phosphorescence emission spectra of 7,8-benzoquinoline in 10^{-2} M β-cyclodextrin. Even though this compound showed significant fluorescence, the RTP

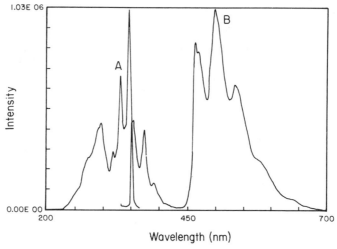

Figure 10.4. Uncorrected excitation (A) and corrected emission (B) spectra of 5×10^{-5} M 7,8-benzoquinoline in 10^{-2} M β-cyclodextrin: (A) emission wavelength, 502 nm; slits, 3.6 nm EX, 14.4 nm EM; (B) excitation wavelength, 298 nm; slits, 14.4 nm EX, 3.6 nm EM; scan rate, 1 nm s^{-1}; 1,2-dibromoethane concentration, 0.58 M. (Reprinted with permission from S. Scypinski and L.J. Cline Love, *Anal. Chem.*, <u>56</u>, 331. Copyright 1984 American Chemical Society.)

emission dominated the spectrum.   Based on the luminescence
spectral data for quinoline, isoquinoline, and several derivatives, it
was postulated that axial inclusion of quinoline and isoquinoline
occurred with cyclodextrin and 1,2-dibromoethane.   Axial inclusion
of the quinolines is illustrated below.   Of the two-ring compounds

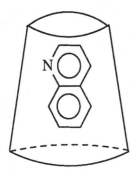
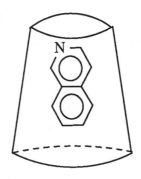

containing two nitrogens, only quinazoline gave RTP.   Three-ring
azaphenanthrenes and diazaphenanthrenes both gave unquenched
fluorescence and RTP.   The RTP of biphenyl and its bridged
derivatives provided details on the inclusion mechanism for these
compounds. Biphenyl only gave a weak RTP signal because of the
flexible nature of the biphenyl molecule.   However, bridging the
two rings by the formation of a five-membered ring between the
two benzene rings forced the molecule to be planar, and more
intense RTP was observed. For example, fluorene, whose two rings
are connected by a methylene group, gave almost an order of
magnitude higher phosphorescence intensity than for biphenyl at the
same concentrations.

DeLuccia and Cline Love (38) reported the effect of cyclodextrin cavity size on the sensitization of the RTP of biacetyl. (See
Chapter 9 for more information on sensitized phosphorescence.)
They found that the triplet-triplet energy-transfer reactions of
several polynuclear aromatic compounds were enhanced to different
extents in α-, β-, and γ-cyclodextrin media when biacetyl was
employed as an acceptor species.   The RTP intensity of biacetyl
increased as the size of the cyclodextrin cavity increased. Because
cyclodextrin cavities have specific dimensions, they place
constraints on the number of donor and acceptor molecules that can
be included in the cavity. The larger the cavity the more effectively
it is able to include the reactants and, most likely, to position them
to increase the rate of triplet-triplet energy transfer. Chrysene and

Table 10.3. Phosphorescence Excitation and Emission Wavelengths and Sensitized Limits of Detection (LODs) for Selected Polynuclear Aromatic Compounds in α-, β- and γ-CD Solutions

Compound	LOD, M^a		
	α-CD	β-CD	γ-CD
Biphenyl	2.3×10^{-6}	5.4×10^{-7}	4.1×10^{-7}
Naphthalene	5.6×10^{-6}	3.6×10^{-7}	2.9×10^{-7}
Chrysene	not detected	not detected	2.6×10^{-6}
Triphenylene	not detected	not detected	2.9×10^{-6}
Phenanthrene	3.2×10^{-6}	9.5×10^{-7}	5.0×10^{-7}

[a]Measured at 522 nm with a typical precision of 0.2×10^{-X} M.

Reprinted with permission from F.J. DeLuccia and L.J. Cline Love, *Talanta* 1985, 32, 665.

triphenylene, which were unresponsive as biacetyl sensitizers in α- and β-cyclodextrin, were excellent sensitizers of biacetyl in γ-cyclodextrin. γ-Cyclodextrin has the largest cavity of the three cyclodextrins. Table 10.3 gives the limits of detection for five polynuclear aromatic compounds. The limits of detection increased in the order γ-cyclodextrin < β-cyclodextrin < α-cyclodextrin. In related work with cyclodextrins, Jules et al. (39) considered the fluorescence enhancement of several hallucinogenic drugs in cyclodextrin media, and Femia et al. (40) discussed the fluorescence characteristics of polychlorinated biphenyl isomers in cyclodextrin media.

The RTP spectra and excited-state lifetimes were reported for several polynuclear aromatic hydrocarbons included within the cavity of heptakis(6-bromo-6-deoxy-β-cyclodextrin) (Br-β-CD) by Femia and Cline Love (41). They synthesized Br-β-CD by replacing the primary hydroxyl groups with bromine. By having the heavy atoms in fixed positions on the cyclodextrin molecule, the phosphorescence and dynamic interactions depended mainly on the entrance rate and exit rate constants of the lumiphor from the cyclodextrin. This simplified the kinetic scheme and data interpretation. The researchers used a mixture of N,N-dimethyl-formamide (DMF) and water, due to the extreme water insolubility of Br-β-CD, to determine the degree of bulk solvent hydrophobicity on the inclusion process. The previous effect was investigated by differences in the Br-β-CD spectral profile of phenanthrene relative to that in aqueous unbrominated β-cyclodextrin/1,2-dibromoethane and also by the fluorescence peak ratio variations as a function of the DMF/water ratio. Indirect evidence was provided by luminescence lifetime values on the relative exit-to-entrance rate constant ratios. They found that the optimum DMF/water ratio for maximum phosphorescence intensity was 4/1, and the optimum concentration of brominated β-cyclodextrin in the 4/1 DMF/water was 0.01 M. Their results showed that the heavy-atom effect was capable of exerting influence either through attachment of bromine to the fixed carbon skeleton of the β-cyclodextrin complexed to a lumiphor or from the close approach of the lumiphor to the bromo-substituted, narrow end of the cyclodextrin molecule.

Sushe et al. (42) reported the determination of α-naphthaleneacetic acid by β-cyclodextrin induced RTP with 1,2-dibromoethane as a heavy-atom material in fluid solution. They studied the experimental conditions for the formation of the inclusion complex and discussed the optimum experimental conditions for the determination of α-naphthaleneacetic acid. Sushe et al. (43) further studied the determination of α-naphthalene-acetic acid by β-cyclodextrin induced RTP with 1,2-dibromopropane as the heavy-atom species. Experimental conditions, calibration curves, and precision data were presented. Calibration curves were linear over 0-10^{-7} M, 0-10^{-7} M, 0-10^{-5} M, depending on the experimental condition used.

10.4. Micellar-Cyclodextrin Solutions

Cline Love et al. (44) considered the ability of microscopically organized media, in the form of surfactant micelles and α- and β-cyclodextrins, to augment the luminescence phenomena of several licit and illicit drugs. Because many physiological samples cannot be measured by direct spectrometric measurement prior to pretreatment, liquid chromatography was also investigated. The fluorescence enhancement of various hallucinogenic drugs such as N,N-dimethyltryptamine, mescaline, and ibogaine was observed in cyclodextrin media compared to homogeneous solutions. They found that heavy-atom-substituted sodium dodecyl sulfate micelles yielded phosphorescence from cationic and/or hydrophobic drugs at room temperature in fluid solution. Drugs such as propranolol, diflunisal, naphalozine, and some quinoline derivatives could be determined easily. For several drugs such as brethine, cocaine, didrate, estradiol, meprobarbital, methaqualone, phenobarbital, and sulfanilamide, sensitized phosphorescence was observed.   The sensitized phosphorescence could be enhanced significantly when micellar solutions were used as the solvent.   The energy-transfer step was favored by the organizing ability of the micelle, and limits of detection could be decreased by over two orders of magnitude compared to homogeneous solvents.   In addition, sensitized phosphorescence could be measured in cyclodextrin solutions, but the sensitivity was inferior to that in micellar media. The authors concluded that for drug determinations, micelles were more versatile than cyclodextrins and were generally a more applicable medium for both enhanced luminescence signals and chromatography. Micelles offered charge selection by the choice of anionic, cationic, neutral, or zwitterionic surfactants, and polarity selection through the choice of normal (aqueous) micelles or reversed (apolar) micelles.

Femia and Cline Love (45) experimented with mixed organized media and reported the effects of micellar-cyclodextrin solutions on the phosphorescence of phenanthrene.   Intense RTP from phenanthrene is observed when it is included in the β-cyclodextrin in the presence of 1,2-dibromoethane. The phosphorescence profile of phenanthrene in β-cyclodextrin shows considerably higher vibronic resolution than that observed in 70:30 Na/Tl dodecyl sulfate (Na/TlDS) micellar solutions.   Below the critical micelle concentration, in a mixed system of β-cyclodextrin:1,2-dibromoethane and Na/TlDS, phosphorescence was observed which was characteristic of that induced by cyclodextrin. By adding more surfactant, but still below the critical micelle concentration, the phosphorescence intensity from the β-cyclodextrin microenvironment was enhanced.   The authors interpreted this as favorable conditions produced by aggregation of the surfactant monomers at the open ends of the β-cyclodextrin torus and/or partial inclusion that minimized the phenanthrene- water contact. This condition increased the hydrophobicity of the phosphor's microenvironment and shielded it from quenching effects. With surfactant concentrations above the critical micelle concentration, phenanthrene was transferred from the β-cyclodextrin to a micellar environment. The experimental evidence for this was the sharp changes in the phosphorescence spectral profile, wavelength maximum, and sensitivity to oxygen quenching.   Their results also indicated that 1,2-dibromoethane preferred to reside in the micellar aggregate over β-cyclodextrin. Generally, for hydrophobic species like phenanthrene, the micelle provided a preferential environment over that of cyclodextrin.

Femia and Cline Love (46) compared the application of synchronous wavelength scanning to two fluid-phase RTP techniques by using β-cyclodextrin and micelles. They discussed the selectivity, the sensitivity, the susceptibility to light scattering interference, and the classes of compounds that could be determined by the two RTP approaches. The synchronous wavelength scanning with cyclodextrin focused on its advantages for identification of nitrogen heterocycle phosphors. They reported improved spectral resolution for both RTP techniques by second-derivative manipulation of digitally stored synchronous spectra. In general, micelle-induced RTP was found to be superior to cyclodextrin-induced RTP. This conclusion was reached based on the analysis of various mixtures of twelve heterocycles and fifteen carbocyclics and for a mixture of the drug, propranolol, and its 4-hydroxy metabolite. Cyclodextrin RTP is a very effective technique for producing intense emission from many compounds, but had limited usefulness with synchronous wavelength scanning because of the frequent significant scatter interference from the heavy-atom-species:-cyclodextrin complex. It is also limited to compounds that physically form inclusion complexes. Thus, there is a dependence on size and steric factors for the solute. Spectral scatter interference, analyte size, and steric hinderance are not important in the micelle-RTP approach. The authors concluded that it remains the method of choice in fluid RTP spectroscopic techniques for the majority of applications. See Chapters 6-8 for informaiton on solid-surface luminescence with cyclodextrins.

10.5. Phosphorescence from Colloidal and Microcrystalline Suspensions

Weinberger and Cline Love (47) described the luminescence properties of polycyclic aromatic hydrocarbons in colloidal or microcrystalline suspensions. It is well known that polycyclic aromatic hydrocarbons are very insoluble in aqueous media. Weinberger and Cline Love added these solutes as a concentrate dissolved in a water-miscible solvent, and a rapid and uniform dispersion was obtained. The aqueous dispersion was very stable. In glass containers, the suspensions were stable for several weeks. Observation under a light-polarizing microscope revealed substantial Brownian motion, which accounted partly for their stability. A significant Tyndall effect was also observed. Brownian motion precluded photomicroscopy of the suspensions with available equipment, and thus, the suspension was air-dried prior to photography. Anthracene microcrystals were needlelike with a maximum size of less than 100 μm in length. The pyrene crystals were much smaller, and the shape was not readily discernible. In general, the suspensions meet the criteria for colloids, except the mean particle size for some of the compounds was larger than that generally recognized for such a classification. Weinberger and Cline Love (47) considered various aspects of self-absorption, excimer formation, quantum yields, and RTP that was oxygen insensitive. All of these effects were constant down to the insolubility limit of each compound in water, and this aspect could be used to differentiate dissolved or suspended matter without manipulation of the matrix.

Weinberger and Cline Love (48) considered in more detail the means of producing RTP from colloidal or microcrystalline suspensions of aromatic molecules in aqueous media. They found that the relative ease of generating RTP was somewhat surprising given the fact that with other RTP techniques deaeration of the solution is normally needed. Because the solute is present as a solid suspension, diffusion of molecular oxygen through the solute is not possible, thus giving it immunity to oxygen quenching. Not all the compounds investigated gave RTP. For example, phenanthrene gave RTP, but no RTP was observed from chrysene. The wavelengths for the compounds that gave RTP are given in Table 10.4 and compared with those found using micelle-stabilized RTP. As shown in Table 10.4, the colloidal phosphorescence is blue-shifted significantly from the micelle-stabilized RTP for all of the examples given. The reason for the blue shift is still unclear. The absence of RTP for several of the polynuclear aromatic compounds was somewhat puzzling. However, by using temporal discrimination and comparison of the spectra of chrysene in homogeneous solution and in the suspension, delayed fluorescence was resolved. For the compounds that did not give RTP, it was postulated that delayed fluorescence was a process that could be competitive with RTP. Weinberger and Cline Love (48) emphasized that unlike other RTP techniques, the analytical usefulness of this phenomena appears to be limited. However, the approach could be used for the accurate determination of the solubility of polycyclic aromatic hydrocarbons and to distinguish

Table 10.4. Comparison of Phosphorescence Wavelengths of Selected Molecules in Microcrystalline Suspensions and in Micellar Media at Room Temperature

	Phosphorescence wavelengths (nm)	
Compound	Microcrystals	Micelles[a]
Biphenyl	386, 409, 434, 461	452, 482, 505
4-Bromobiphenyl	378, 389, 410, 442	498, 530, 560
2-Bromonaphthalene	394, 414, 440	495, 525, 563, 613
Dibenzofuran	400, 422, 444, 471	424, 450, 480, 516
Fluorene	330, 346, 362, 383	437, 465, 491, 533
Phenanthrene	402, 412, 447, 480	464, 516, 553, 609
Naphthalene	387, 410, 436, 462	486, 521, 557, 601

[a] Species dissolved in aqueous 0.15 M Tl/Na dodecyl sulfate micelles (30:70) and deaerated.

Reprinted with permission from R. Weinberger and L.J. Cline Love, *Appl. Spectrosc.* 1985 39, 516.

between suspended and dissolved matter.

10.6. Concluding Comments

A common aspect of the RTP techniques discussed in this chapter is the necessity of some form of molecular association. Thus, molecular organization is needed to stabilize the phosphorescence emission of the individual solute molecules. The microcrystalline RTP approach is similar to immobilization techniques because the matrix is fixed at the time of preparation. RTP with cyclodextrins results from a more dynamic system compared to the microcrystal technique because the solute can exit the cavity and is less vibrationally constrained. The micellar system represents the most dynamic of the RTP techniques discussed in the chapter. The spectra are less resolved than with the previous two RTP techniques, and the exit and entrance rates of solutes are rapid. For all three of the RTP phenomena, some form of immobilization, protection, or both, is needed, which permits the observation of RTP (49).

In a variety of analytical situations, the RTP methodologies discussed can give improved selectivity, and both enhanced fluorescence and/or phosphorescence can be obtained. More useful analytical results can be acquired by a liquid chromatographic/-micelle-stabilized phosphorescence detector, and synchronous wavelength scanning/second-derivative/micelle-stabilized phosphorescence. Also, by the use of several combinations of methodologies, more data from complex systems can be obtained. Thus, from a single experiment on a complex mixture, it may be possible to obtain an equivalent amount of information that now takes several experiments to acquire (1).

References

1. Cline Love, L.J.; Weinberger, R. *Spectrochim. Acta* 1983, 38B, 1421.
2. Turro, N.J.; Gratzel, M.; Braun, A.M. *Angew. Chem. Int. Ed. Engl.* 1980, 19, 675.
3. Thomas, J.K. *Chem. Rev.* 1980, 80, 283.
4. Kalyanasundaram, K. In *Photochemistry in Microheterogeneous Systems*; Academic Press: New York, 1987; Chapter 3.
5. Cline Love, L.J.; Habarta, J.G.; Dorsey, J.G. *Anal. Chem.* 1984, 56, 1132A.
6. Cordes, E.H. *Pure Appl. Chem.* 1978, 50, 617.
7. Cline Love, L.J.; Skrilec, M. In *Solution Behavior of Surfactants*; Mittal, K.L.; Fendler, E.J., Eds.; Plenum Press: New York, 1982, Vol. 2, pp 1065-1082.
8. Cline Love, L.J.; Skrilec, M.; Habarta, J.G. *Anal. Chem.* 1980, 52, 754.
9. Skrilec, M.; Cline Love, L.J. *Anal. Chem.* 1980, 52, 1559.
10. Cline Love, L.J.; Habarta, J.G.; Skrilec, M. *Anal. Chem.* 1981 53, 437.
11. Almgren, M.; Grieser, F.; Thomas, J.K. *J. Am. Chem. Soc.* 1979, 101, 279.
12. Cline Live, L.J.; Skrilec, M. *Anal. Chem.* 1981, 53, 1872.
13. Skrilec, M.; Cline Love, L.J. *J. Phys. Chem.* 1981, 85, 2047.
14. Woods, R.; Cline Love, L.J. *Spectrochim. Acta* 1984, 40A, 643.
15. Weinberger, R.; Yarmchuk, P.; Cline Love, L.J. *Anal. Chem.* 1982, 54, 1552.
16. Arunyanart, M.; Cline Love, L.J. *Anal. Chem.* 1984, 56,

1557.

17. Femia, R.A.; Cline Love, L.J. *Anal. Chem.* 1984, 56, 327.
18. Garcia, M.E.D.; Sanz-Medel, A. *Anal. Chem.* 1986, 58, 1436.
19. Nugara, N.E.; King, A.D. *Anal. Chem.* 1989, 61, 1431.
20. Sanz-Medel, A.; Garcia, P.L.M.; Garcia, M.E.D. *Anal. Chem.* 1987, 59, 774.
21. Fernandez de la Campa, M.R.; Diaz Garcia, M.E.; Sanz-Medel, A. *Anal. Chim. Acta* 1988, 212, 235.
22. Garcia, M.E.D.; Fernandez de la Campa, M.R.; Hinze, W.L.; Sanz-Medel, A. *Mikrochim Acta* 1988, III, 269.
23. Sanz-Medel, A.; Fernandez de la Campa, R.; Alonso, J.I.G. *Analyst* 1987, 112, 493.
24. O'Reilly, A.M.; Winefordner, J.D. *Spectrochim Acta* 1988, 44A, 1395.
25. Riehl, T.E.; Malehorn, C.L.; Hinze, W.L. *Analyst* 1986, 111, 931.
26. Hinze, W.L. In *Ordered Media in Chemical Separations*; Hinze, W.L.; Armstrong, D.W., Eds.; ACS Sympsoium Series 342; American Chemical Society: Washington, DC, 1987; Chapter 1.
27. Patonay, G.; Fowler, K.; Shapira, A.; Nelson, G.; Warner, I.M. *J. Inc. Phenom.* 1987, 5, 717.
28. Nelson, G.; Patonay, G.; Warner, I.M. *J. Incl. Phenom.* 1988, 6, 277.
29. Bender, M.L.; Komiyama, M. *Cyclodextrin Chemistry*; Springer-Verlag: New York, 1978.
30. Szejtli, J. *Cyclodextrins and Their Inclusion Complexes*; Akademiai Kiado: Budapest, 1982.
31. Szejtli, J. *Cyclodextrin Technology*; Kluwer Academic Publishers: Boston,MA, 1988.
32. Turro, N.J.; Cox, G.S.; Li, X. *Photochem. Photobiol.* 1983, 37, 149.
33. Scypinski, S.; Cline Love, L.J. *Anal. Chem.* 1984, 56, 322.
34. Turro, N.J.; Bolt, J.D.; Kuroda, Y.; Tabushi, I. *Photochem. Photobiol.* 1982, 35, 69.
35. Turro, N.J.; Okubo, T.; Chung, C.J. *J. Am. Chem. Soc.* 1982, 104, 1789.
36. Casal, H.L.; Netto-Ferreira, J.C.; Scaiano, J.C. *J. Incl. Phenom.* 1985, 3, 395.
37. Scypinski, S.; Cline Love, L.J. *Anal. Chem.* 1984, 56, 331.
38. DeLuccia, F.J.; Cline Love, L.J. *Talanta* 1985, 32, 665.
39. Jules, O.; Scypinski, S.; Cline Love, L.J. *Anal. Chim. Acta*

1985, <u>169</u>, 355.

40. Femia, R.A.; Scypinski, S.; Cline Love, L.J. *Environ. Sci. Technol.* 1985, <u>19</u>, 155.

41. Femia, R.A.; Cline Love, L.J. *J. Phys. Chem.* 1985, <u>89</u>, 1897.

42. Sushe, Z.; Changsong, L.; Yulong, B. *Anal. Chem.* (P.R. China) 1988, <u>16</u>, 494.

43. Sushe, Z.; Changsong, L.; Yulong, B. *Anal. Chem.* (P.R. China) 1988, <u>16</u>, 682.

44. Cline Love, L.J.; Grayeski, M.L.; Noroski, J. *Anal. Chim. Acta* 1985, <u>170</u>, 3.

45. Femia, R.A.; Cline Love, L.J. *J. Colloid Interfac. Sci.* 1985, <u>108</u>, 271.

46. Femia, R.A.; Cline Love, L.J. *Spectrochim. Acta* 1986, <u>42A</u>, 1239.

47. Weinberger, R.; Cline Love, L.J. *Spectrochim. Acta* 1984, <u>40A</u>, 49.

48. Weinberger, R.; Cline Love, L.J. *Appl. Spectrosc.* 1985, <u>39</u>, 516.

49. Weinberger, R.; Rembish, K.; Cline Love, L.J. In *Advances in Luminescence Spectroscopy*; Cline Love, L.J.; Eastwood, D., Eds.; ASTM Publication 863: Philadelphia, PA, 1985; pp 40-51.

CHAPTER 11

PHOSPHORESCENCE OF PROTEINS, POLYPEPTIDES, AND PEPTIDES

11.1   Introduction

This chapter will present only a brief survey of the application of phosphorescence spectrometry in protein research. By far, fluorescence spectrometry has been used more extensively in the characterization of protein. However, phosphorescence has some unique advantages relative to fluorescence. For example, the lifetime of phosphorescence is much greater than the lifetime of fluorescence. This permits one to pursue various structural aspects of proteins that cannot be investigated by fluorescence spectrometry. In addition, room-temperature phosphorescence (RTP) can be obtained from a considerable number of proteins, which makes the application of the technique easier to use than low-temperature phosphorimetry.

McCarthy and Winefordner (1) have emphasized that it has been known for some time that the low temperature emission of proteins is due to tyrosine, tryptophan, and phenylalanine residues. However, all three residues do not always emit phosphorescence. For example, ribonuclease and insulin yield tyrosine emission but no tryptophan emission. Also, detailed studies have been performed on the relationship of the phosphorescence of single amino acids to protein phosphorescence (1,2). Konev (2) has given an extensive discussion of the low-temperature phosphorescence of proteins including phosphorescence emission wavelengths of several proteins, decay times of proteins, and the absorption and emission phosphorescence polarization of proteins. In the remaining part of this chapter, several examples will be discussed that illustrate the applicability of phosphorescence spectrometry to the study of proteins, polypeptides, and peptides.

11.2. Applications

Steiner and Kolinski (3) explored the phosphorescence of oligopeptides containing tryptophan and tyrosine. The phosphorescence of tyrosine was greatly suppressed in 50% ethylene glycol, and the ionization of tyrosine produced a major decrease in the ratio of fluorescence to phosphorescence. They concluded that from the set of oligopeptides investigated, the phosphorescence characteristics were primarily those of tryptophan, irrespective of the ionization state of tyrosine. In addition, the excitation spectrum for tryptophan phosphorescence was shifted to longer wavelengths upon ionization of tyrosine.

Several years ago, Weinryb and Steiner (4) pointed out that the interpretation of protein luminescence was hampered by the multiplicity of environments in which the aromatic amino acids could be found because of the wide variations in primary, secondary, and tertiary structures for different proteins. The systematic investigation of simple peptide derivatives of the aromatic amino acids would be expected to bridge the gap between the free amino acids and natural proteins. Thus, they undertook a study of the fluorescence and phosphorescence of peptides and other derivatives of tryptophan and phenylalanine, with the objective of systematically extending knowledge of these residues in a peptide or polypeptide environment. For tryptophan compounds, they found that the luminescence was much more sensitive to chemical modification at 25°C than at liquid nitrogen temperature. Also, the observed phosphorescence lifetimes were invariant at 91 K, and the fluorescence to phosphorescence ratios were fairly insensitive to the chemical nature of the tryptophan derivatives. In contrast to the tryptophan series of compounds, the fluorescence to phosphorescence ratios for phenylalanine peptides at liquid nitrogen temperature showed significant changes according to their chemical nature. The phosphorescence lifetimes were very similar for all the phenylalanine peptides.

Purkey and Galley (5) used low-temperature phosphorescence, observed in 1:1 ethylene glycol:buffer glass at 77 K, to investigate the environmental heterogeneity for tryptophyl residues in proteins. The phosphorescence spectra of horse liver alcohol dehydrogenase, yeast alcohol dehydrogenase, papain, and trypsin revealed heterogeneity for the tryptophan residues. The phosphorescence data suggested that two components arose from trypto-phans either buried in the polarizable protein core or occupying an environment where they were exposed to the polar solvent.

Saviotti and Galley (6) reported RTP of selected proteins in solution. The tryptophan phosphorescence arose from residues which were hindered from interaction with dissolved oxygen by the folding of the polypeptide chains. Measuring the phosphorescence lifetime of horse liver alcohol dehydrogenase as a function of oxygen concentration showed that internal tryptophan residues were periodically exposed to oxygen. From this information, they were able to calculate rate constants for conformational oscillations in the enzyme. Their research showed the feasibility of employing RTP for proteins in solution and the advantage of such experiments in probing the dynamic aspects of the structure of proteins.

The relative contributions of tryptophan and tyrosine to the phosphorescence emission of human serum albumin at low temperature were discussed by Waldmeyer et al. (7). The phospho-rescence emission, excitation spectra, and decay profiles of human serum albumin were investigated in the wavelength regions of the tryptophan and tyrosine absorption and emission spectra in potassium phosphate buffer at 77 K. The emission and excitation spectra were discovered to be linear superpositions of the contributions of the tryptophan and tyrosine residues. Their results indicated that there was no significant tyrosine to tryptophan energy transfer in serum albumin at low temperature. The phosphores-cence decay was multiexponential with lifetime components of 5.95, 2.7, and 1.2 s. The lowest lifetime was characteristic of tryptophan, and the two short components were assigned to two types of tyrosine residues located in different environments within the protein.

Strambini (8) investigated the RTP of alcohol dehydroge-nase from horse liver with continuous excitation to elucidate the quenching role of oxygen. Surprisingly, the emission appeared only after a relatively large amount of excitation radiation was absorbed by the sample. The light-induced phosphorescence was interpreted as arising from photochemical depletion of oxygen from solution. Strambini and Gabellieri (9) reported tryptophan phosphorescence from proteins in the solid state at room temperature. They found the RTP to be very intense and long lived whatever the macromolecule examined or the particular location of the chromophore in it. Figure 11.1 shows the RTP spectra of several protein samples. The RTP emission was extremely sensitive to the degree of hydration of the

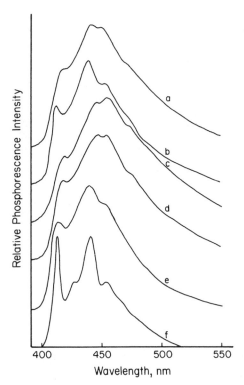

Figure 11.1. Room-temperature phosphorescence spectra of dry protein powders excited at 295 nm. Spectra are normalized at 440 nm and uncorrected for instrumental factors: (a) Ribonuclease A lyophilized in presence of tryptophan in 1:1 molar ratio: (b) *Staphylococcus* nuclease; (c) human serum albumin: (d) lysozyme; (e) horse liver alcohol dehydrogenase; (f) horse liver alcohol dehydrogenase 50 μM in pyrophosphate buffer pH 8.6 (Reprinted with permission from G.B. Strambini and E. Gabellieri, *Photochem. Photobiol.* 1984, <u>39</u>, 725.)

macromolecule, and formation of a water monolayer was sufficient to restore the primary kinetic and spectral features found in dilution solutions. From the triplet- lifetime data, they concluded that the dry macromolecule possessed very little flexibility in structure, and that with a modest degree of hydration, it fully gained features peculiar to the isolated molecule in bulk water. However, the broadness of the spectra and multicomponent decays gave strong evidence that in the solid state a heterogeneous population of macromolecules existed.

Strambini (10) reported on the quenching of alkaline phosphatase phosphorescence by O_2 and NO. The rate constant for the phosphorescence quenching of alkaline phosphatase by molecular oxygen was measured as a function of temperature. The results showed that diffusion of O_2 in the macromolecule was a highly hindered process. With NO as a quencher, similar rate constants were found. Strambini concluded that the proposed mechanism of quenching departed from the usual interpretation of colliding partners. It was postulated that there existed compact regions of protein structure through which even the smallest molecules were unable to penetrate in the micro-millisecond time scale.

Information on the effects of crystallization upon the structure of horse liver alcohol dehydrogenase was obtained by a comparison of the phosphorescence properties of its tryptophan residues in solution and in the crystalline state (11). In the crystalline state, a red-shift in the phosphorescence spectrum of the solvent-exposed tryptophan-15 indicated a decreased polarity of its environment. This was consistent with its shielding away from the aqueous solvent polarity through its involvement in an intermolecular contact. In contast, the triplet-state lifetime of tryptophan-314, which was buried deeply in the coenzyme-binding domain, showed that the flexibility of this region of the macromolecule was unaffected by crystallization. The previous conclusion was also supported by the similarity in the rate of oxygen quenching of its phosphorescence.

Strambini and Gabellieri (12) carried out an investigation of the phosphorescence emission properties of tryptophan in glyceraldehyde-3-phosphate dehydrogenase from pig and rabbit muscle. By using the external heavy-atom effect, the dependence on excitation wavelength, and thermal quenching profiles, they established that the 0,0 vibronic band, peaked at 406 nm in the pig and rabbit proteins, consisted of overlapping contributions from two tryptophan residues. They concluded that in the muscle enzymes all three aromatic side chains were phosphorescent. In addition, when the characteristics of the local environment of each residue were compared to the crystallographic structure of lobster glyceraldehyde-3-phosphate dehydrogenase, a completely new assignment of the individual phosphorescence spectra resulted. With each protein, a single tryptophan, identified as tryptophan-310, was discovered to display a long-lived RTP. The decay of the RTP gave evidence of conformational homogeneity among the subunits of the tetrameric molecule.

Calhoun et al. (13) carried out experiments to measure the ability of oxygen to collisionally quench the phosphorescent and fluorescent tryptophans in alcohol dehydrogenase and alkaline phosphatase. In all the experiments, the luminescence was quenched with rate constants close to 1×10^9 M^{-1}s^{-1}. The rate of oxygen reaching the buried tryptophans was not affected much by solvent viscosity due to added glycerol. Also, the quenching by oxygen was not due to a protein-opening reaction. The quenching appeared to be rate-limited by internal protein diffusion rather than by the entry step. Their results resolved various discrepancies in the literature concerning oxygen penetration in horse liver alcohol dehydrogenase and alkaline phosphatase. They concluded from fluorescence and phosphorescence data that oxygen does reach the most protected tryptophan in these proteins with rate constants that approach the solution diffusion-limited rate. In other work, Calhoun et al. (14) performed experiments to test the ability of a variety of small molecules to quench the phosphorescence and fluorescence of the well-buried and the more accessible tryptophans of horse liver alcohol dehydrogenase and alkaline phosphatase. The quenchers studied included acrylamide derivatives, saturated and unsaturated ketones, and inorganic anions and cations. The RTP of the buried tryptophan in horse liver alcohol dehydrogenase was quenched by several of the chemicals tested, with rate constants grouped in a relatively narrow range 5 to 6 decades below the diffusional limit. Also, the phosphorescent residues in alkaline phosphatase were reached several decades more slowly. The major conclusion from their work was that small molecules, in general, could not enter native protein structures with the exception of oxygen.

As already discussed in the chapter, RTP can be obtained from proteins. However, Vanderkooi et al. (15) tested a large number of proteins for the property of intrinsic phosphorescence in deoxygenated aqueous solution at room temperature. They found that the majority of proteins showed phosphorescence under normal solution conditions. Phosphorescence lifetimes from 0.5 ms to 2 s were observed in three-fourths of the proteins tested. The lifetime appeared to correlate with the relative isolation of the tryptophan indole side chain from the solvent. In their earlier work (13), they showed that oxygen quenched the phosphorescence of the very well protected tryptophans of alcohol dehydrogenase and alkaline phosphatase with a rate constant of 10^9 M^{-1}s^{-1}. Therefore, when oxygen is present at normal solution concentration (250 μM in air-exposed water), phosphorescence with a lifetime of 5 s would be reduced in intensity by a factor of 10^6 and give a lifetime of only 4 μs. The authors emphasized that this explains why the phosphorescence of

Table 11.1. Phosphorescence from Proteins at 20° C

Protein	Lifetime (ms)
Albumin, bovine serum	0.9
Albumin, chicken egg	15
Alcohol dehydrogenase[a]	550
Asparaginase, *Escherichia coli*	50
Azurin, *Pseudomonas aeruginosa*	400
Cellulase, *Aspergillus niger*	200
Edestin, hemp seed	500
Glucose oxidase, *Aspergillus niger*	120
Lactic dehydrogenase, rabbit muscle	25
Beta-lactoglobulin	15
Myosin, rabbit muscle	100
Nuclease, micrococcal	400
Parvalbumin (with calcium), cod	5
Protease (acid), *Aspergillus saitoi*	100
Protease *Streptomyces griseus*	600
Protease (subtilisin Carlsberg), *Bacillus subtilis*	10
Streptokinase, streptococcus	50

[a]D_2O solvent.

Reprinted with permission from J.M. Vanderkooi, D.B. Calhoun, and S.W. Englander, *Science*, 236, 568. Copyright 1987 by the AAAS.

many proteins in solution has escaped detection. Table 11.1 gives a partial listing of the RTP lifetimes of proteins given by Vanderkooi et al. (15). For the data in Table 11.1, the oxygen concentration was reduced to subnanomolar levels, and phosphorescence was measured in the time-resolved mode. The more intense fluorescence emission was removed by gating the data acquisition to start at 1 ms after the excitation flash. The widespread occurrence of protein RTP should provide many opportunities for expanding photoluminescence studies of intra- and intermolecular dynamics into the rather wide phosphorescence time domain.

Korkidis (16) used time-resolved phosphorescence to study protein mobility in normal and abnormal red blood cells. The experiments were designed to study changes in the external membranes of red blood cells. Severe blood diseases such as sickle-cell anemia are linked to abnormalities in red blood cells. Because fluorescence lifetimes are generally measured in nanoseconds, this approach is limited to the determination of short rotational relaxation times.   Time-resolved phosphorescence spectroscopy can be employed into the range in which significantly longer relaxation times can be measured. This type of measurement is important to the study of long-range protein mobility in living cells.   Korkidis showed the effectiveness of this approach in the investigation of the aggregation of intracellular proteins that induce the characteristic deformation of the erythrocyte membrane in sickle-cell anemia.   In one experiment, the phosphorescence decay from normal cells gave a relaxation time that was one-half of the time noted for sickle-cells. The approach was highly sensitive for monitoring changes in minute protein concentrations in red blood cells.

Maniara et al. (17) observed phosphorescence at room temperature for the 2-(p-toluidinyl)naphthalene-6-sulfonate (TNS) and 1-anilinonaphthalene-8-sulfonate (ANS) bovine serum albumin complexes. The TNS and ANS are normally used as fluorescence probes for hydrophobic sites in biological structures.   The phosphorescence lifetimes were in the millisecond time range when these molecules were bound to the hydrophobic site on serum albumin.   The authors concluded that phosphorescence could be used to quantitate rotational and lateral diffusion and to measure distances between chromophores.

Rumsey et al. (18) considered phosphorescence imaging for measuring oxygen distribution in perfused tissue. This approach provided a method for monitoring oxygen distribution within the vascular system of intact tissues. Isolated rat livers were perfused through the portal vein with media containing palladium coproporphyrin.   The previous material phosphoresced and was used to image the liver at various perfusion rates. Because oxygen is a powerful quenching agent for phosphors, the transition from well-perfused liver to anoxia (no flow of oxygen) resulted in a large increase of phosphorescence.   By stepwise restoration of oxygen flow, the phosphorescence images showed prominent heterogeneous patterns of tissue reoxygenation. These results showed that there were regional inequalities in oxygen delivery.

Ghiron et al. (19) reported the decay of the indole triplet of single tryptophan-containing proteins and model compounds at room temperature in aqueous solution by monitoring the triplet-triplet absorption or phosphorescence emission following a 265 nm exciting laser pulse.   They studied the quenching action of acrylamide on the triplet excited state of indole side chains and discussed several factors affecting the triplet quenching efficiency.

Rousslang et al. (20) described the phosphorescence emission and decay times of oxidized nicotinamide adenine dinucleotide (NAD^+) and its fluorescent etheno derivative, ϵ-NAD^+. NAD^+ is nonfluorescent, but is strongly phosphorescent. The reduced dinucleotide (NADH), which is strongly fluorescent, is only weakly phosphorescent. In the systems studied, the phosphorescence was due to the adenine ring. All their measurements were performed at liquid nitrogen temperature. Phosphorescence decay data were analyzed by a nonlinear least squares program. The main purpose of their work was to establish the ability to employ pyridine nucleotide coenzymes to monitor enzyme-coenzyme interactions. They showed that ϵ-NAD^+ was a more useful phosphorescence probe than NAD^+ for their investigations. In the future, they hope to establish the origin of the phosphorescence quenching of tryptophan and of the coenzymes in the ternary pyrazole complexes of NAD^+ and ϵ-NAD^+.

11.3. Concluding Comments

Although this chapter only presented a brief coverage of phosphorescence in protein research, it does show the applicability of phosphorescence in this important area of research. With the recent results showing that RTP can be obtained from numerous proteins, phosphorescence will be used more extensively in protein research. Also, the time scales for fluorescence and phosphorescence are very different. Thus, various molecular motions can be pursued with phosphorescence spectrometry that cannot be studied by fluorescence spectrometry. Also, fluorescence and phosphorescence spectrometry complement one another in protein investigations. The combined use of these two luminescence techniques provides powerful tools in exploring the structures of proteins.

References

1. McCarthy, W.J.; Winefordner, J.D. In *Fluorescence: Theory, Instrumentation, and Practice*; Guilbault, G.G., Ed.; Marcel Dekker: New York, 1967; Chapter 10.
2. Konev, S.V. *Fluorescence and Phosphorescence of Proteins and Nucleic Acids*; Plenum Press: New York, 1967.

3.   Steiner, R.F.; Kolinski, R. *Biochem.* 1968, 7, 1014.
4.   Weinryb, I.; Steiner, R.F. *Biochem.* 1968, 7, 2488.
5.   Purkey, R.M.; Galley, W.C. *Biochem.* 1970, 9, 3569.
6.   Saviotti, M.L.; Galley, W.C. *Proc. Nat. Acad. Sci.* 1974, 71, 4154.
7.   Waldmeyer, J.; Korkidis, K.; Geacintov, N.E. *Photochem. Photobiol.* 1982, 35, 299.
8.   Strambini, G.B. *Biophys. J.* 1983, 43, 127.
9.   Strambini, G.B.; Gabellieri, E. *Photochem. Photobiol.* 1984, 39, 725.
10. Strambini, G.B. *Biophys. J.* 1987, 52, 23.
11. Gabellieri, E.; Strambini, G.B.; Gualtieri, P. *Biophys. J.* 1988, 30, 61.
12. Strambini, G.B.; Gabellieri, E. *Biochem.* 1989, 28, 160.
13. Calhoun, D.B.; Vanderkooi, J.M.; Woodrow, G.V.; Englander, S.W. *Biochem.* 1983, 22, 1526.
14. Calhoun, D.B.; Vanderkooi, J.M.; Englander, S.W. *Biochem.* 1983, 22, 1533.
15. Vanderkooi, J.M.; Calhoun, D.B.; Englander, S.W. *Science* 1987, 236, 568.
16. Korkidis, K. *Spectrosc.* 1987, 2, 44.
17. Maniara, G.; Vanderkooi, J.M.; Bloomgarden, D.C.; Koloczek, H. *Photochem. Photobiol.* 1988, 47, 207.
18. Rumsey, W.L.; Vanderkooi, J.M.; Wilson, D.F. *Science* 1988, 241, 1649.
19. Ghiron, C.; Bazin, M.; Santus, R. *Photochem. Photobiol.* 1988, 48, 539.
20. Rousslang, K.; Allen, L.; Ross, J.B.A. *Photochem. Photobiol.* 1989, 49, 137.

CHAPTER 12

PHOSPHORESCENCE IN POLYMER RESEARCH

12.1. Introduction

In a similar fashion, as in the previous chapter on the phosphorescence of proteins, this chapter will present a survey of the applications of phosphorescence spectrometry in polymer research. Fluorescence spectrometry has been employed more extensively in polymer research, but phosphorescence has some unique properties that permit important structural information to be obtained from polymers. For example, because the lifetime of phosphorescence is considerably greater than the lifetime of fluorescence, various molecular motions can be pursued by phosphorimetry that cannot be undertaken with fluorescence spectrometry. Several authors have discussed the use of phosphorescence in polymer research in various chapters. Guillet (1) considered the use of phosphorescence and photochemistry of polymers. Webber (2) gave a rather detailed consideration of phosphorescence and other delayed emissions of polymers. Horie (3) focused on the phosphorescence decay and dynamics in polymer solids. Toynbee and Soutar (4) use both fluorescence and phosphorescence to study molecular motions in poly(n-butylacrylate). In the remaining part of the chapter, selective examples will be discussed that illustrate the importance and usefulness of phosphorescence in polymer research.

12.2. Applications

Somersall and Guillet (5) investigated triplet-triplet annihilation and excimer fluorescence in poly(naphthyl methacrylate). The delayed emission spectrum of this material was measured in

H$_4$furan-ether glass at 77 K. The bands at 495, 528, and 550 nm corresponded to the normal phosphorescence of naphthalene and had a mean lifetime of 0.7 s. However, the band at 340 nm had a mean lifetime of 0.1 s and corresponded to the fluorescence of the naphthalene chromophore. The position and lifetime of this band was consistent with a delayed fluorescence emission, similar to that observed for poly(1-vinyl-naphthalene). The delayed fluorescence intensity was shown to depend on the square of the intensity of excitation. This was consistent with a bimolecular triplet-triplet annihilation process for delayed fluorescence. Evidence for triplet energy migration along the polymer chain was obtained from phosphorescence quenching experiments with piperylene and 1,3-cyclooctadiene interaction with poly(naphthyl methacrylate).

Somersall et al. (6) obtained the phosphorescence of a wide variety of polymer films containing ketone and/or naphthalene groups in the temperature range 77-300 K. As the samples were warmed up, the emission decreased, and Arrhenius curves for the temperature dependence indicated distinct linear regions with changes in slope at the temperature corresponding to the onset of characteristic subgroup motion in polymers. The specific relaxation processes, involving small segments and side groups of the polymer chains, were important in affecting the local accessibility of oxygen to quenching the triplet state. They obtained phosphorescence data for styrene, ethylene, methacrylate, vinyl chloride, acrylonitrile, and polyketone polymers. One of the most important aspects of their work was that the change in slopes appearing in the graphs of phosphorescence as a function of temperature were directly related to the local relaxation in the polymers.

Kilp and Guillet (7) studied the triplet energy migration in poly(acrylophenone). Several experiments were performed. Of particular interest were phosphorescence quenching experiments in 77 K solid solutions. The quenching of phosphorescence was achieved by the addition of naphthalene to the polymeric solutions. Naphthalene is an effective quencher of aryl alkyl ketone triplets and was not directly excited by the 340-nm radiation that was used in the work. They applied the well-known Stern-Volmer equation to their data and found that the resulting graph gave a distinctly upward curvature. This implied that simple Stern-Volmer kinetics were not applicable to the quenching behavior of their systems. However, the active-sphere model of Perrin (8) better described the data. With their results and with this model, they concluded that intramolecular energy migration in polymers was occurring.

Burkhart et al. (9) investigated the triplet photophysical properties of the alternating copolymer of 2-vinylnaphthalene with methyl methacrylate. The delayed luminescence spectra of the alternating copolymer was obtained in frozen glassy solutions at 77 K, solid films at 77 K, and fluid solutions at ambient temperature. Delayed fluorescence was observed in all three media. In the solid films at 77 K and in fluid solutions at ambient temperature, the delayed fluorescence emission consisted of both monomeric and excimeric components. Both the phosphorescence and delayed fluorescence emissions were monomeric in character in frozen glassy solutions. Essentially no phosphorescence was observed at 77 K from solid films of the copolymer which had been treated previously by heating under vacuum.

Starzyk and Burkhart (10) discussed a number of aspects of triplet emission from poly(3,6-dibromo-N-vinylcarbazole) (PdBVK). The triplet delayed emission of PdBVK in a 2-methyltetrahydrofuran frozen solution at 77 K was investigated within the spectral and time regimes of 400-570 nm and 0.2-50 ms, respectively. The influence on optical absorption of bromine substitution into the carbazole ring (3- and 6-positions) of poly(N-vinylcarbazole), delayed triplet emission spectra, and decay curves for triplet emission were monitored. At different delay times after excitation, the phosphorescence spectra were recorded. The phosphorescence decays were primarily exponential in the wings of phosphorescence band but showed unusual nonexponential behavior in the range 460-475 nm, which was the range of maximum phosphorescence intensity. The phosphorescence band was resolved into Gaussian components, which yielded excellent fits by using three components at delay times less than 2 ms and two components for spectra taken at longer delay times. With an average of 12 different spectra recorded at various times after excitation, the calculated wavelengths at maximum intensity for the Gaussian components were found to be 448 ± 2 nm, 473 ± 4 nm, and 501 ± 8 nm. The 501-nm component was somewhat broad and was the least intense of the three component bands. This band also disappeared at delay times longer than 2 ms after the excitation pulse. They postulated that the 448-nm component originated from independent chromophoric groups or was possibly a trapped species not very different in energy from the triplet state of individual chromophores. The 473-nm component was possibly associated with deeper energy trap sites. The origin of the 501-nm component was uncertain, but might have been associated with a trap site or with a low-energy vibronic component of the independent chromophore emission.

The effect of varying the polymer matrices and gases on the singlet and triplet behavior of 2-piperidinoanthraquinone was reported by Natarajan and Becker (11). Although these authors employed primarily laser flash photolysis and fluorescence spectrometry, their work is of interest because the triplet state in the polymer samples was considered. The photophysical behavior of 2-piperidinoanthraquinone (2-PAQ) was investigated by embedding the solute dye in several polymer matrices like poly(methyl methacrylate), poly(ethyl methacrylate), polycarbonates, polystyrene, and épóxy resins at room temperature. Polymers containing the dye were cast in the form of blocks and as very thin films. They employed radiation of high input laser energy to demonstrate the occurrence of cavitation in fluid solvents like benzene or toluene, whereas in thin polymer films such cavitation did not occur and only smooth triplet decay curves were observed. The exponentiality of the triplet decay in thin films was energy dependent. Only a single exponential decay was observed at low laser energy.

Horie and Mita (12) studied the effects of molecular weight and solvent on the rate constants k_q for quenching of phosphorescence of polystyrylbenzil by polystyrylanthracene in benzene, cyclohexane, and butanone at 20-40°C by using a 10-ns dye laser or a nitrogen laser pulse. The k_q in benzene was inversely proportional to the 0.29 power of the degree of polymerization, P, for the range of P = 23-3900. The reduced rate constants $k_q\eta_o/T$ in poor solvents (cyclohexane and butanone) were larger than $k_q\eta_o/T$ in good solvents for $P < 10^3$ and were opposite for $P > 10^3$.

Encinas et al. (13) considered the triplet benzophenone quenching by poly(α-naphthyl methacrylate) and copolymers of α-naphthyl methacrylate and methyl methacrylate as a function of the copolymer composition, molecular weight, and thermodynamic character of the solvent. Equation (12.1) was used in the interpretation of the data. Equation (12.1) is the Stern-Volmer equation.

$$I^o{}_{Ph}/I_{Ph} = 1 + k_Q\tau_T[Q] \tag{12.1}$$

In Equation (12.1), $I^o{}_{Ph}$ is the intensity of phosphorescence without quencher; I_{Ph} is the intensity of phosphorescence with quencher; [Q] is the quencher concentration in terms of the molar concentration of naphthyl groups; k_Q is the rate constant for quenching; and τ_T is the lifetime of the triplet state. They found that the quenching rate constant was determined only by the volume of the macromolecular coil.

Yu and Torkelson (14) considered the phosphorescence quenching of benzil by polystyrene in dilute and semidilute solutions. For benzil phosphorescence quenching by polystyrene, the interaction is a non-diffusion-controlled process. That is, the polymer and benzil molecules may encounter each other several times before quenching takes place. Stern-Volmer kinetics was employed to analyze the data. Figure 12.1 shows the benzil phosphorescence lifetime as a function of polystyrene concentration for several molecular weights of polystyrene in cyclohexane and in toluene solutions. As shown in this figure, the lifetime drops

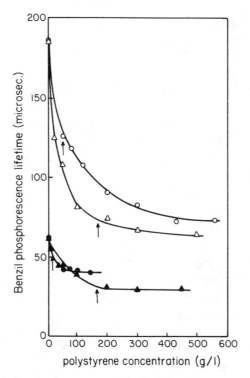

Figure 12.1. Benzil phosphorescence quenching in polystyrene solutions (arrows designate $c = 1 / [\eta]$ for the cyclohexane and toluene solutions); τ versus concentration. Polymer solutions: (O) PS with MW 47500 in cyclohexane; (Δ) PS with MW 4000 in cyclohexane; (•) PS with MW 670000 in toluene; (▲) PS with MW 4000 in toluene. (Reprinted with permission from D.H.S. Yu and J.M. Torkelson, *Macromolecules*, <u>21</u>, 852. Copyright 1988 American Chemical Society.)

significantly with increasing polymer concentration in the dilute concentration region and approaches a plateau at higher concentrations. The authors also measured phosphorescence intensities as a function of polymer concentration. By comparing the intensity values and lifetime values, they concluded that significant static quenching was present in addition to dynamic quenching.

In later work, Yu and Torkelson (15) considered, in detail, diffusion-limited phosphorescence quenching interactions in polymer solutions. Diffusion-limited interactions between benzil and anthracene were investigated by phosphorescence quenching in polystyrene-cyclohexane, polystyrene-toluene, poly(methyl methacrylate)-toluene, and polybutadiene-cyclohexane solutions. The values of the bimolecular diffusion-limited quenching rate constant, k_q, were obtained by measuring benzil phosphorescence lifetime as a function of anthracene concentration and applying Stern-Volmer analysis. In addition to polymer species and solvent, k_q was measured as a function of polymer molecular weight and concentration, up to 560 g/L. The rate constant k_q was determined to be independent of polymer molecular weight in polystyrene-cyclohexane solutions and showed a slight molecular weight dependence in polystyrene-toluene solutions. The polymer concentration dependence of k_q in polystyrene-cyclohexane and polystyrene-toluene solutions imitate the polymer concentration dependence of the solvent self-diffusion coefficient. This result was consistent with the concept that $k_q/k_{q_0} \sim D_s/s_0$, where D_s is the solvent self-diffusion coefficient, and the subscript "o" refers to the value at zero polymer concentration. In this particular work, Yu and Torkelson (15) showed how the phosphorescence technique could be used successfully to determine the effects of polymer species, molecular weight, concentration, and solvent diffusion-limited interactions in polymer solutions.

Miller and North (16) reported the first measurements of phosphorescence depolarization as a way of probing molecular motion in the millisecond time scale in polymer systems. Because phosphorescence lifetimes range from fractions of a millisecond to seconds, this range gives a wide spectrum of time scales for polymer molecular motions that can't be studied with the short time scale for fluorescence. The primary purpose of their work was to investigate the possible use of phosphorescence depolarization in aromatic ketones as a way of studying molecular motions in polymer systems. They employed the phosphorescence polarization of anthrone and benzophenone at room temperature in solutions that

contained various polymers. From the polarization data, they
calculated the rotational time constants for the aromatic ketones.
Then, they commented on the possible polymeric motions that
caused the depolarization of the phosphorescence.

Rutherford and Soutar (17) first reported the phosphorescence depolarization of labeled macromolecules and related the data
to macromolecular segmental relaxations. Steady-state phosphorescence polarization experiments were carried out on samples of
poly(methyl acrylate) which contained 0.5 wt% copolymerized
acenaphthylene or 1-vinyl naphthalene as phosphorescent probes
over the temperature range 77 to 310 K. Depolarization of
phosphorescence occurred with the onset of segmental motion of
the polymer at about 278 K. They were able to characterize the
motion of either probe by calculating an activation energy from the
polarization data. The activation energy was in fair agreement with
the activation energy estimated for the segmental relaxation of
poly(methyl acrylate) by dielectric and mechanical relaxation
techniques. The results of their work show that phosphorescence
depolarization is capable of measuring segmental relaxation times
in bulk polymers. Rutherford and Soutar (17) also commented that
phosphorescence intensity and lifetime data can be used to detect
transitions in polymers, but this data yields less detailed information
regarding the mechanism of relaxation in polymers.

Rutherford and Soutar (18) further considered the phosphorescence depolarization measurements of segmental relaxation
processes in polymers. They used the Perrin equation, which is
given by Equation (12.2), to interpret their data. In Equation (12.2),

$$\left(\frac{1}{p} - \frac{1}{3}\right) = \left(\frac{1}{p_o} - \frac{1}{3}\right)\left(1 + \frac{3\tau}{\rho}\right) \qquad (12.2)$$

p is the degree of polarization; p_o is the degree of polarization in the
absence of external depolarizing factors; τ is the excited-state
lifetime; and ρ is the rotational relaxation time of the luminescent
molecule. Disubstituted acenaphthene was employed as a phosphorescent label in poly(methyl acrylate) (PMA) and poly(methyl
methacrylate) (PMMA). Rotational relaxation times were calculated from observed p and τ values by means of Equation (12.2).
Figure 12.2 gives the graphs of log of ρ as a function of T^{-1} for
PMMA and PMA. The graph for PMA shows a single relaxation
region, whereas the graph for PMMA gives evidence for two

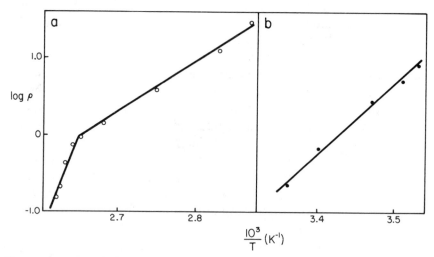

Figure 12.2. Thermal dependence of segmental relaxation time of (a) PMMA and (b) PMA. (Reprinted with permission from H. Rutherford and I.Soutar, *J. Polym. Sci. Lett. Ed.* 1978, <u>16</u>, 131.

relaxation regions. The graphs in Figure 12.2 illustrate the type of information that can be obtained from the polymers by the phosphorescence depolarization method.

Rutherford and Soutar (19) reported on several aspects of relaxation effects in bulk polymers using phosphorescence spectrometry. They were primarily interested in phosphorescence emission depolarization; however, phosphorescence intensity and phosphorescence lifetime data were also discussed. Phosphorescence depolarization measurements, under steady-state polarized excitation, were used to examine the relaxation behavior of bulk poly(methyl methacrylate). Phosphorescent labels were synthesized by copolymerization of small quantities of acenaphthylene, 1-vinylnaphthalene, 2-vinylnaphthalene, 1-naphthyl methacrylate, and 2-naphthyl methacrylate, respectively, with methyl methacrylate. Also samples of pure poly(methyl methacrylate) containing dispersed traces of naphthalene and acenaphthene were prepared so a comparison of the emission characteristics with those of the labeled systems could be made. From the phosphorescence depolarization data, activation energies were calculated and related to various relaxation phenomena in the polymers. For example, the rotational motions of labels and probes on a scale sufficient to cause depolarization were absent at temperatures below that of the β relaxation in the poly(methyl methacrylate). One of the disadvantages of the phosphorescence approach in the work by Rutherford and Soutar (19) was the low phosphorescence intensities. However, they emphasized that the application of single photon and transient excitation techniques should improve the study of relaxation processes in polymers by phosphorescence spectrometry. In other work, Toynbee and Soutar (4) extended previous work with phosphorescence depolarization to poly(n-butyl methacrylate). Also, fluorescence depolarization measurements were made to provide complementary information regarding the higher frequency behavior of the polymer. In addition, luminescence intensity and lifetime data were obtained from the labeled polymer and the emission of dispersed naphthalene, acenaphthene, and 1,1-dinaphthyl-1,3-propane. These measurements provided information not only on the relaxation behavior of the polymer matrix but also on the photophysics of the polymer system.

12.3. Concluding Comments

Although this chapter was not comprehensive in its treatment of phosphorimetry in polymer research, it does illustrate how phosphorescence can be used to investigate various molecular motions in polymers and a variety of structural features associated with polymers. Because the time scales for fluorescence and phosphorescence are quite different, these two techniques complement one another in polymer research. The low intensity of phosphorescence has been a problem with some molecular probes used to study the structural aspects of polymers. However, using probes with a higher phosphorescence quantum yield or using more sensitive detector systems should minimize this problem. The types of fundamental phosphorescence information that can be obtained from polymers are excitation and emission spectra, lifetimes, polarization data, and intensity data. From these types of information, many photophysical parameters have been calculated that are very helpful in elucidating properties of polymers that could only be achieved by using phosphorescence spectrometry. Based on the research performed with phosphorimetry in polymer characterization, one should continue to see this technique used in polymer research in obtaining unique structural information on polymers (1-4).

References

1. Guillet, J. *Polymer Photophysics and Photochemistry*; Cambridge University Press: New York, 1985; Chapter 8.
2. Webber, S.C. In *Polymer Photophysics*; Phillips, D., Ed.; Chapman and Hall: New York, 1985; Chapter 2.
3. Horie, K. In *Photophysics of Polymers*; Hoyle, C.E.; Torkelson, Eds.; ACS Symposium Series 358; American Chemical Society: Washington, DC, 1987; Chapter 8.
4. Toynbee, J.; Soutar, I. In *Photophysics of Polymers*; Hoyle, C.E.; Torkelson, Eds.; ACS Symposium Series 358; American Chemical Society: Washington, DC, 1987; Chapter 11.
5. Somersall, A.C.; Guillet, J.E. *Macromolecules* 1973, 6, 218.
6. Somersall, A.C.; Dan, E.; Guillet, J.E. *Macromolecules* 1974, 7, 233.
7. Kilp, T.; Guillet, J.E. *Macromolecules* 1981, 14, 1680.
8. Turro, N.J. *Modern Molecular Photochemistry*; Benjamin-/Cummings: Menlo Park, CA, 1978; pp 317-319.
9. Burkhart, R.D.; Haggquist, G.W.; Webber, S.E. *Macromolecules* 1987; 20, 3012.
10. Starzyk, F.C.; Burkhart, R.D. *Macromolecules* 1989, 22, 782.
11. Natarajan, L.V.; Becker, R.S. *Macromolecules* 1988, 21, 73.
12. Horie, K.; Mita, I. *Macromolecules* 1978, 11, 1175.
13. Encinas, M.V.; Lissi, E.A.; Gargallo, L.; Radic, D.; Olea, A.F. *Macromolecules* 1984, 17, 2261.
14. Yu, D.H.S.; Torkelson, J.M. *Macromolecules* 1988, 21, 852.
15. Yu, D.H.S.; Torkelson, J.M. *Macromolecules* 1988, 21, 1033.
16. Miller, L.J.; North, A.M. *J. Chem. Soc. Faraday Trans. II* 1975, 71, 1233.
17. Rutherford, H.; Soutar, I. *J. Polym. Sci. Polym. Phys. Ed.* 1977, 15, 2213.
18. Rutherford, H.; Soutar, I. *J. Polym. Sci. Lett. Ed.* 1978, 16, 131.
19. Rutherford, H.; Soutar, I. *J. Polym. Sci. Polym. Phys. Ed.* 1980, 18, 1021.

CHAPTER 13

FINAL COMMENTS AND FUTURE TRENDS IN PHOSPHORIMETRY

Phosphorimetry has found its place in the analytical tools available for organic trace analysis. It is continuing to evolve in the areas of theory, instrumentation, and applications. Phosphorimetry is complementary to fluorimetry in many respects. However, the use of phosphorescence in many analytical problems may be the only reasonable analytical approach to achieve the desired sensitivity and selectivity. In addition, the lifetime of phosphorescence is such that it can be used to study molecular motions in materials such as biological samples and polymers that cannot be studied by fluorimetry because of the very short lifetime of fluorescence. There are several new ways to implement analytical phosphorescence compared to low-temperature solution phosphorimetry. For instance, sensitized room-temperature solution phosphorescence, micelle-stabilized room-temperature phosphorescence, and solid-surface room-temperature phosphorescence have been successfully employed in many analytical situations. These techniques will find continued use in the future for obtaining low limits of detection and high selectivity. As an example, with solid-surface luminescence techniques, picogram detection limits have been obtained, and several compounds in multicomponent mixtures have been readily characterized and quantitated without the need to perform any separation steps. In Table 13.1 are listed some of the types of phosphorimetry that have been employed in chemical analysis. Which phosphorimetric approach would be used for research on an analytical problem would depend on the type of sample and overall objectives of the project. Low-temperature solution phosphorimetry was the original phosphorimetric approach and has been employed for years in chemical analysis and will continue to be used in the future. The other phosphorimetric

Table 13.1. Types of Phosphorimetry

Low-temperature solution
Room-temperature solid surface
Room-temperature micelle
Sensitized
Quenched
Room-temperature solution cyclodextrin
Room-temperature colloidal/microcyrstalline
Electron-impact-induced
Collisional induced

techniques listed in Table 13.1 are rather new and have been applied to various extents. Room-temperature solid-surface phosphorimetry has been used very extensively over the past several years. Given the speed, sensitivity, and moderate cost, and the possibility of automating solid-surface phosphorescence, many applications of this technique should continue to appear. Room-temperature micelle, sensitized, quenched, and room-temperature solution cyclodextrin phosphorimetry have not been employed in chemical analysis as much as room-temperature solid-surface phosphorimetry partly because they are relatively new techniques, but partly because the approaches are not applicable to as many phosphors as room-temperature solid-surface phosphorescence. Room-temperature colloidal/microcrystalline, electron-impact, and collisional phosphorimetry have seen limited use and most likely will be used in specialized ways in the future.

With the advent of the new phosphorimetric methodology, one might wonder if phosphorimetry will supersede fluorimetry. This most likely will not occur in the future because more compounds give fluorescence than those that give phosphorescence. Also, in general, it is easier experimentally to obtain fluorescence spectra and fluorescence intensity data for a larger number of compounds in solution at room temperature than with the phosphorimetric techniques in Table 13.1. However, with room-temperature solid-surface phosphorimetry, one adsorbs a tiny aliquot (~ 1 μL or less) of a solution on a solid surface, dries the solid matrix, and then measures the phosphorescence. Also, as mentioned earlier, the measurement step can be automated quite easily. The equipment needed to measure fluorescence lifetime ($\sim 10^{-8}$ s) is more sophisticated and costly than the equipment required to obtain phosphorescence lifetime ($\sim 10^{-4}$ to 10 s). Thus, phosphorimetry has the edge in lifetime measurements. In addition, because the phosphorescence emission wavelengths are generally far removed from the excitation wavelengths, phosphorescence intensity and spectral measurements normally are not subject to interference from scattered excitation radiation. For fluorimetry, scattered radiation can be more of a problem because the emitted fluorescence wavelength region is closer to excitation wavelength region. Phosphorimetry should be somewhat more selective than fluorimetry in a practical way because of the ease of measuring phosphorescence lifetimes. Also experimental conditions are less critical for fluorescence compared to phosphorescence. Thus, the potential exists for more compounds to give fluorescence rather than phosphorescence. This fact would actually favor selectivity for phosphorescence. For example, in some experimental situations, it is possible to adjust conditions so that phosphorescence from unwanted phosphorescent components can be eliminated or minimized. However, the same experimental conditions would not affect the fluorescence signals.

Because the possibility exists to measure both fluorescence and phosphorescence at room temperature, the analyst can have both fluorescence phenomena and phosphorescence phenomena readily available to work with in dealing with some analytical problems or research problems. This feature would be particularly useful in dealing with complex mixtures of organic compounds and in the characterization of complex materials such as certain polymers.

What about the future of phosphorimetry? As mentioned earlier, it is still unfolding in the areas of theory, instrumentation, and applications. Of the types of phosphorimetry listed in Table 13.1, the theory of low-temperature solution phosphorescence is the most developed, but advances in theory of low-temperature solution phosphorescence are still being made today and will continue to be made in the future. Of the remaining areas of phosphorimetry in Table 13.1, various aspects of theory have been developed, but much more development in theory is needed. This is especially true for room-temperature solid-surface phosphorescence. With advances in the theoretical aspects in the various areas of phosphorimetry, it may be possible in the future to develop a unified theory for the different types of phosphorimetry that would be more comprehensive than present day phosphorescence theory. With present day instrumentation, there is an impressive array of accessories, instruments, and instrumental systems available. These instruments or instrumental systems are used for simple collection of routine phosphorescence data to sophisticated data processing and handling. In the future there should be more extensive use of laser sources and various detectors such as diode arrays. Because fluorescence and phosphorescence data can be obtained at room temperature, and with the availability of relatively inexpensive computers and instrumental components, it should be possible to design a luminescence instrument to simultaneously collect fluorescence and phosphorescence lifetime data, intensity data, and spectral data. With a system of this type, the collection of fluorescence and phosphorescence information would be more rapid than with today's instruments, and the amount of fluorescence and phosphorescence information on a given sample would be very substantial. With advances in theory and instrumentation, one will see many more applications of phosphorimetry in the future. This would result partly because of improved sensitivity and selectivity, but also because of the ease of handling samples and in refining the techniques and methodology for obtaining phosphorescence from materials and compounds via the phosphorimetric approaches in Table 13.1.